*Advances in
Acoustics Technology*

# WILEY – EC AERONAUTICS RESEARCH SERIES

# Advances in
# Acoustics Technology

*Edited by*
**J. M. Martin Hernandez**

European Commission Aeronautics Research Series

JOHN WILEY & SONS
Chichester • New York • Brisbane • Toronto • Singapore

Publication EUR 15495 EN of the
European Commission
Directorate-General XII for Science, Research and Development, Brussels

Published in 1995 by John Wiley & Sons Ltd,
              Baffins Lane, Chichester,
              West Sussex PO19 1UD, England

        Telephone (+44) 243 779777

*Other Wiley Editorial Offices*

John Wiley & Sons, Inc., 605 Third Avenue,
New York, NY 10158-0012, USA

Jacaranda Wiley Ltd, 33 Park Road, Milton,
Queensland 4064, Australia

John Wiley & Sons (Canada) Ltd, 22 Worcester Road,
Rexdale, Ontario M9W 1L1, Canada

John Wiley & Sons (SEA) Pte Ltd, 37 Jalan Pemimpin #05-04,
Block B, Union Industrial Building, Singapore 2057

*British Library Cataloguing in Publication Data*

A catalogue record for this book is available from the British Library

ISBN 0471 95149 8

Typeset in 10/12pt Palatino by Keytec Typesetting Ltd, Bridport, Dorset, UK
Printed and bound in Great Britain by Bookcraft (Bath) Ltd

# Contents

# Foreword

Over the last thirty years the European aeronautical industry has achieved a respected and internationally successful position. Aeronautical products made in Europe are able to secure large market shares, and in some instances have even become dominant. Nevertheless, the level of global competition continues to be set by the United States of America, which is traditionally committed to preeminence in this field.

Despite the importance of the European dimension, it is only in the comparatively recent past that the European Community has started to play a significant role in the technological challenge in aeronautics.

In 1988, the major European aircraft manufacturers presented the EURO-MART study (European Cooperative Measures for Aeronautical Research and Technology) to the European Commission. The study identified areas of research which were considered to be critical to the future competitiveness of the industry in world markets. Separate reports, conveying views on the content of a European research activity in aeronautics, were submitted by representative groups of the European Aero-Engine Manufacturers and the European Aerospace Equipment Manufacturers and Systems Suppliers.

Following these extensive consultations with industry, the Commission launched its first dedicated aeronautical research programme in March 1989. This initiative became known as the aeronautics 'pilot phase' and was in fact Area 5: Specific Activities Relating to Aeronautics, of the BRITE/EURAM-Programme, itself a Specific Programme within the Community Second Framework Programme for Research and Technology Development (1989–1992). This action is being continued under the 3rd EC Framework Programme (1991–1994) and will be pursued under the 4th EC Framework Programme (1994–1998).

The pilot phase comprised 28 projects, covering four technology areas— aerodynamics, acoustics, airborne systems and equipment, and propulsion systems.

Despite a relatively small budget (35 MECU over 2 years), this exploratory action achieved considerable success in stimulating wide-ranging cooperation between all types of companies. These included airframe, engine and equipment manufacturers, small and medium-sized enterprises, research centres and universities from the Member States of the Community and also from EFTA countries.

This EC-Aeronautics Research Series provides the opportunity to present the results of these research projects which have reached completion and to illustrate the achievements of this pilot action.

This volume presents the three projects which, under this pilot phase, carried out work in three areas of noise and acoustics technologies: exterior noise, cabin noise and acoustic loads in relation to structural fatigue.

The ASANCA project was a comprehensive effort to demonstrate the feasibility of active noise control technologies for reduction of cabin noise levels in propeller and jet aircraft. It included not only analytical modelling and laboratory testing, but also extensive ground and flight tests of two experimental demonstrator systems. Seven aircraft manufacturers worked together with universities and research institutes from eleven European countries in this project, which also covered, to a lesser extent, work on interior noise in helicopters.

In ACOUFAT, fourteen partners, including five aircraft manufacturers, have combined their efforts to improve the understanding of the fatigue behaviour of advanced aeronautical materials and structural designs subjected to acoustic excitations. The project highlights an integrated approach to three research aspects: analytical modelling of acoustic loads, enhanced computational methods for structural dynamic response and experimental generation of acoustics fatigue strength data on selected materials and designs.

The European helicopter manufacturers, supported by a number of research establishments, universities and other industrial enterprises, undertook in the HELINOISE project the acquisition of a most comprehensive aeroacoustics experimental data on a 40% dynamically scaled wind tunnel rotor model. The purpose was to develop and validate improved aerodynamics and noise prediction codes with the ultimate goal of providing the needed design tools for development of quiet helicopters.

The notable technical successes achieved in these projects illustrate the added value of collaboration at European level.

Thanks are due to all partners of the three consortia, particularly to the scientists, engineers and managers who contributed to the success of these aeronautical projects, reflecting the high level of collaboration attained under a common interest of advancing the European Union technology base in this crucial sector.

Thanks are especially due to the authors of the three technical project reports.

**J. M. Martin Hernandez**
*Brussels*
*January 1994*

# 1 *Advanced study of active noise control in aircraft (ASANCA)*

**I. U. Borchers*** *et al.*

This report, for the period July 1990 to September 1992, covers the activities carried out under the BRITE/EURAM Area 5 'Aeronautics' Research Contract No. AERO-CT90-0028 (Project AERO-P1067) between the Commission of the European Communities and the following:

Dornier Luftfahrt GmbH* (coordinator)
Instituto Superior Tecnico, Lisboa
University of Patras
Fokker Aircraft B.V.
SAAB Scania AB
Matra Sep Imagerie et Informatique
Centre National de la Recherche Scientifique, CNRS
TNO Institute of Applied Physics
Metravib R.D.S.
Leuven Measurements and Systems
Katholieke Universiteit Leuven, KUL
Alenia S.p.A.
Trinity College
Captec
Reson System ApS.
Agusta S.p.A.
CIRA
Eurocopter Deutschland GmbH
Brüel & Kjaer
Westland Helicopter Ltd
Institute of Sound and Vibration Research, ISVR-Southampton
Defence Research Agency, DRA

Contact: I. U. Borchers, Dornier Luftfahrt GmbH, Dept. EV53, D-88039 Friedrichshafen, Germany

*Advances in Acoustics Technology* Edited by J.M. Martin Hernandez. © ECSC-EEC-EAEC, Brussels–Luxembourg, 1994. Published in 1995 by John Wiley & Sons Ltd.

## Abstract

A broad cooperative study of active noise control in aircraft interiors has been performed involving 22 organisations located in 11 European countries. The main tasks and goals of the program are described and important results presented here. In particular, information on feasible optimum active noise control systems for future practical applications were identified and demonstrated successfully during flight testing. Detailed baseline information for further improving this technology was established.

The major part of the study related to fixed-wing aircraft and included interior noise and vibration measurements in flight and on the ground in different aircraft. Related initial noise calculations without and with active noise control were conducted. Based on this, two systems for laboratory and flight testing were developed. Studies on alternative active noise control approaches and development work on advanced actuators and sensors were also performed.

The flight and ground tests were performed in the Dornier 228, Saab 340, ATR 42 and Fokker 100. These aircraft were selected in order to show that active noise control is feasible in aircraft with large differences in acoustical character.

The noise control calculations were based primarily on finite-element analysis and advanced analytical methods. In addition, numerical modelling analyses were conducted based on models with reduced numbers of degrees of freedom. The calculations were performed, for example, for the same test aircraft as were used in the experiments, and they should provide supporting theoretical inputs for selecting optimum active noise control approaches, including the optimum number and locations of actuators and sensors for given control systems. The development of the demonstrator systems included system arrangement optimisation and selection and testing of feasible control algorithms. Based on this, the control unit architecture was designed and corresponding hardware realised. The control units were assembled and integrated with selected actuators and sensors into complete control systems.

A smaller work program related to rotary-wing aircraft. It included the design, manufacture and testing of a model helicopter fuselage, as well as assessments of prediction codes and comparison with experimental results. A laboratory validation of a new noise transmission path identification method and a review and an appraisal of active noise control methods for helicopters were conducted. The model fuselage was excited by a mechanical shaker and plane acoustic waves. Prediction code assessments considered, in particular, finite-element, boundary-element and statistical energy analysis methods.

Active noise control may be an effective means for reducing the critical low frequency aircraft interior noise. Full-scale ground data evaluation and flight tests indicated possible average noise reductions at the fundamental propeller tone of up to 15–20 dB with local noise reductions up to even 27 dB. The study identified that further research is required on improved control system definition, and system weight and size minimisation.

## 1.1   Introduction

Initial assessments have indicated that active noise control—the introduction of 'anti-noise' to cancel the original noise—has potential for solving the critical low-frequency interior noise problem of fixed-wing and rotary-wing aircraft. The problem is caused by the intense low-frequency noise of, for example, aircraft propellers or helicopter rotors, which is difficult if not impossible to reduce by applying lightweight and standard noise reduction measures.

The active noise-control system may consist, for example, of different sensors providing synchronous noise signals and of measured residual interior noise data. The main system component is a computerised control unit which uses these data as inputs and which derives optimised output control signals. The latter are fed to a set of actuators generating the required anti-noise fields for primary noise field reduction (Figure 1.1). Based on the low-frequency content and the deterministic character of the exterior noise, very high noise reductions (above 10 dB) appear possible using this technique.

In order to meet this goal, extensive advanced research is required into, for example, aircraft interior acoustics including vibro/acoustic interactions, the development of advanced actuators and sensors, and the optimisation of control-unit algorithms and related hardware components. This is needed to meet the constraints set by the complex aircraft environment; for example, minimum control-unit weight and size, minimum number of actuators and sensors, easy implementation of these components into the aircraft interior, adaptability to stationary and in-flight conditions, and various safety aspects.

The major part of the present research program related to fixed-wing aircraft. The main goal was to initiate this research, taking the various aircraft constraints into account. Feasible optimum active noise-control systems for future practical applications were to be identified and demonstrated during first flight testing. In addition, detailed baseline information for further improving this

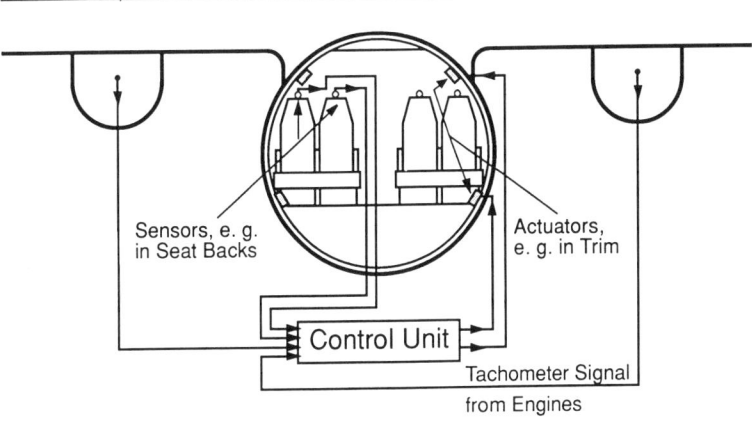

**Figure 1.1**   Scheme of possible active noise-control system

technology needed to be established. A smaller program on rotary-wing aircraft aimed to identify, for example, the most promising research methods that are available for helicopters.

## 1.2   Research objectives

### 1.2.1   Fixed-wing aircraft

The main goal was to investigate techniques for propeller-aircraft interior noise control, and to identify and demonstrate feasible control-system configurations. In particular, this research should identify the practical potential of noise reduction using this technique and obtain first estimates of weight/noise-reduction ratios, the costs of practical systems and on the cabin space required.

New information was wanted on the efficiency of different anti-noise actuator methods and approaches (e.g. speakers, shakers), and on reduction of the complexity of promising actuator methods (e.g. the required number of actuators and sensors).

With regard to the control-system design, the aim was to identify optimum noise-control algorithms and hardware components and to identify aircraft integration requirements and system performance limitations.

Finally, detailed information was wanted on the prediction of optimum noise-control systems in the early aircraft design stage, and basic information established for further improving this promising noise-control technology.

### The tasks

In order to reach these goals, the study included systematic and advanced experimental as well as theoretical work. It was subdivided into nine well-defined main tasks:

1.  Summary and evaluation of existing and pertinent active noise-control work.

2.  Detailed experimental surveys of the interior sound fields in partners' aircraft and in a fuselage test section (flight and ground tests).

3.  Theoretical description of the interior sound fields for selected aircraft and a fuselage test section.

4.  Performance of tests in aircraft and in a fuselage test section to determine the optimum active noise-control approach and optimum control systems.

5.  Preparation of theoretical methods, and performance of supporting calculations, for selecting the optimum active noise-control approach and optimum control systems.

6.  Data evaluation, and preselection of the optimum noise-control approach and system configurations.

7.  Development of prototype control units, including hardware and software realisation.

8.  Integration of complete prototype control systems and establishment of system descriptions.

9.  Evaluation of results and recommendations for prototype control-system improvements.

The work of *Task 1* is presented in Section 1.3.1, 'Literature'. The main testing of *Tasks 2* and *4* included real aircraft tests, which in most cases were performed together; they are thus summarised in Section 1.3.2, 'Testing'. The theoretical work of *Tasks 3* and *5* was conducted on a similar basis and is thus given in Section 1.3.3, 'Analyses'. Work on development of the control-system units was conducted successively in *Tasks 6, 7* and *8* and is summarised and presented in Section 1.3.4, 'Control systems'. In addition, in *Task 4* important initial work on alternative noise-control approaches and advanced actuators and sensors was performed, which is given separately in Section 1.3.5, 'Alternative approaches' and Section 1.3.6, 'Advanced actuators'. *Task 8* included also the laboratory and flight testing of prototype systems; because of its importance this is presented separately in Section 1.3.7, 'Laboratory and flight testing'. The work of the final *Task 9* is described as appropriate throughout the chapter.

## 1.2.2 Rotary-wing aircraft

The main goal was, having identified the most promising research methods that are available for helicopters, to provide a basis for a more extensive future helicopter interior noise-control project.

In more detail, the basic objectives were a review by the helicopter manufacturers of their existing internal noise data and the validation of a novel experimental method for noise-path identification. In addition, a review of the current situation and the future potential for active noise-reduction techniques for helicopters was to be performed and a preliminary first comparative look at theoretical prediction methods conducted. Additional research requirements aimed at lightweight palliative design guides should be identified.

### The tasks

Based on these goals, the related study included the following five main tasks:

1.  Critical review of available relevant knowledge and interior noise-source identification.

2.  Design, manufacture and testing of a model helicopter fuselage.

3.  Prediction codes assessment, and comparison with experimental results.

4.  Laboratory validation of the path identification method.

5.  Review and appraisal of active noise-control methods.

A detailed description of the work performed in these tasks, and selected results, are given in Section 1.4.


## 1.3   Research activities and results for fixed-wing aircraft

In this section the work performed in the various tasks and selected important results are described on a task-by-task basis. However, wherever appropriate two or three tasks are considered together because of their closely related work content.


### 1.3.1   Literature

*Task 1* was to summarise and review the existing pertinent active noise-control work. Important interior noise studies and control-unit hardware and software-related work were considered separately.

The object was to provide detailed baseline information on the problem to all partners at an early stage of the program. In addition, updated information for the preselection of an optimum noise-control approach and system configurations was supplied, and information during the course of the project for possible further updates of the noise control approach, control systems and objectives provided as required.


*Examination of relevant aircraft and control-system related work*

Based on the open literature and the internal know-how of the partners, a review was made of experimental and theoretical studies related to the noise fields inside the cabins of passenger aeroplanes with and without active noise control. In addition, a critical survey was performed of existing and planned anti-noise control systems, including both hardware and software aspects. The results were summarised in a report which was forwarded to all partners for review.


*Updates of study evaluation reports, and recommendations*

Throughout the course of the project, new developments in the state of the art were closely followed. The results were used to update the reports of the previous sub-tasks on a four-monthly basis. The updated reports were also forwarded to all partners for review.

On the basis of the previously established results, recommendations for preselection of a noise-control approach and control system were provided. Recommendation updates and/or updates for the project objectives were derived as appropriate.

### Conclusions

The work performed in this task turned out to be very beneficial to the overall project and helped significantly to make the total problem transparent to all partners. It resulted in six reports, including updates, and some recommendations were made for the final control system demonstrators. The demonstrators were tested successfully in the later phase of the project as described in Sections 1.3.4 and 1.3.7.

## 1.3.2 Testing

In this section, the aircraft testing of *Tasks* 2 and 4 and important results are described and discussed. Software development was the largest part of these tasks. Further activities of *Task 4* covered initial research on alternative noise-control approaches and advanced actuators, which are presented separately in Sections 1.3.5 and 1.3.6.

Aircraft testing played a primary role in the current project, as it was considered and selected as the most direct and best method to arrive at feasible control-system demonstrators, including initial flight evaluation, in the limited project time of two years. To define the demonstrators, and with this the selected noise-control approach, reliable information on the primary interior sound fields of selected partner aircraft and of secondary sound fields of anti-noise sources installed in these aircraft were of great importance. The main objective of aircraft testing was to establish and evaluate this information as early as possible so that control-system work could proceed in time. Furthermore, an experimental database was to be provided to validate theoretical methods, including related modal analyses, response analyses and first active noise control calculations.

### The selected aircraft and the tests

Aircraft testing was performed in four aircraft: the Dornier 228, Saab 340, Alenia ATR 42 and Fokker 100. These were selected in order to identify general noise control information, and to show that active noise control is feasible in aircraft with large differences in acoustical character. The Dornier 228 is, for example, a propeller aircraft with an unpressurised fuselage and with a rectangular cross-section. In contrast, the Saab 340 and ATR 42 are pressurised propeller aircraft with a circular fuselage, and the Fokker 100 is a jet aircraft with aft-mounted engines.

**Table 1.1**   Typical noise frequencies in the selected aircraft

| Aircraft | Noise source for active control | Fundamental frequency (Hz) | Number of important harmonics | Highest frequency to be controlled (Hz) |
|---|---|---|---|---|
| Dornier 228 | Propeller | 102–106 | 4 | 424 |
| Saab 340 | Propeller | 85 | 2 | 170 |
| ATR 42 | Propeller | 68.8 | 4 | 275.2 |
| Fokker 100 | Jet engine N1 spool | ± 115 | 2 | 250 |
| | N2 spool | ± 180 | 1 | |

In all the selected aircraft, low-frequency tones are present which contribute to the overall interior noise. Table 1.1 lists the frequencies of the critical fundamental tones and the highest frequencies at which active noise control was considered to be desirable. The fundamental frequencies range between about 70 Hz and 180 Hz, so these and similar aircraft are especially suitable for active noise control.

The measurements performed were in two parts. First, in *Task 2* the in-flight primary sound fields were measured in detail to identify clearly the noise distribution to be controlled. Not only the amplitudes but also the phase distributions of the various tones of interest were determined throughout the three-dimensional cabin space, including in the plane of the passengers' heads. Special synchronisation signals which were coherent with the propellers or engines were recorded.

Secondly, in *Task 4* extensive ground tests were performed for various positions of loudspeakers and shakers to determine effective noise-cancelling of secondary sound fields. Typically in each aircraft around 45 loudspeaker and shaker positions were individually tested. The responses of these were measured for about 60 microphone positions, including in the plane of the passengers' heads. All transfer functions between the individual actuator locations and microphone positions were determined. From these data, the amplitude and phase distribution of the secondary field of each actuator was extracted, at the frequency of the tones of interest.

*Test preparation*

**Test procedures**   Flight and ground test procedures were agreed in detail by all the partners. For instance, in the Dornier 228 and Saab 340 the complete aircraft was surveyed and the secondary-source transfer functions were determined for all seat positions. In the other aircraft the effort was concentrated on a limited number of seat rows; e.g. the seven aft seat rows in the Fokker 100 and the six forward seat rows in the ATR 42. As a consequence the resulting four test procedures were very similar and enabled the partners Metravib and LMS, which primarily performed the measurements, to use basically the same measurement set-up in all four aircraft.

**Test equipment** For the measurements and data analysis, special test equipment had to be prepared. The aircraft companies, for example, developed high-quality synchronisation-signal pickups to be attached at the propellers or engines. For this, Dornier and Saab used an acoustic sensor which detected optical pulses from a disc mounted on the propeller axis. Alenia used the existing tachometer signal. Fokker made special electronic equipment to translate the available tachometer signals of the jet engine's N1 and N2 spools into the required synchronisation signals. In addition, the aircraft companies supplied converters for power supply to the measurement equipment.

Metravib and LMS used their extensive data-acquisition equipment. This was mounted in instrumentation racks which were flexibly mounted to a steel plate which was itself attached to the aircraft floor. Two racks were needed: one consisted of all data-acquisition devices; the other consisted of a complete computer system to load the data on hard disks and to enable preliminary checking of the validity of the data during the measurements.

In addition, Metravib constructed two antennas each to hold 24 microphones (Figures 1.2 and 1.3). The microphones were selected from a very large number to minimise their phase differences; the resulting maximum phase error was 0.3 degrees. All microphones were individually calibrated prior to the tests.

**Figure 1.2** Microphone antenna for in-flight interior noise surveys in the ATR 42

**Figure 1.3**   Horizontal microphone antenna for actuator ground tests in the Dornier 228

The array was mounted on a rail and could be moved throughout the aircraft cabin to cover all measurement locations. Typically 20 measurement sections were needed for a complete three-dimensional survey. An example of the measurement locations is given in Figure 1.4. All 24 microphone signals could be acquired simultaneously. In addition, a reference microphone signal was recorded continuously.

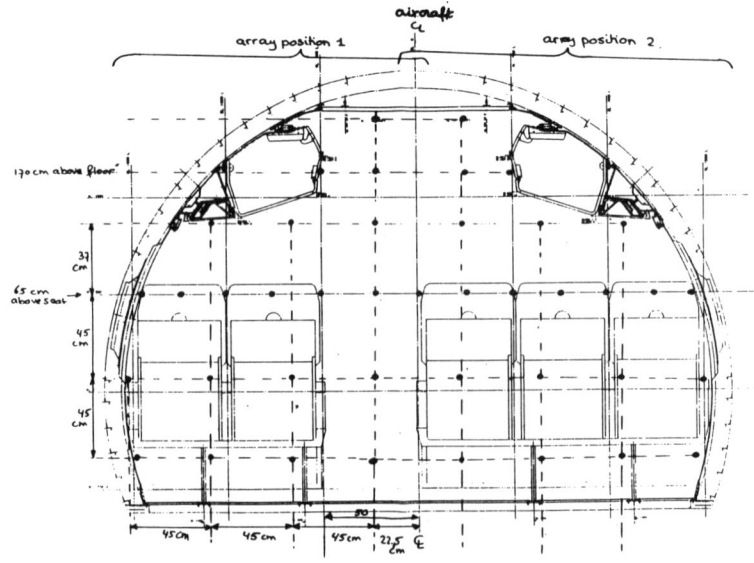

**Figure 1.4**   Measurement positions for an in-flight noise survey at 14 axial positions in the Fokker 100

LMS also provided and mounted 80 vibration pickups, which were all individually calibrated. Recordings of these pickup signals could be made in three batches. In addition, specific acquisition command software and data-processing software was developed by this partner.

*Results of in-flight measurements*

**Amplitude and phase of the primary sound field**   The complete three-dimensional sound fields within the aircraft were surveyed. The non-coherent parts of the sound fields gave an estimation of the maximum noise reduction which could be obtained with active noise control, as the prevailing tones to be suppressed are masked by the non-coherent background noise.

In addition, the contributions of the left-hand and right-hand propellers or engines could be separated out to get an impression of their relative contributions and to enable calculations of synchrophaser effects. With this information, the synchrophase angle could in principle be tuned to the optimum value, even in combination with active noise control. An example of a measured primary field for the fundamental tone at the blade passing frequency (1 × BPF) is given in Figure 1.5.

**Vibration patterns of aircraft structure and trim panels**   Vibration levels were recorded, to be used for verification of finite-element calculations and for general understanding of the noise transmission path. The locations of the

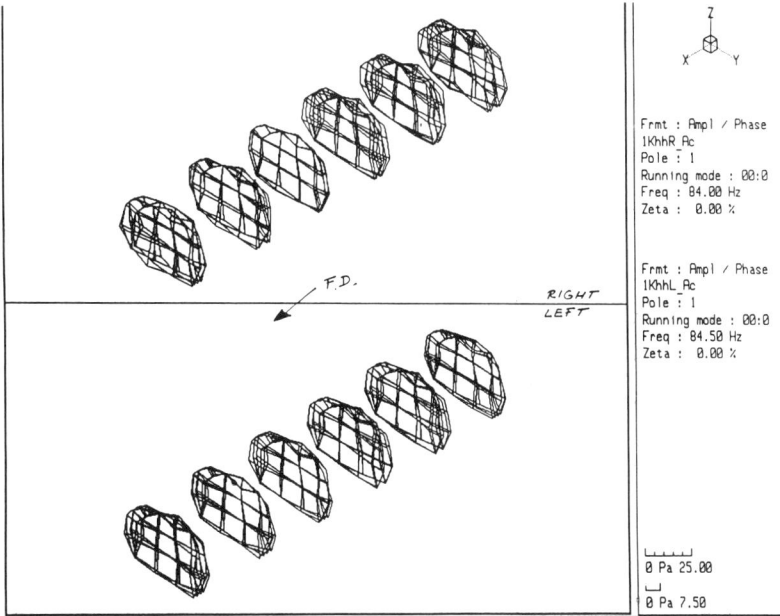

**Figure 1.5**   Measured three-dimensional primary sound field in the Saab 340 (1 × BPF, left and right propeller parts separated)

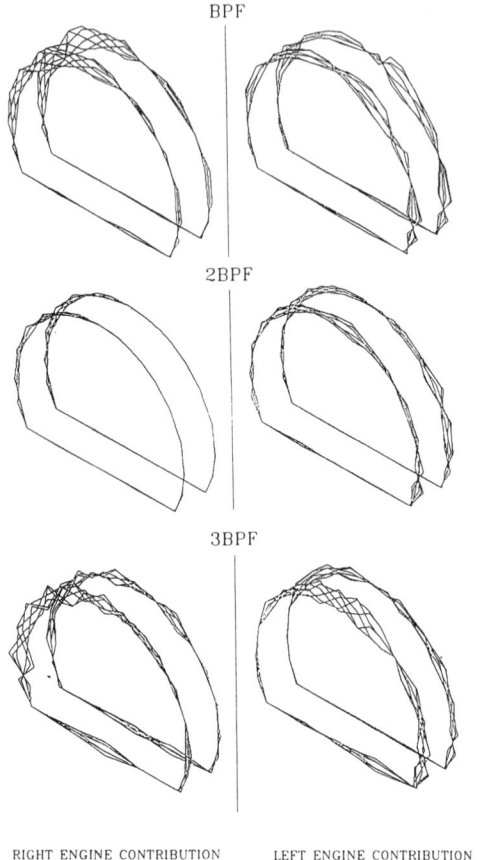

BPF

2BPF

3BPF

RIGHT ENGINE CONTRIBUTION        LEFT ENGINE CONTRIBUTION

**Figure 1.6**   Results of vibration measurements in the ATR 42

vibration pickups were adapted to the specific circumstances. In the Dornier 228, the pickups were located on the fuselage frames. In the Alenia ATR 42, the pickups were positioned in two fuselage frames and on the trim panel between these frames. In the Fokker 100, selected trim panel vibrations were measured. In the Saab 340, vibration measurements were performed on some trim panels and on the cabin floor. One example of the results is given in Figure 1.6.

**Data analysis**   For the general data analysis and presentation, existing LMS software and Metravib acoustical imaging software was used. The results of the data analysis were transferred to computer files and provided to partners for optimisation of the positions and number of secondary sources and sensors, algorithm simulations etc. A special software code was developed within the project by CNRS. Details of this software are given in the later subsection, 'Data analysis for optimum active noise control'.

**Stability of the propeller or engine rotational frequency**   To permit design of the control units, information was needed on the stability of the tones within the aircraft. The data obtained would be used in simulations to be performed by Matra and TNO.

In addition, recordings were made with a very high sampling frequency at typically five locations in each aircraft, in order to enable further data analysis if required for the design of the control systems.

An example of a very stable propeller tone is shown in Figure 1.7. In other aircraft, however, the stability was somewhat less.

## Results of ground measurements

**Reverberation time at the frequencies of interest** The purpose of this simple test was to check whether the coherence measurements would give correct results for the chosen block length. The reverberation time in the various aircraft appeared to be very similar, although it depended on the exact location of the exciting loudspeaker and recording microphone. Typical measured values are given in Table 1.2.

**Acoustic-cavity modal analysis** One way to obtain a global noise reduction with active noise control may be based on so-called 'modal control'. In order to use this technique, the acoustic modes of the aircraft cabin have to be determined. With these data, the in-flight sound field can be decomposed to

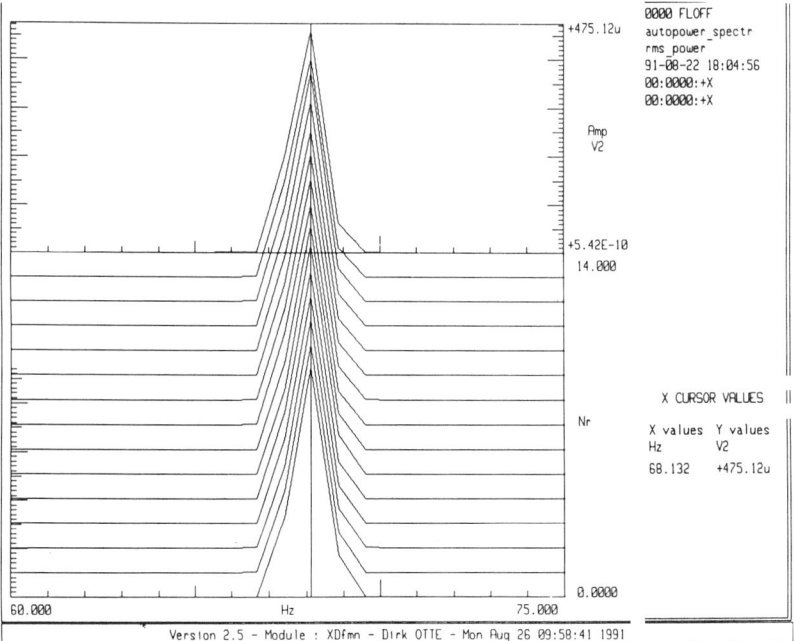

**Figure 1.7** Measured autopowers of the right tacho signal during a flight test with the ATR 42

Table 1.2   Measured reverberation times

| Aircraft | Reverberation time $T_{60}$ at fundamental tone (seconds) |
| --- | --- |
| Dornier 228 | 0.2–0.6 |
| Saab 340 | ~0.25 |
| ATR 42 | 0.4–0.6 |
| Fokker 100 | 0.2–0.5 |

select the most important modes. If these are known, selection of secondary source positions is more straightforward. For acoustic-cavity modal analysis, a complete survey of the aircraft volume has to be made with four excitation sources.

From the measurement results it appeared, however, that the acoustic volumes (which included all cabin trim, seats, etc.) were highly damped, which made the planned modal analysis far from straightforward. In addition, it was concluded from FEM calculations that 'modal sources' are not easily identified. Thus, this approach was not considered any further.

**Secondary-source sound fields and related impulse responses**   The secondary sound fields in the plane of the passengers' heads generated by individual secondary sources (loudspeakers) located at a large number of positions, were determined in amplitude and phase. The measurements were carried out using the standard method (Dornier 228 and Fokker 100) or the reciprocity technique (Saab 340 and ATR 42). In the latter method the loudspeakers were located on the seats and the microphones were placed on the walls of the aircraft. In this way, measurement time could be somewhat reduced, after reference tests had shown that this technique could be used without loss of accuracy. However, in the Fokker 100 the reciprocity tests gave negative results, so in this aircraft the straightforward method was used.

In the Dornier 228, the loudspeakers were incorporated as much as possible into the cabin trim of the prototype aircraft. The complete impulse response between the current of each loudspeaker and the pressure response at each measurement microphone was needed for algorithm simulations to be performed by Matra and TNO. Typically 45 secondary-source positions were measured in each aircraft. Owing to the large number, it was decided to measure the response only in the plane of the passengers' heads. Also, it was considered that it would not be worthwhile to aim for a noise reduction at other locations (for instance, under the seats).

Some of the secondary fields showed typical modal properties (Figure 1.8), while others showed a typical running wave behaviour (Figure 1.9). A sample of the results obtained by reciprocity measurements is given in Figure 1.10.

In addition to loudspeakers, shakers located at a limited number of positions were tested as secondary sources. However, from the very limited tests which could be performed it appeared that it was not possible to sufficiently excite the acoustic volume to get reliable results.

**Flight**

**Direction**

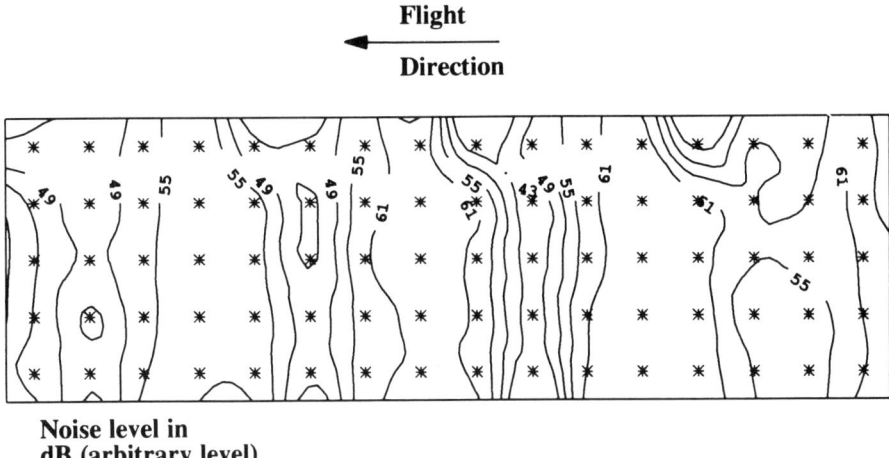

**Noise level in
dB (arbitrary level)**

**Figure 1.8** Measured secondary sound field of a loudspeaker located at the rear cabin wall of the Dornier 228

**Flight**

**Direction**

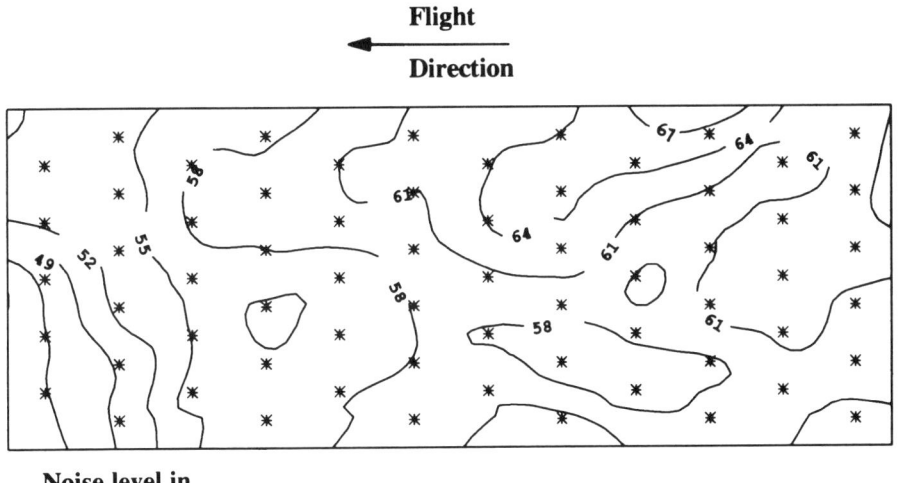

**Noise level in
dB (arbitrary level)**

**Figure 1.9** Measured secondary sound field of a loudspeaker below a seat at the right-hand cabin side of the Fokker 100

**Repeatability of the transfer function signals** The stability of the transfer functions from some secondary sources to some microphones was determined. For instance, it was tested whether a walking person would considerably influence the transfer function. This information was needed to determine the required control-unit convergence rate and to enable a correct choice of algorithm.

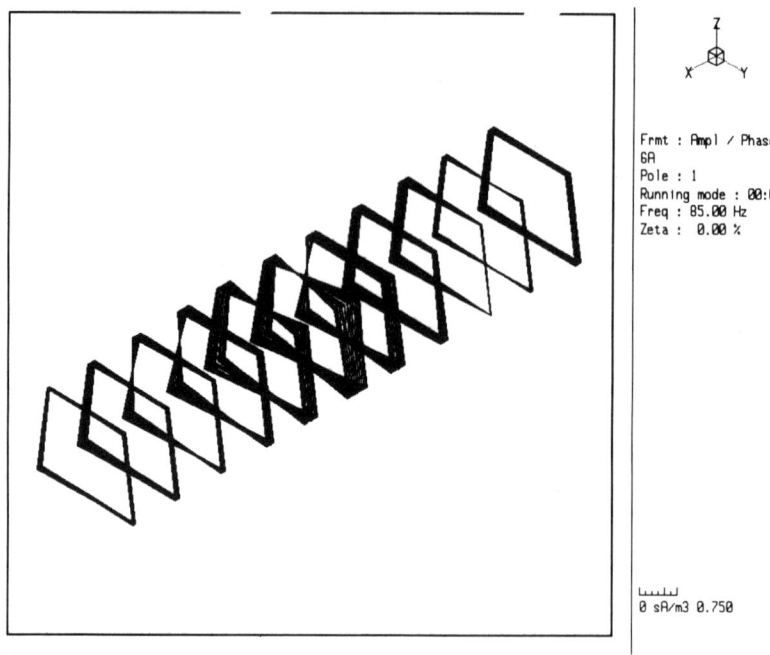

Frmt : Ampl / Phase
6A
Pole : 1
Running mode : 00:0
Freq : 85.00 Hz
Zeta :   0.00 %

0 sA/m3 0.750

**Figure 1.10**   Measured reciprocal secondary sound field of a loudspeaker in the middle of the cabin of the Saab 340

*Data evaluation for optimum active noise control*

From the ground tests, the microphone and source signals and the related transfer functions were determined. From these data, the amplitude and phase distributions of the secondary sound fields in the plane of the passengers' heads were extracted for the frequencies of interest. From the flight tests, the amplitude and phase of the prevailing sound fields were also available. Both of these data sets were used for the optimisation of actuator loudspeaker and sensor microphone positions.

A special computer code was developed by CNRS to be used by Trinity College and the aircraft companies. Within this code, the effect of a given active noise-control configuration, consisting of a specific set of loudspeakers and microphones, could be estimated. The noise reduction was calculated by superposition of the measured primary sound field and the computed secondary sound field, while the amplitude and phase of the selected secondary sources were optimised in the same manner as would be done in the actual active noise-control systems. In this direct manner of calculation, all types of secondary sound fields (with a modal or running-wave character) could be used without any constraints.

As the number of possible ways of combining all secondary fields was virtually infinite, the true optimum actuator loudspeaker configuration for active noise control could not be determined. However, several practical approaches were applied by CNRS, Trinity College, Dornier, Fokker and Saab to find configurations evidently rather near to this optimum.

One such approach started with no sources and added, step-by-step, the source with the best noise reduction. This method could be extended by adding two or more sources at each step and gave the best results for selecting a small control system, consisting of a few loudspeakers and microphones. Another method started with all measured sources and deleted, step-by-step, the source(s) with the smallest additional noise reduction. This method gave the best results for selecting a large control system. Another approach selected the sources which had the best fit or spatial coherence with the prevailing sound field, which had the advantage that the power consumption of the loudspeakers was minimised. In addition, a genetic algorithm approach which could find a good configuration with an extremely small calculation effort was tested.

The sensor microphone locations could likewise be optimised using the computer code.

In Figure 1.11, the potential noise reduction using active noise control, obtained by applying this evaluation, is plotted against the number of sources for the different aircraft and fundamental tones. The data correspond to average values of 48 or 36 sensor microphones and show that active noise control can be very effective. Figure 1.12 gives an example of the noise

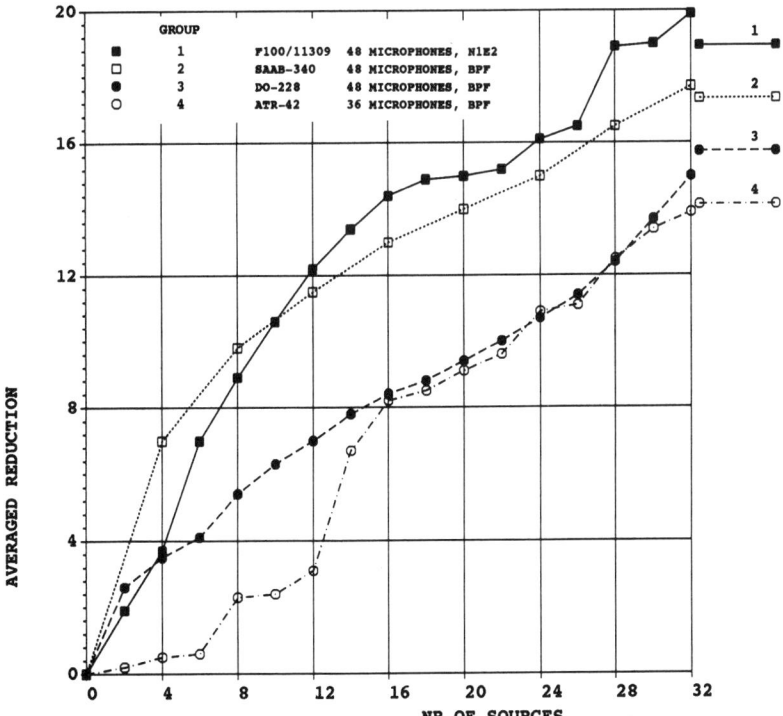

**Figure 1.11** Determinated noise reduction for active noise control based on actuator ground testing

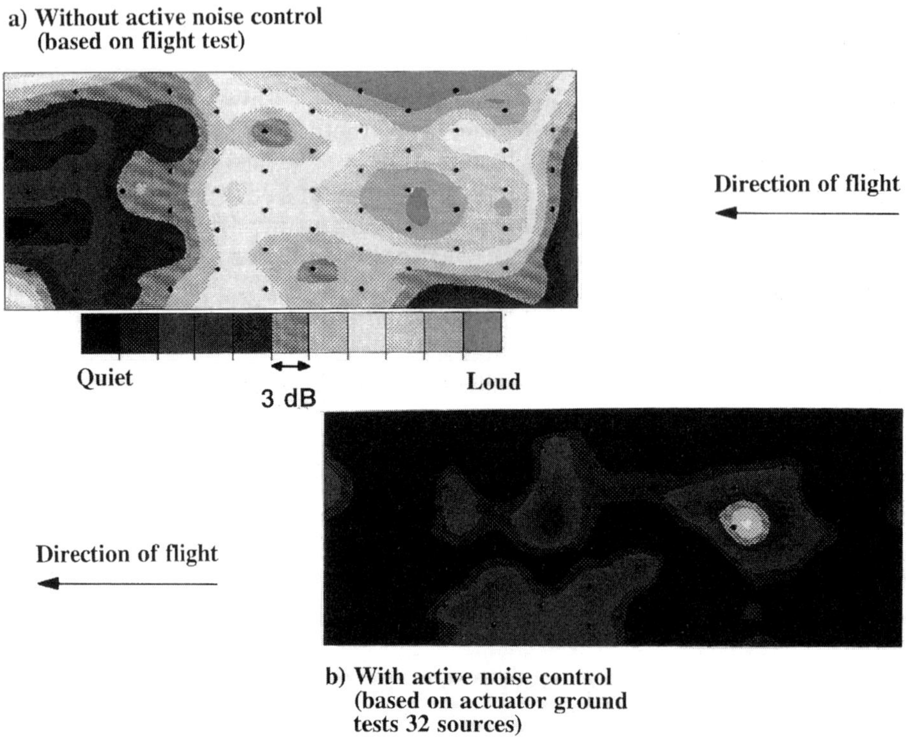

a) Without active noise control
   (based on flight test)

Direction of flight
←————————

Quiet        **3 dB**        Loud

Direction of flight
←————————

b) With active noise control
   (based on actuator ground
   tests 32 sources)

**Figure 1.12**  Determined noise levels in the plane of the passengers' heads at the rear part of the Fokker 100, without and with active noise control

reduction which could be reached locally using a control system with 32 loudspeakers and 48 sensors in the Fokker 100. The noise reduction predicted for the Dornier 228 was virtually confirmed by later actual measurements, which are described in Section 1.3.7.

## Conclusions

Owing to the high damping properties of an aircraft cabin, the dominant modes could not be determined experimentally. Thus, from the measurements it appeared at first that a full global control approach taking the entire cabin space into account would not be possible. However, a purely local control approach, using several independent one-channel systems, was considered to be not at all appropriate. Therefore, it was decided that at least a global localised control approach in the complete plane of the passengers' heads should be reached. This approach appeared to be possible, even with a limited number of actuator loudspeakers and sensor microphones.

With these and other results, the aircraft testing phase was successfully concluded. The work overall was described in a total of 54 reports, and provided all inputs required for the control system definitions and theoretical work as planned.

### 1.3.3  Analysis

This section describes important work performed in *Tasks* 3 and 5. The first task was concerned with primary interior noise predictions (primary response), while the other was concerned with active noise-control simulations (secondary response) and initial studies on advanced active damping measures.

The longer-term objective of this work was to initiate research which in the future could permit predictions of optimum control systems early in the aircraft design stage and/or which could provide effective future control methods with minimum weight. Primarily, supporting information for selecting and/or improving active noise-control configurations was to be supplied, and a database established for identifying possible general trends for most simple control systems. Theoretical results for improved interpretation of test data were also required.

The primary response analyses performed in *Task* 3 consisted of the following:

- Description of the external excitation field.

- Aircraft fuselage and cavity finite-element modelling.

- Eigenvalue analyses of a complete coupled system.

- Response calculations for external field excitation.

The simulations of *Task* 5 included finite-element calculations on the following subjects:

- Active noise control of a propeller-aircraft interior noise field.

- Active noise control to improve the transmission loss of a double panel.

*External excitation fields*

For the Dornier 228, the external acoustic pressure field was measured during flight tests with 10 outside microphones at each side of the aircraft for a given favourable syncrophase setting. Based on these data, which were dominated by propeller noise, related left- and right-hand side pressure and phase distributions were determined as inputs for the planned response calculation. As an example, Figure 1.13 shows the pressure field of the first blade passage frequency of 102.0 Hz (1 × BPF) on the right side of the aircraft. The pressure has a strong peak near the plane of the propeller. This and the left-hand-side pressure field were used with the corresponding phase distributions as the excitation force field for the 1 × BPF response calculations, and as an approximate force field for frequencies below 1 × BPF and up to about 1.5 × BPF.

For the Saab 340, too, the external acoustic pressure field is dominated by propeller noise. The pressure field was predicted using an existing propeller noise prediction code and taking installation effects into account. The funda-

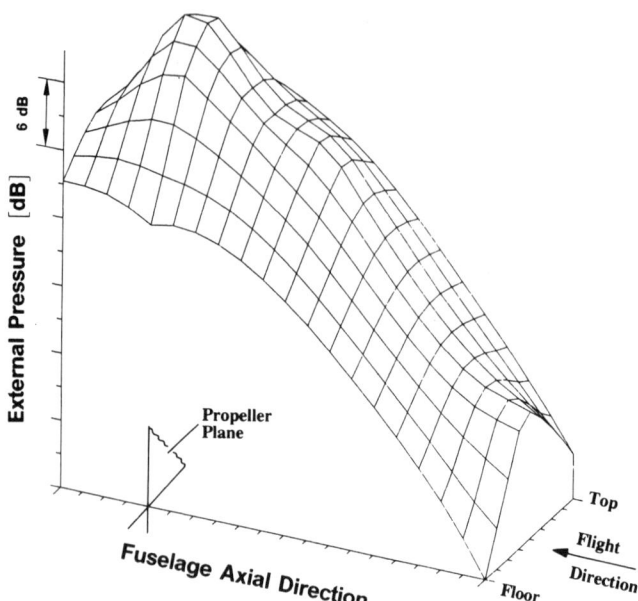

**Figure 1.13**  Applied external pressure field of the Dornier 228 based on test data (fuselage right side, 1 × BPF)

mental propeller tone had a frequency of 84.7 Hz. For the initial response calculations, the pressure field was calculated in the plane of the propellers around the circumference of the aircraft. It was then assumed to be constant in axial direction over a width of about 0.5 m including the propeller plane, based on the strong peak of the pressure field around the plane of rotation. Since the pressure field of each propeller is not stationary in space, both real and imaginary parts were calculated. The structural responses were analysed correspondingly and thus consist also of these parts.

*Finite-element models*

For both aircraft, the finite-element idealisation consisted of separate models for the fuselage structure and the enclosed air cavity, which for the analyses were correspondingly coupled.

For the Dornier 228, the full passenger compartment was idealised. For the structural model, detailed design influences were taken into account such as local stiffeners, wall panels, windows etc., in addition to standard frame, stringer and floor constructions. A view of this model is shown in the upper part of Figure 1.14. The model was set up to predict structural modes up to 350 Hz. The acoustic cavity was idealised in a way to represent especially the plane of the passengers' heads and to get accurate eigenvalue results up to about 400 Hz. A view of this model is shown in the lower part of Figure 1.14.

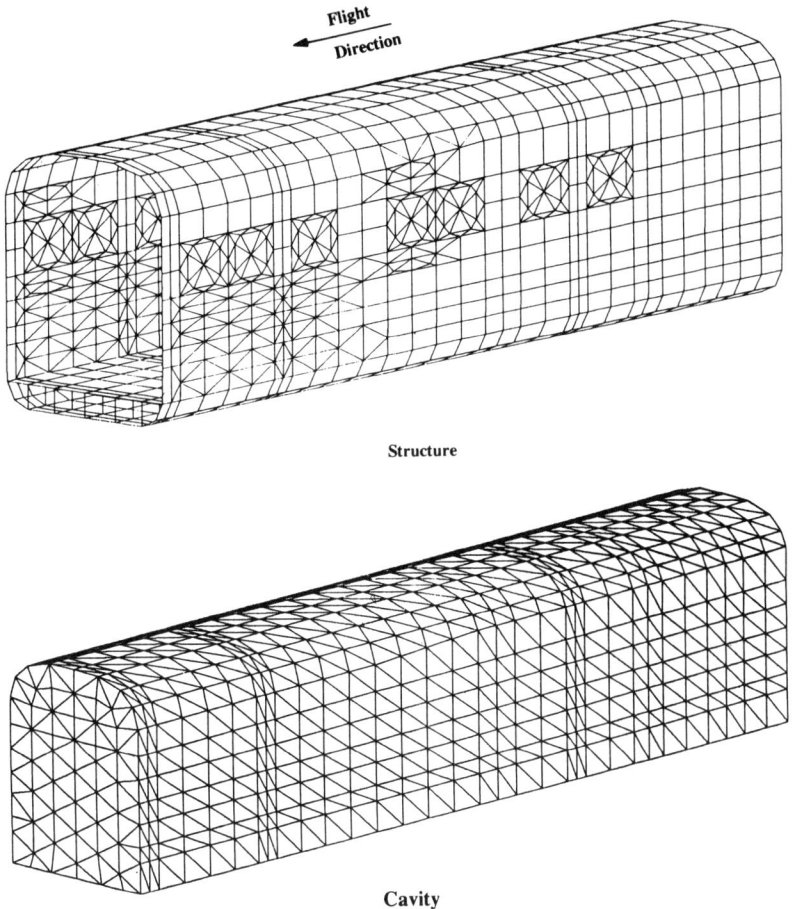

Flight Direction

Structure

Cavity

**Figure 1.14**  Fuselage and cavity finite-element idealisation of the Dornier 228

The structural model of the Saab 340 was made up of two sub-nets each representing one half of a fuselage section of 3 m length located in the main propeller noise excitation region. A picture of this model, including design details as in the model of the Dornier 228, is given in the upper part of Figure 1.15. The model of the acoustic cavity included four sub-nets and covered the entire cabin length. A view of the total cavity model is shown in the lower part of Figure 1.15. Using sub-nets for the structure and cavity, the calculation time for each step could be reduced to a reasonable amount.

*Eigenvalue analyses*

For the Dornier 228, uncoupled eigenvalue analyses for the structure and cavity were performed. The models for the structure and cavity had about

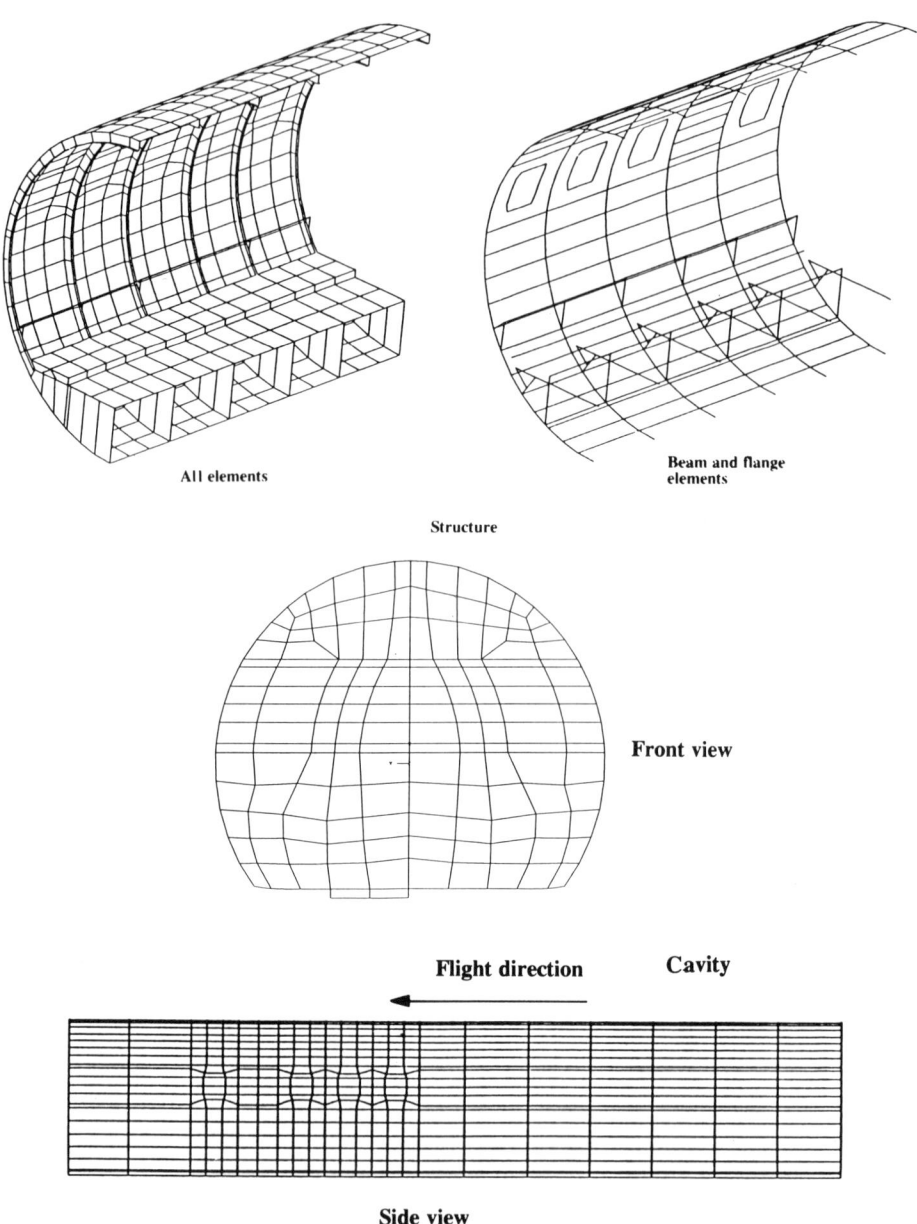

**All elements**

**Beam and flange elements**

**Structure**

**Front view**

**Flight direction**          **Cavity**

**Side view**

**Figure 1.15**   Fuselage and cavity finite-element idealisation of the Saab 340

22 000 and 14 000 degrees of freedom respectively, which were reduced to about 2400 and 2100. The analyses provided the modal basis for the coupling procedure and yielded important uncoupled mode information necessary for interpreting the coupled results. About 300 structural and 250 cavity frequencies were calculated for frequencies up to 409 Hz and 470 Hz, respectively. Two

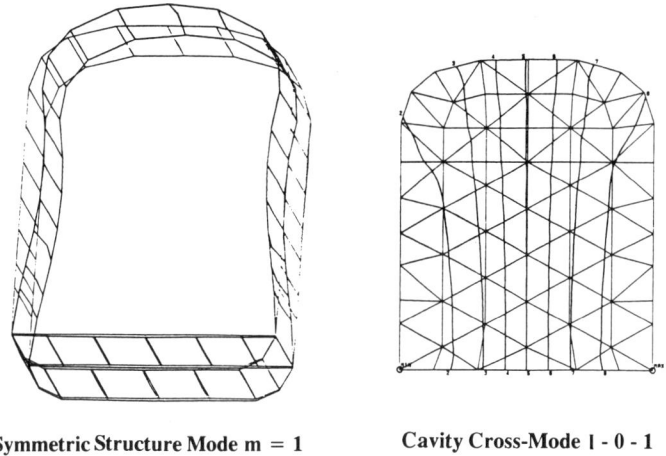

Symmetric Structure Mode m = 1          Cavity Cross-Mode 1 - 0 - 1

**Figure 1.16** Selected uncoupled structure and cavity modes of the Dornier 228

typical results are shown in Figure 1.16. The coupling procedure used was a modal synthesis of the uncoupled structure and cavity modes. Both systems were connected at their common interface area at 950 points. The related coupled analyses yielded coupled system modes as linear combinations of all uncoupled modes with different weightings and covered eigenvalues up to 470 Hz. The resulting mode shapes were classified as structure-dominated modes, cavity-dominated modes, or simply as coupled modes if both structure and cavity uncoupled modes were approximately equally involved.

The eigenvalue analyses of the Saab 340 followed similar steps. First the eigenmodes of each of the substructures were extracted. Since the frequency considered was up to about 400 Hz, up to about 150 eigenmodes were obtained. These eigenmodes then entered the equations at the next higher level of substructure hierarchy as the degrees of freedom of the substructures. In the last step, the problem was solved at the main net level and the lowest 255 coupled system eigenmodes were calculated. With this modal synthesis the total number of degrees of freedom of about 60 000 was reduced to about 5000. The eigenmodes obtained were classified as described above. The structural part of an identified important coupled mode at the frequency $f = 85.5$ Hz is shown as an example in Figure 1.17.

*Response analyses*

Response analyses were performed for the exterior pressure fields described earlier. For the Dornier 228, corresponding calculations were done for frequencies between 50 Hz and 150 Hz to show all critical resonances which may be excited in this frequency range. An example of the results obtained is given in Figure 1.18, which shows predicted interior noise spectra averaged over all

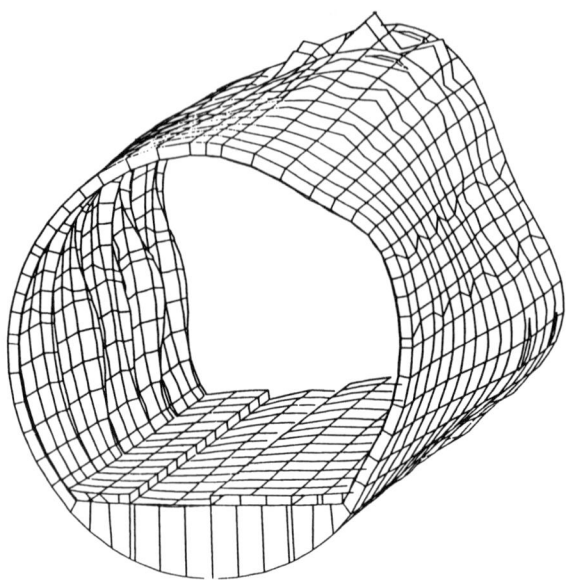

**Figure 1.17**   Structural part of coupled mode 17 of the Saab 340 ($f = 85.5$ Hz)

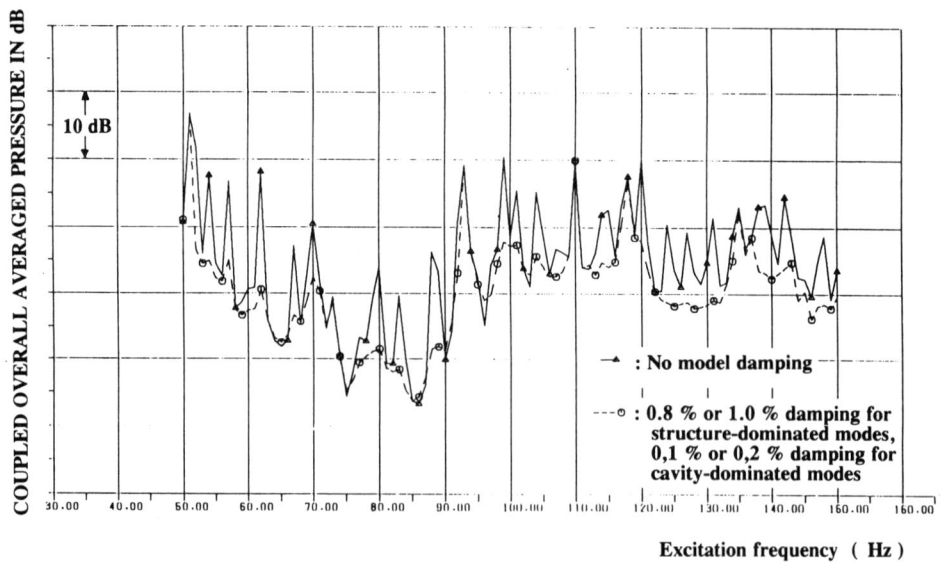

**Figure 1.18**   Predicted overall average interior noise of the Dornier 228 *versus* frequency (coupled response)

points of the whole cavity. For this, no modal damping, or a damping of 0.8% or 1.0% for structure-dominated and of 0.1% or 0.2% for cavity-dominated modes, was assumed.

Figure 1.19 presents a three-dimensional plot of individual spectra averaged over all points in 33 different cross-sectional planes in the axial direction.

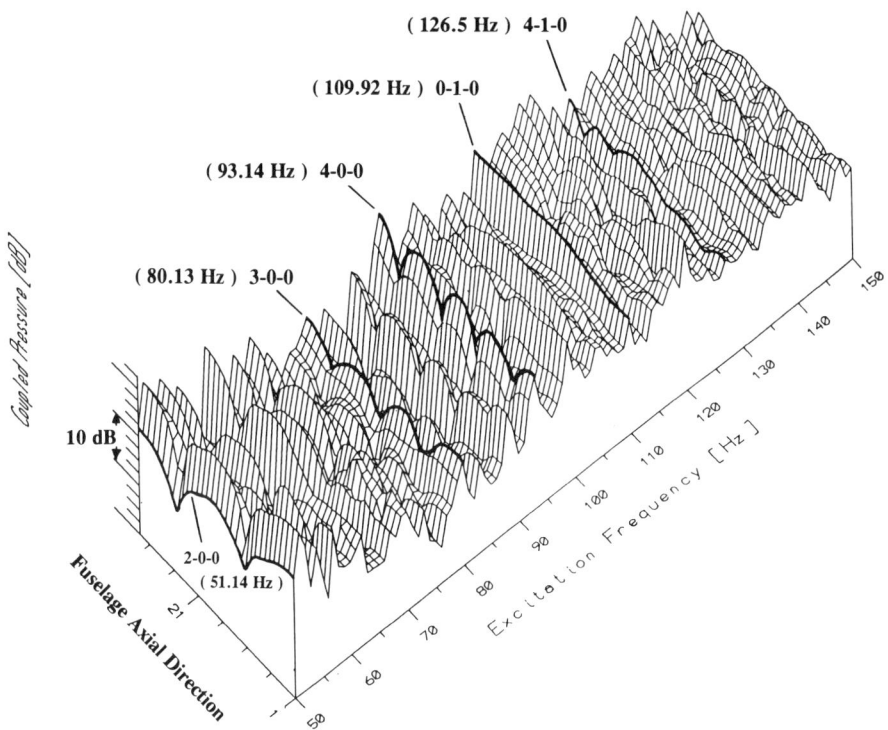

**Figure 1.19** Predicted frequency-dependent average noise levels inside the Dornier 228 *versus* fuselage axial direction

Figure 1.20 shows predicted pressure distributions for 102 Hz and 103 Hz excitation for the plane of the passengers' heads. From these and other results, indications were found that the structure may strongly influence the cavity response. Also, the tendency was identified that transverse modes especially govern the resulting overall sound pressure level in the front cabin section.

For the Saab 340, generalised responses for the different modes were calculated for 1 × BPF and 2 × BPF. Furthermore, acoustic and structural responses for these excitation frequencies were predicted in the form of sound pressure contours and deflection shapes. Corresponding results are given in Figures 1.21–1.23. From the generalised response diagram in Figure 1.21 it follows that the 1 × BPF response is dominated by only a few modes. Most of these were classified as strongly coupled modes with the exception of mode 20 which was deemed to be a cavity-dominated mode. Since all the interesting modes showed a more or less clear side–side behaviour, the related response in the cavity can be expected to look also like a side–side pressure variation. The 1 × BPF response due to left propeller excitation is dominated by the same modes as the one due to right propeller excitation, with one addition, mode 22. This, which is a symmetrical structural mode of the S4 type (four circumferential wavelengths), gives a local pressure increase close to the top of the cavity which might result in a distortion of the otherwise dominating side–side

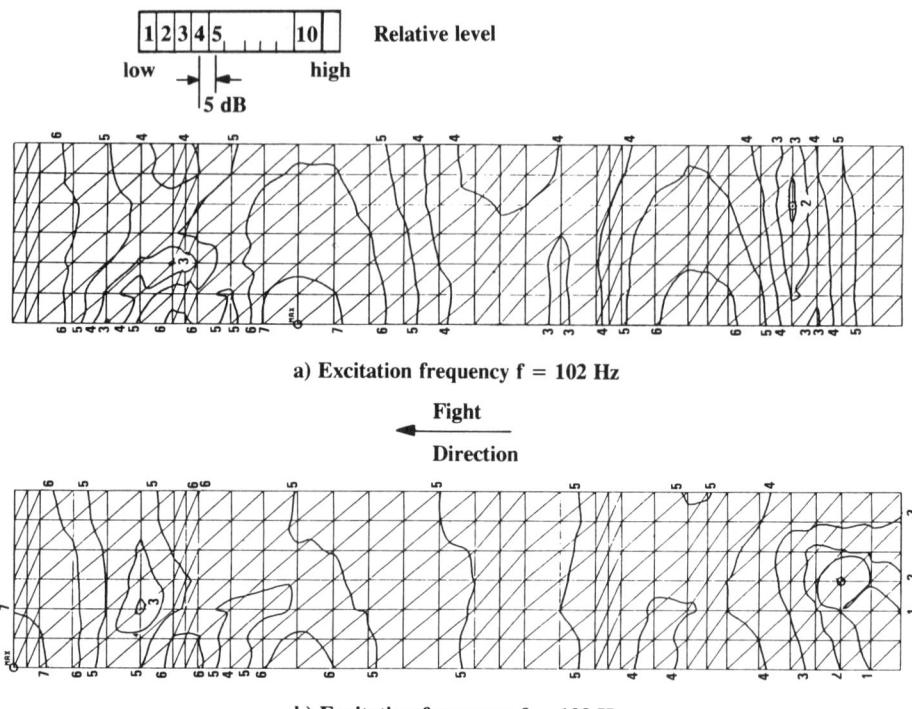

a) Excitation frequency f = 102 Hz

**Fight**
◄━━━━━━
**Direction**

b) Excitation frequency f = 103 Hz

**Figure 1.20**   Calculated noise distribution in the plane of the passengers' heads for the Dornier 228

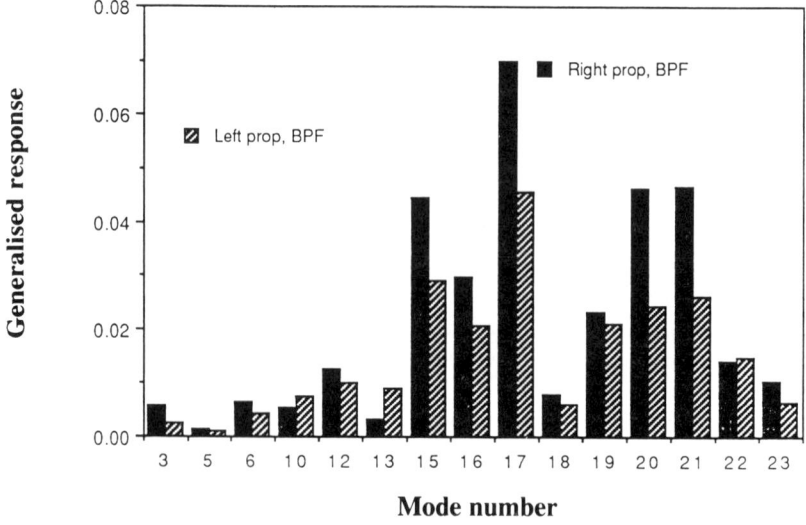

**Figure 1.21**   Generalised response diagram of the Saab 340 for left and right propeller (1 × BPF)

behaviour. Figure 1.22 confirms that the acoustic response at $1 \times BPF$ is clearly one of the side–side type; i.e. high pressure at the sidewalls, with a tendency to a decaying pressure towards the rear of the cabin. The related structural response of the frames in the propeller plane is a combined A4/S4 response in the real part and a pure A4 response (antisymmetric) in the imaginary part (see Figure 1.23).

### Finite-element simulation of active noise control

In this task, active control was applied to theoretical finite-element models of an aircraft structure and a double panel. Based on evaluations of the calculated primary fields, secondary source configurations were devised and their amplitude and phase settings calculated using standard optimisation procedures.

The approach taken was based on an assumed modal behaviour; i.e. a system with low, proportional damping subjected to low-frequency excitation.

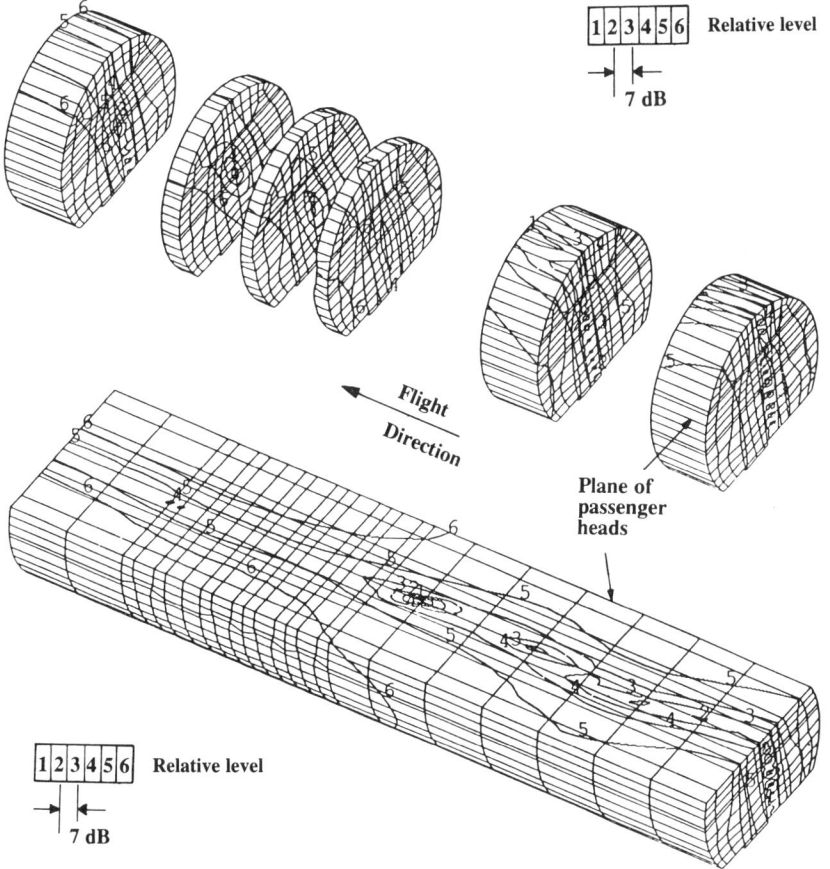

**Figure 1.22** Predicted acoustic part of coupled response of the Saab 340 for $1 \times BPF$ due to right propeller

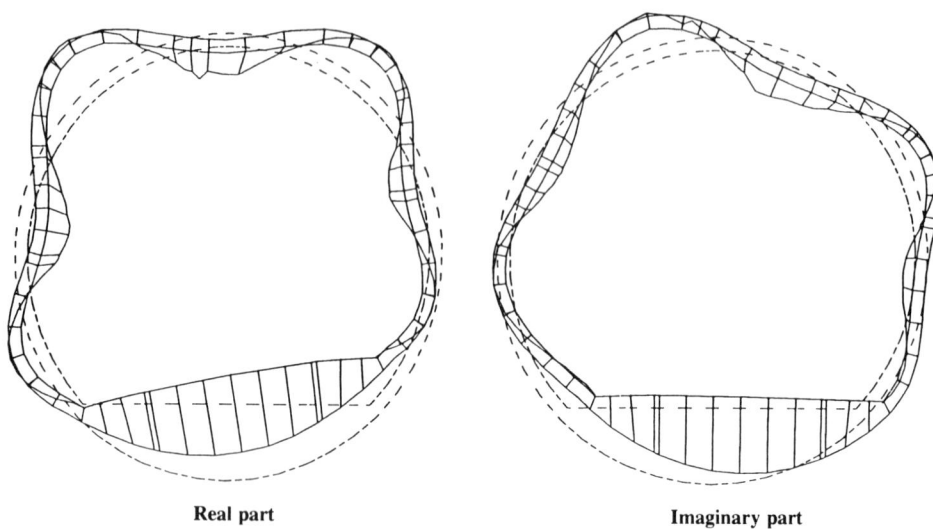

Real part                                    Imaginary part

**Figure 1.23**  Calculated structural part of coupled response of the Saab 340 for 1 × BPF due to right propeller

The response was determined by calculating the coupled modal frequency response functions.

For the aircraft simulation the goal was to reduce the noise levels in the cabin. The way of achieving this was to reduce the modal response in the most dominating modes, by devising secondary source configurations which matched the global pressure patterns of the modes. The secondary sources, in the form of loudspeakers, were applied at the relevant locations inside the aircraft cabin.

For the double-panel system the goal was to increase the transmission loss by application of secondary sources in the cavity between the two panels. Tests were performed with microphones both inside and outside the cavity.

**Control of propeller-aircraft interior noise field**   A total of 255 eigenmodes were calculated for the coupled structure fluid system. Using the information about which modes contribute most to noise levels in the cabin, the six most important modes were selected as candidates for the noise reduction simulation. A loudspeaker was then placed at the pronounced local maxima and minima for each mode, which gave the following configuration:

| Mode | 15 | 16 | 17 | 19 | 20 | 21 |
|---|---|---|---|---|---|---|
| Number of sources | 2 | 2 | 4 | 4 | 6 | 4 |

For the simulation, 22 loudspeakers were thus used as secondary sources. The responses from the loudspeakers (with unit input) belonging to a specific mode were then summed to create a source function, resulting in six source functions, which are the unknowns in the system.

The same theoretically based transfer function matrix as for the primary sources was used to calculate the generalised response referring to the unit secondary sources. An optimisation problem was then solved as described above and the resulting optimum secondary responses for the different modes were obtained.

The pressure fields before and after secondary sources were applied are shown in Figures 1.24 and 1.25, with the difference shown in Figure 1.26. The average pressure reduction obtained in this case was around 4–5 dB, starting from average levels around 65–69 dB, for a syncrophasing angle (phase angle between the two propellers) of +30 degrees. Inspection of the pictures of the pressure fields before and after secondary sources were applied shows that the reductions were global.

Of particular interest also is the influence of the syncrophasing angle, which has a significant impact on the theoretically calculated average sound pressure inside the aircraft using all (about 16 000) points in the finite-element model (Figure 1.27). As may be seen, the synchrophasing angle strongly influences the achievable minimum pressure, and the optimum angle setting will also be slightly different with and without the secondary sources.

It should be noted that the reason why a global noise reduction may be achieved by using just six parameters is that the response from the primary field is dominated by only a few modes.

**Control to improve the transmission loss of a double panel**   For the double-panel simulation, secondary sources in the form of two loudspeakers were

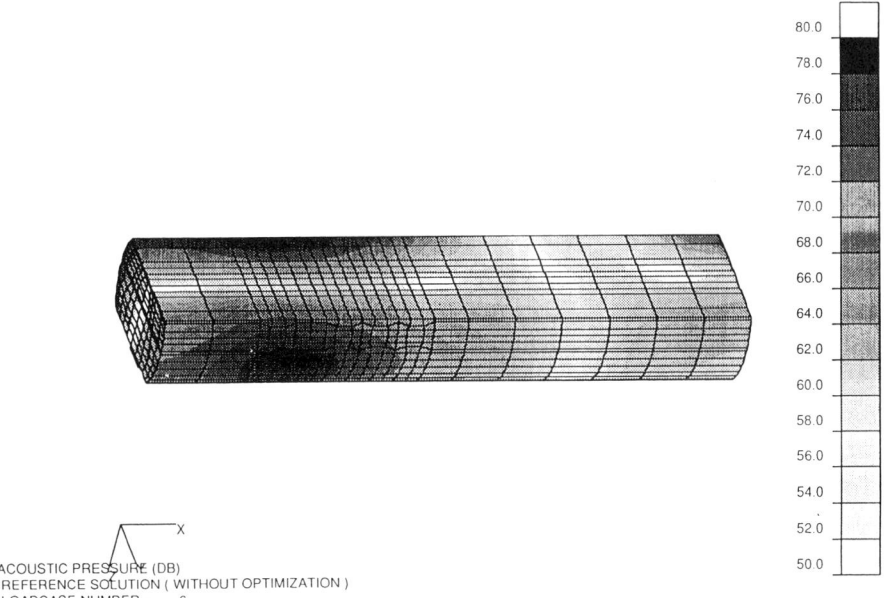

ACOUSTIC PRESSURE (DB)
REFERENCE SOLUTION ( WITHOUT OPTIMIZATION )
LOADCASE NUMBER =    6

**Figure 1.24**  Predicted pressure field *without* secondary sources

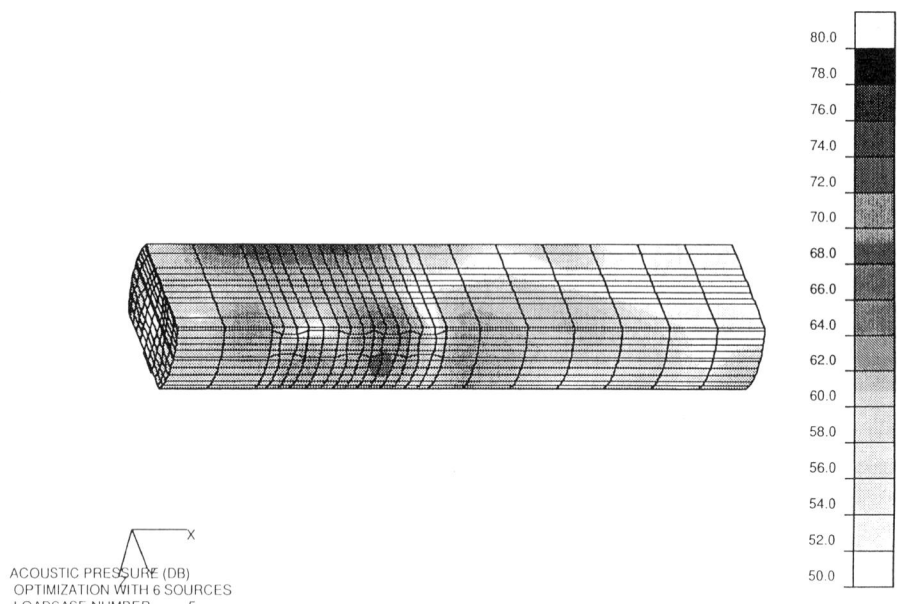

ACOUSTIC PRESSURE (DB)
OPTIMIZATION WITH 6 SOURCES
LOADCASE NUMBER =    5

**Figure 1.25**   Predicted pressure field *with* secondary sources

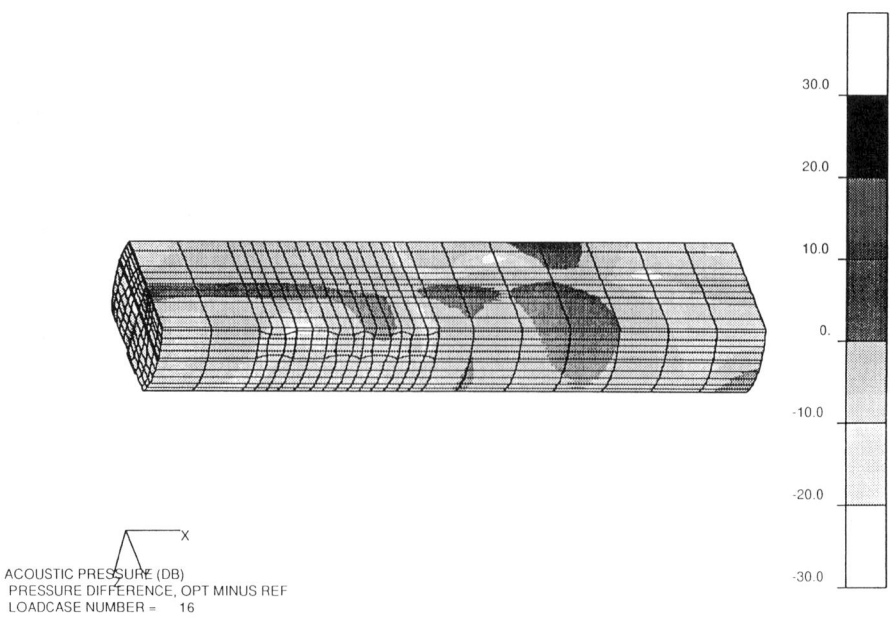

ACOUSTIC PRESSURE (DB)
PRESSURE DIFFERENCE, OPT MINUS REF
LOADCASE NUMBER =    16

**Figure 1.26**   Calculated noise reduction due to secondary sources

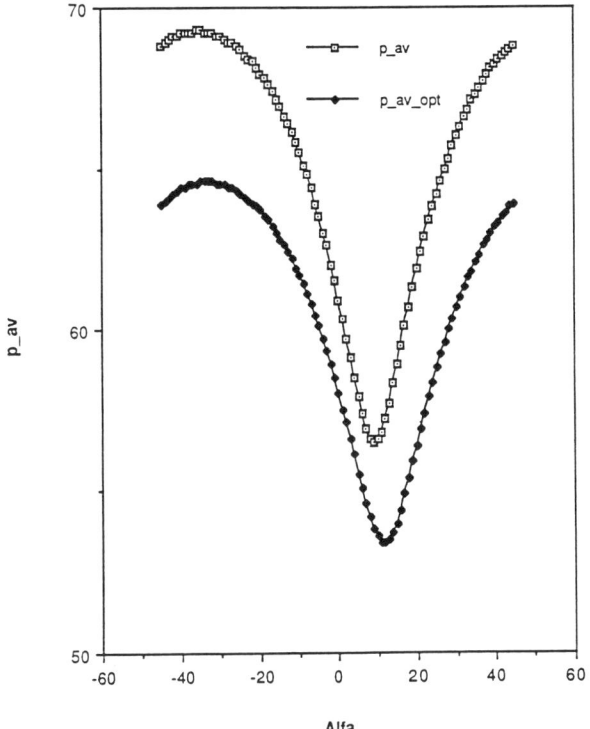

**Figure 1.27** Influence of the phase angle between the two propellers on the interior noise level

placed between the panels. The criterion used for the optimum active noise-control configuration was then the one giving the highest overall sound pressure level reduction above the test box.

A particularly interesting part of the investigation was to find the best location for the control microphones. In the present study two different cases were investigated; i.e. inside the cavity and above the plates. It was found that placing the control microphones above the plates was much the most efficient in terms of achieving the highest reduction in transmitted sound pressure level. For the best configuration, an average of 12 dB was achieved as measured over all the available control points (Figures 1.28 and 1.29). An interesting observation was that the sound pressure level inside the double panel was substantially higher with the control microphones above the plates.

*Conclusions*

During the program all theoretical work was performed as planned. The results provided important detailed information on aircraft interior noise and related active noise control. The information was used to a large extent to define and

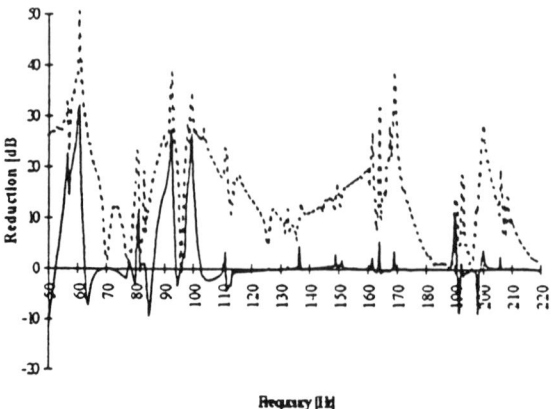

**Figure 1.28**   Attenuation curves with control microphones between the plates

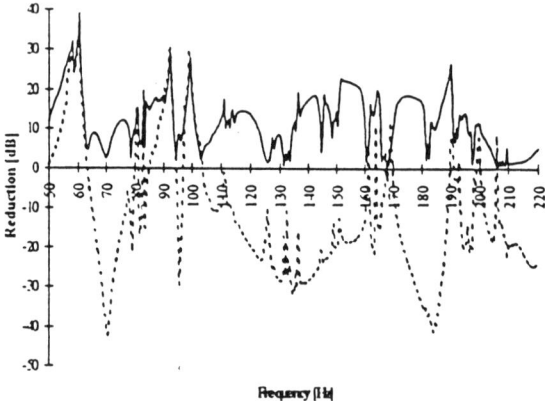

**Figure 1.29**   Attenuation curves with control microphones above the upper plate

update the control system development and support interpretation of test data. The work, documented in 27 reports, provides an excellent basis for more extensive studies in the future.

### 1.3.4   *Control system development*

In this section, work on development of the all-important digital control-system units is described, which was successively performed in *Tasks 6, 7* and *8*. In particular, information on the selected general development approach is given, and the adaptive control algorithms are explained. The various control-unit architectures are described, and some selected controller-related results achieved during initial ground and flight tests are presented. More detailed results of laboratory and flight testing with the control units are given in Section 1.3.7.

*Development approach*

Since one of the main objectives of the project was to demonstrate the potential of active noise control under realistic flight conditions, the design and realisation of suitable digital control units was certainly one of the most critical study tasks. A large number of input and output channels is needed, so the required computation power is important and multiprocessing is definitely required. Architecture design was thus an important part of controller development.

During the first development phase, feasible control algorithms were selected based upon the ground and in-flight measurements and later computer simulations. The control algorithm is responsible for continuously adapting the signals sent to the secondary sources in order to minimise the average sound level in the cabin; i.e. the global energy measured by the selected set of sensor microphones. In order to obtain broad information, control algorithms were implemented in the frequency domain and in the time domain and were tested in detail.

The second development phase started after the control algorithms had been defined, and included the control-unit architecture design. Two different approaches were tried, so two different control units had to be built. This was done primarily by Matra and TNO.

In the third and final development phase, the control units were assembled and integrated with the selected amplifiers, actuator loudspeakers and sensor microphones. Laboratory and flight tests with these systems were then performed.

*Time-domain algorithm with offline system identification*

**Identification part** The purpose of offline identification is to evaluate the transfer functions between the $N$ actuators and the $M$ sensors. Each transfer function is modelled by an FIR filter $B_{ij}$, where $i = 1, \ldots, N$ and $j = 1, \ldots, M$ are the sensor and actuator indexes, respectively. This identification is realised on the ground without primary noise.

The identification process is done sequentially, source-by-source. At the $i$th step, the reference signal $X$ generated by the controller is sent to the $i$th actuator. The filter coefficients for each $B_{ij}$ are optimized in order to match the paths between the $M$ sensors and the $i$th actuator (see Figure 1.30). The corresponding recursive equation is:

$$B_{ij}^{(n+1)} = B_{ij}^{(n)} + 2\mu\varepsilon_{ij}^{(n)}X^{(n)}$$

where

$X^{(n)}$ = the state vector including $Bd$ last samples of the reference signals ($Bd$ is the depth of the $B$ filter)

$\varepsilon_{ij}$ = difference between the model and the physical signals; i.e. $\varepsilon_{ij} = d_{ij} - y_{ij}$

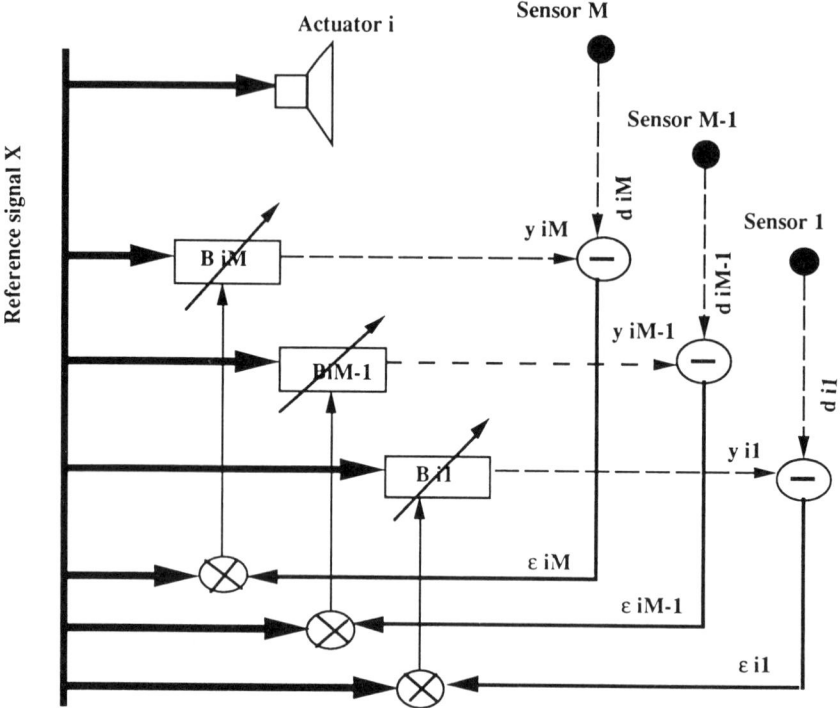

**Figure 1.30**　Block diagram of the time-domain identification algorithm

$d_{ij}$ = signal from sensor $j$ excited by the $i$th actuator

$y_{ij}$ = output of the FIR filter $B_{ij}$

$\mu$ = convergence factor.

**Control part**　One reference signal which contains all the harmonic components to be controlled is acquired and filtered by $B_{i,j}$, the impulse response of
actuator $i$ on sensor $j$. At each time step, filters $A_i$ are updated using outputs of
$B_{i,j}$ filters and outputs of sensors. Afterwards, the reference signal is filtered by
$A_i$ to generate the command for the $i$th actuator. The functional scheme is
shown in Figure 1.31.

At each step of the algorithm, one must compute:

$$A_{ik}(n) = A_{ik}(n-1) - \mu_i \sum_{j=0}^{N-1} e_j(n) \sum_{m=0}^{Bd-1} B_{ijm} \cdot x(n-m-k)$$

where

$A_{ik}(n)$ = coefficient $k$ of actuator filter $i$ at time $n$

$e_j(n)$ = output of sensor $j$ at time $n$

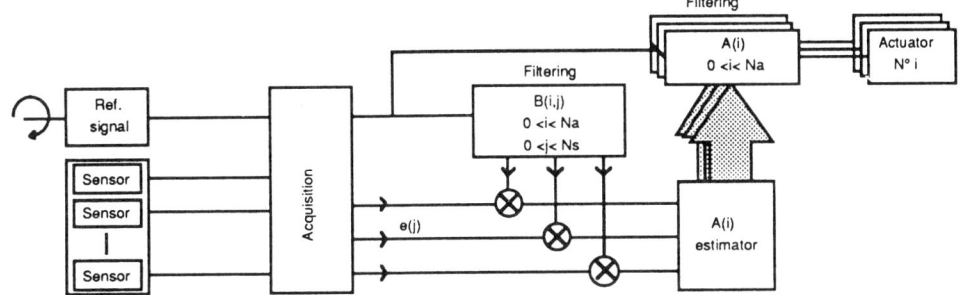

**Figure 1.31** Functional scheme of the time-domain algorithm

$Bd$ = depth of filters $B_{ij}$

$\mu_i$ = convergence factor for actuator $i$

$B_{ijm}$ = $m$th coefficient of filter $B_{ij}$

$x(n)$ = $n$th reference signal sample.

## Time-domain algorithm with online system identification

**Identification part** The purpose of online identification is to evaluate the transfer functions between the $N$ actuators and the $M$ sensors while the primary sound field is present; that means identification during flight with engines running. Each transfer function is modelled by an FIR filter $h_{ij}$.

The identification process is done sequentially, first for the primary noise field, thereafter source-by-source. For the sources, a harmonic reference signal is sent to the respective source. The filter coefficients for each $h_{ij}$ are estimated using the RLS algorithm:

$$t(n) = \lambda^{-1} P(n-1)\Phi(n)$$

$$k(n) = [1 + \Phi^H(n)t(n)]^{-1}t(n)$$

$$\alpha[n] = \alpha[n] - \theta^H(n-1)\Phi(n)$$

$$\theta(n) = \theta(n-1) + k(n)\alpha(n)$$

$$P(n) = \lambda^{-1} P(n-1) - k(n)t^H(n)$$

where

$$M = \text{number of FIR model coefficients}$$

$$\theta(n) = [\hat{h}_{1,j}^T(n)\hat{h}_{2,j}^T(n) \ldots \hat{h}_{i,j}^T(n)\hat{h}_j^{pT}(n)]^T \text{ vector } 33M \times 1$$

$$\Phi(n) = [s_1^T(n)s_2^T(n) \ldots s_i^T(n)x^T(n)]^T \text{ vector 33M} \times 1$$

$$\alpha[n] = \theta_j[n] - \sum_{j=0}^{L-1} \theta(1)\Phi(n-1) \quad (L = 33M)$$

$$t(n) = \text{helpvector } 33M \times 1$$

$$P(n) = \text{matrix } 33M \times 33M$$

$$\lambda = \text{forgetting factor } (0 < \lambda \leqslant 1).$$

It may be noted that an LMS algorithm was used with the offline system identification. An RLS algorithm can make the identification more accurately using fewer data samples (shorter test time required), at the cost of considerably more calculations. This procedure can also be used for offline operation (the primary sound field will be estimated as zero).

**Control part**    This is identical to the control part described for offline system identification, using the same kind of LMS algorithm.

### Frequency-domain algorithm

This algorithm is recursive. It deals with measured data (sound pressure or vibration level on each sensor for the frequency concerned), commands (voltage applied to each actuator) and estimated data. The purpose of the algorithm is to identify both the matrix of influence between actuators and sensors for each frequency and the primary sound field in the cabin. The functional scheme is given in Figure 1.32.

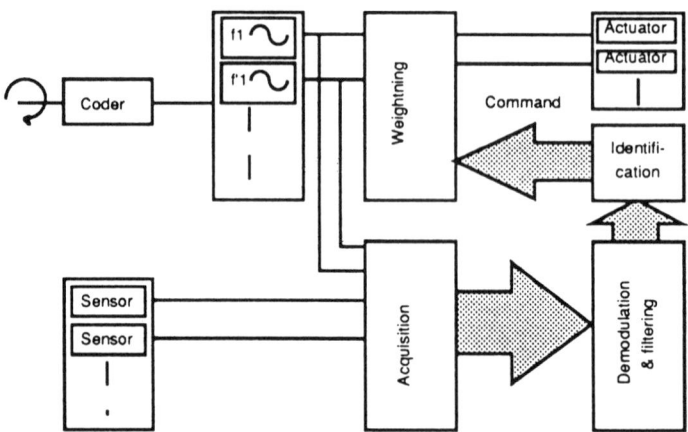

**Figure 1.32**    Functional scheme of the frequency domain algorithm

To describe the algorithm the following convention may be used:

$$Ns = \text{number of sensors}$$

$$Na = \text{number of actuators } (Na \leqslant Ns)$$

$$Fc = \text{controlled frequency}$$

$$A[2.Ns, 2.(Na + 1)] = \text{normalised transfer matrix at } Fc \ (A = [HE])$$

$$Y[2.Ns] = \text{demodulated values of sensors at } Fc$$

$$X[2.(Na + 1)] = \text{normalised vector of applied commands at } Fc \ (X = [F, 1, 0])$$

$$E[2.Ns] = \text{primary sound field vector at } Fc$$

$$F[2.Na] = \text{vector of applied commands at } Fc$$

$$P[2.(Na + 1), 2.(Na + 1)] = \text{matrix of least-squares computation}$$

$$G[2.(Na + 1)] = \text{gain vector}$$

$$H[2.Ns, 2.Na] = \text{influence matrix at } Fc$$

$$\gamma = \text{factor such as } \gamma^n = 0.5, \text{ where } n \text{ is the forgetting factor } (\gamma \leqslant 1).$$

The control algorithm proceeds in an iterative way, in which estimated data are evaluated at each control step by using the following formula for each of the controlled frequencies:

$Y$ is measured by demodulation of sensor outputs at $Fc$

. . . . . . . . . . . . . . . . . . . . . . . . . . . *Identification with real and imaginary parts*

$$G^+ = P.X/(X^T P.X + \gamma)$$

$$P^+ = (P.G^+.(P.X)^T)/\gamma$$

$$A^+ = A + (Y - A.X).G^{+T} \ \dots\dots\dots\dots\dots\dots\dots\dots\dots\dots (A = [HE])$$

. . . . . . . . . . . *New command evaluation*

. . . . . . . . . . . ($E^+$ is $A^+$ before the last column; $H^+$ is $2.Na$ first $A^+$ columns)

$$F^+ = -(H^{+T}.H^+)^{-1}.H^{+T}.E^+$$

$$X^+ = [F^+, 1, 0]$$

At the first algorithmic step, variables are initialised as follows (the $p$ value is given by the user):

$$F^0 = [0, 0, \ldots, \ldots, \ldots, 0]$$

$$X^0 = [0, 0, \ldots, \ldots, \ldots, 0, 1, 0]$$

$$A^0 = [1, 0, \ldots, 0, 0, 1, \ldots, 0, \ldots, \ldots, \ldots, \ldots, 0, 0, \ldots, 1, 1, 0, \ldots, 0, 0, 1,$$
$$\ldots, 0]$$

$$P^0 = [p, 0, \ldots, 0, 0, p, \ldots, 0, \ldots, \ldots, \ldots, \ldots, 0, 0, \ldots, p]$$

### Control-unit architecture

**The Matra approach**    The control unit developed by Matra has 32 output and 48 input channels and is based on a VME bus architecture. The selection of a standard bus and the maximum use of commercial boards allow easy control-unit integration (for instance, debugging tools are available on the market) and low development cost. Indeed, low cost and short development times were critical issues in the feasibility demonstration. A general sketch of the controller is given in Figure 1.33.

The control unit uses asynchronous sampling for the adaptive FIR filters (only the waveform synthesis is synchronous with the noise fundamental

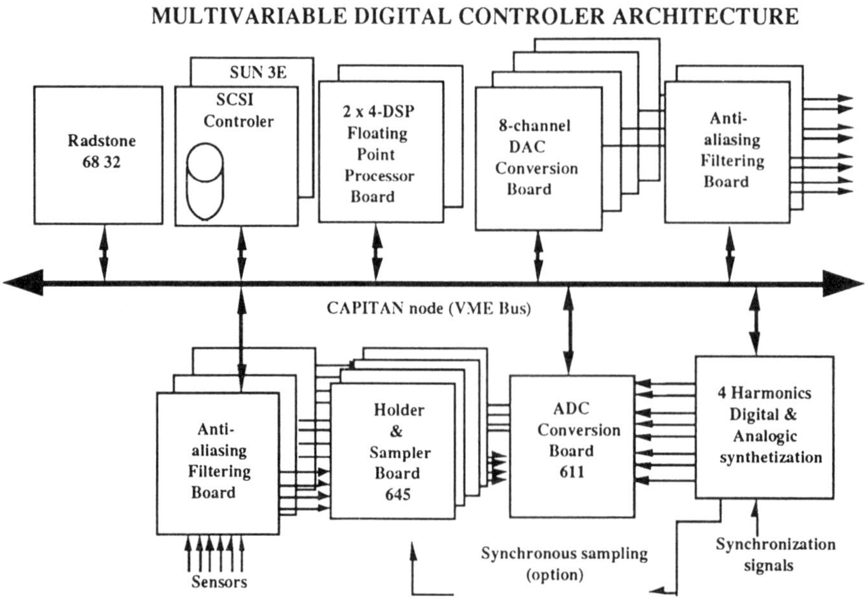

MULTIVARIABLE DIGITAL CONTROLER ARCHITECTURE

**Figure 1.33**    Multivariable digital controller architecture of the Matra control unit

period). With the time-domain algorithm, this feature leads to a significant increase in the algorithm speed and robustness.

**The TNO approach**   The TNO control unit also had 32 output and 48 input channels. The architecture of the control unit was matched with the selected algorithm. The architecture allows for later VLSI (very large scale integration), while in this project standard components were used. All processors in the control unit were TMS 320 C30 DSPs (digital signal processors) for which extensive development and debugging tools are available. This made it possible for a complete system to be designed and realised from a component level up within the time-frame of the project. Although standard components were used, the design philosophy also allowed for smaller dimensions, lower weight and lower power consumption.

A general sketch of the control unit is given in Figure 1.34. It uses propeller-speed synchronous sampling throughout the whole system. Although the speed of the algorithm thus becomes coupled to the sampling speed, the intrinsic higher frequency resolution leads to a significant increase of convergence speed (time to reach iteratively the optimal setting; or, in other words, better tracking performance).

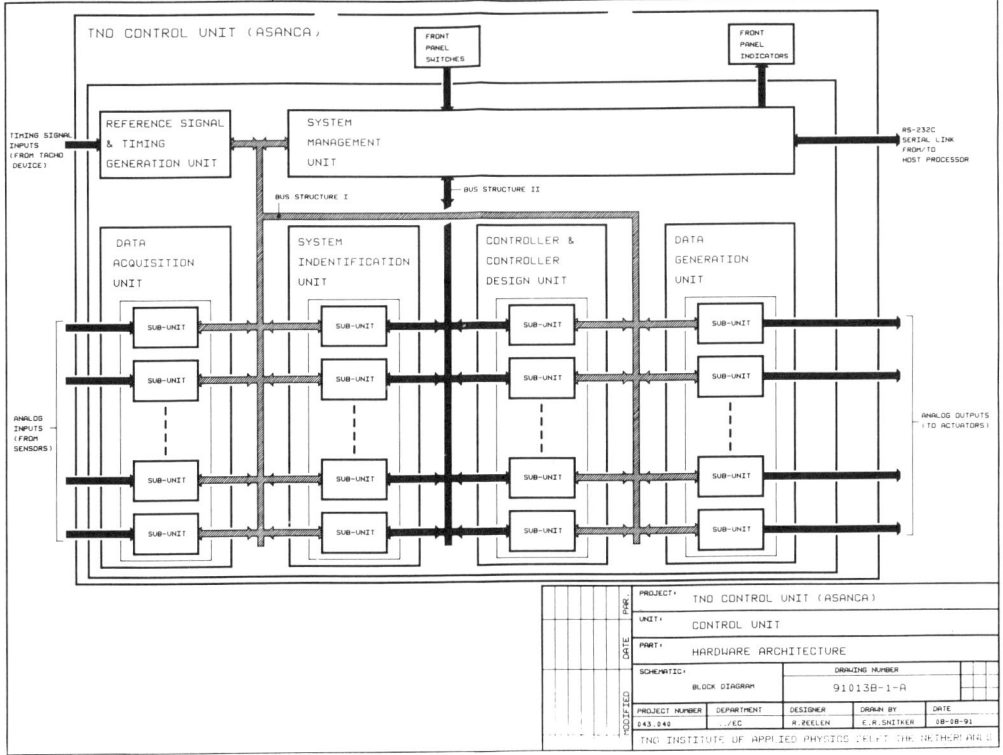

**Figure 1.34**   General sketch of the TNO control unit

*Selected results*

After extensive laboratory testing, both control units were tested in a Dornier 228, on the ground and in flight. The test configuration consisted of 32 loudspeakers distributed along the walls, floor and top panels, and 48 microphones located in the plane of the passengers' heads.

For the ground tests, an artificial source (a multi-pole loudspeaker) was used inside the aircraft to generate the primary sound field. The generated harmonics and their strengths were, as far as possible, the same as for the real propeller. A typical result of these tests corresponding to average levels of all microphones is given in Table 1.3.

During the flight tests, an attempt was made to control the first three harmonics of the propeller noise. For the Matra control unit this worked very well. However, for the TNO unit the control for the third harmonic failed: the offline system identification on the ground for this third harmonic proved to be wrong during flight, causing the noise level to increase some 8 dB at this harmonic. An important conclusion from this is that, for higher harmonics, the identification on the ground might not be a good enough estimate for the actual system in flight. An attempt to perform an online identification in flight was not successful owing to a combination of poor signal-to-noise ratio and a lack of time to adjust the signals during flight. A last measurement was carried out with the control of the third harmonic turned off. Typical results of these tests, corresponding again to average levels of all microphones, are given in Table 1.4. For the Matra control unit, similarly good results were obtained (also at the third harmonic) which are explained in more details in Section 1.3.7.

Table 1.3   Result of a ground test of the TNO control unit in the Dornier 228

|          | 102 Hz   | 204 Hz   | 306 Hz  | 408 Hz     | 0–500 Hz     |
|----------|----------|----------|---------|------------|--------------|
| Primary  | 94.6 dB  | 78.5 dB  | 69.0 dB | 67.3 dB    | Not measured |
| Residual | 71.4 dB  | 64.7 dB  | 59.2 dB | No control | Not measured |
| Reduction| −23.2 dB | −13.8 dB | −9.8 dB | No control | Not measured |

Table 1.4   Selected results of flight tests of the TNO control unit in the Dornier 228

|          | 102 Hz   | 204 Hz  | 306 Hz     | 408 Hz     | 0–500 Hz     |
|----------|----------|---------|------------|------------|--------------|
| Primary  | 105.1 dB | 87.0 dB | 82.0 dB    | 79.1 dB    | Not measured |
| Residual | 90.5 dB  | 78.8 dB | 90.4 dB    | 78.5 dB    | Not measured |
| Reduction| −14.6 dB | −8.2 dB | +8.4 dB    | No control | ~ 12 dB      |
| Primary  | 105.1 dB | 87.0 dB | 82.0 dB    | 79.1 dB    | Not measured |
| Residual | 91.5 dB  | 80.9 dB | 82.1 dB    | 79.3 dB    | Not measured |
| Reduction| −13.8 dB | −6.1 dB | No control | No control | Not measured |

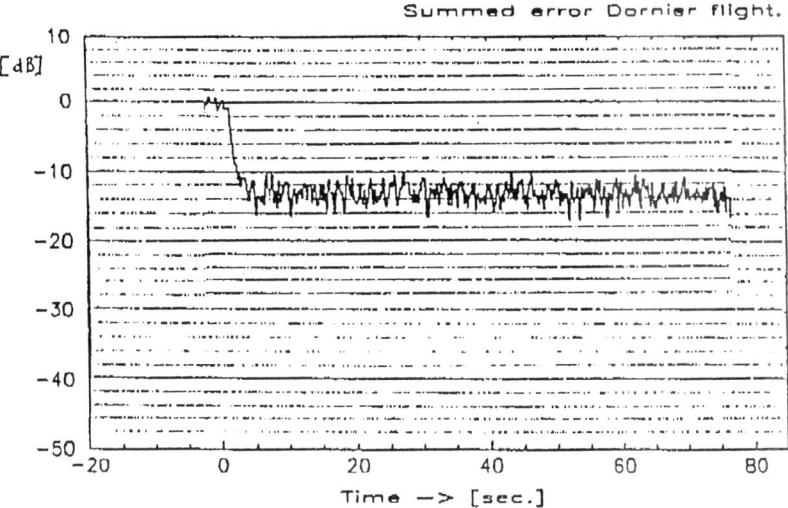

**Figure 1.35**  Dynamic behaviour of the TNO control unit during Dornier 228 tests

In addition to the steady-state performance, the dynamic behaviour of the control units was measured. A typical example, corresponding to the first flight test example of Table 1.4, given in Figure 1.35. At $t = 0$ the unit started its control. As can be seen from the plot, optimum control leaving just background noise is reached within 3–4 seconds.

Finally it may be noted that, for slowly varying conditions, the overall 0–500 Hz performance hardly changed. Tested variations included, for example, changing r.p.m. (3%), air-conditioning on/off, climb and descent, and sharp turns.

*Conclusions*

Both the control units met the project objectives. Testing showed the feasibility of active noise control in a real flight test under stable cruise conditions. In addition, the control units performed well under slowly varying conditions. Finally, the control units have many facilities for future experiments and on-site analysis. The work on *Tasks 6, 7 and 8*, documented in 31 reports, provides an excellent basis for future research.

### 1.3.5  An alternative approach

As mentioned earlier, part of the work of *Task 4* was exploratory investigations of an alternative noise control approach. These are reported in this section. The alternative approach aimed at an improvement of the insertion loss (IL) of the aircraft shell by means of small acoustic secondary sources placed in air gaps between fuselage and trim panels. The approach differs from previous ones [1] in that it uses acoustic inputs as the means of control instead of vibration inputs. The basic idea is straightforward: the IL of a double-panel partition can be considerably improved if the sound field in the air gap of the partition, assumed to be the main coupling element between the two panels, can be successfully cancelled by means of appropriately controlled secondary sources.

In order to investigate this idea in greater detail, the following activities were performed during the course of the project:

• Construction of a dedicated test box (ASANCA box).

• Analytical and numerical calculations.

• Laboratory experiments.

### Construction of a dedicated test box (ASANCA box)

The laboratory test setup was called an ASANCA box and is shown in Figure 1.36.

The main structure was a box with a triple wall insulation. The inner and middle boxes were made with a special heavyweight double-layer gypsum board material, fixed to a wooden frame. The outer one was fabricated from steel, covered with a viscous damping material. The specific masses of the layers and the air gaps between them were dimensioned so that the expected resonance frequencies were below 50 Hz. In order to avoid reflections and to increase damping between the layers, sound absorbing material of 50 mm thick mineral wool was installed. All these measures were taken to make sure that sound inside the box could not transmit to the outside through the side walls

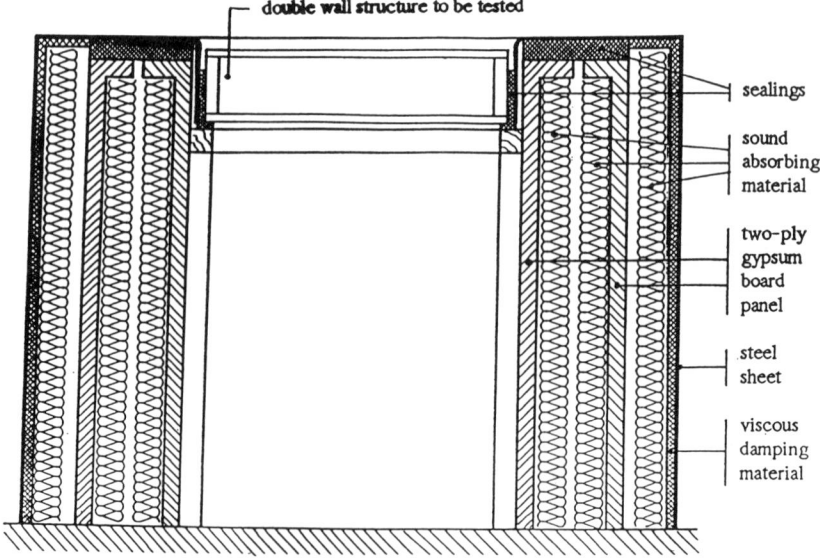

**Figure 1.36**   Cross-section of the ASANCA test box

but only through the upper opening of the box, where the double-panel object to be tested was mounted.

The object was a simplified double-panel partition, formed by two plane and parallel aluminium plates of 1.5 mm thickness clamped to a 10 mm thick and 150 mm high rectangular steel framework. The free dimensions of the plates were 1140 × 730 mm. The partition was instrumented by an array of 42 microphones in the medium plane of its cavity as well as 36 accelerometers on both plates. The microphone spacing was 120 × 127 mm in a 7 × 6 mesh, aimed at mapping the sound pressure distributions in the cavity as closely as possible. The accelerometers were placed in a 6 × 6 mesh with 60 × 60 mm spacing in the central part of the plates to identify the vibration mode shapes. Two small closed-box type commercial loudspeakers were fixed resiliently inside the cavity as the secondary sources.

## Analytical calculations

To gain insight into the physical phenomena underlying the behaviour of a coupled vibroacoustic system controlled by active noise control methods, a theoretical analysis of the double-wall configuration was performed. The method chosen was based on the modal expansion method [2] as interpreted by Fahy [3]. The detailed analysis of the testing object and its results have been described by Sas and colleagues. [4, 5]. The results obtained from the analysis are listed below:

1. Within the frequency range of interest (60–220 Hz), when subject to an acoustic excitation at the incident side of the partition, the sound field inside the cavity of the partition is often composed of only a few uncoupled modes, in several cases dominated by just a single one.

2. In order to control the sound field effectively inside the cavity of the partition, and hence to control sound transmission, the number of secondary sources should be at least equal to the number of coupled acoustic modes of the cavity.

3. For a given number of control sources, one can anticipate a better performance from an active noise control system when the forced primary field is dominated by a few uncoupled modes of the cavity only. On the other hand, only a poor sound attenuation can be achieved at those frequencies where the coupled sound field in the cavity is composed of a large number of uncoupled modes.

## Numerical calculations

The rapid development of ever more powerful and yet cheaper computers has made numerical methods available as an attractive engineering approach.

Finite-element (FE) and boundary-element (BE) codes have already proven their usefulness in a number of applications. Also in vibroacoustic problems, numerical methods become more and more popular in a design environment where the prediction of acoustic behaviour should be taken into account in the early design stage. Therefore, a combined FE–BE model was developed to predict the coupled structural acoustic behaviour of the ASANCA test box. This numerical model was used to optimise the alternative approach.

The numerical model was based on the program SYSNOISE. In the calculation procedure, the structure was modelled with finite elements while the fluid was modelled with boundary elements, following an indirect formulation with a variational approach. The displacements of the structure were projected on a previously calculated modal base of uncoupled structural mode shapes, resulting in a smaller system of equations. With this model, the forced-response calculations were performed at all frequencies of interest. It was found that the numerical calculations agreed quite well with the experimental results with respect to the vibroacoustic behaviour of the partition. The calculations predicted a considerable improvement of IL of the partition by the alternative approach, which was confirmed by the experiments. Moreover, a useful guideline for the optimal active noise control configurations was also provided by the calculations.

## The applied active noise control system

A multichannel digital active noise controller was used in the experimental investigation of the alternative approach. The control strategy employed was feedforward adaptive control or, more specifically, adaptive filtering. The working principle of the controller can be described as follows. Firstly the incident noise is detected. This measured signal is usually called the 'reference input' in the literature and is fed to the controller. The controller itself is formed by a vector of adaptive filters, where the reference input is processed to become the so-called 'anti-noise signal' sent to the secondary loudspeakers. Transformation from the reference signals to the anti-noise signals is accomplished by appropriate changes in the strengths and phases of the former signals so that, when the latter signals are sent out by the secondary speakers, they interfere destructively with the incident noise and result in noise attenuation. Those changes in strengths and phases are realised by means of proper choice of the coefficients of the adaptive filters. To be able to adjust the coefficients to their right values automatically, a time-domain least-mean-square (LMS) type algorithm was developed [6].

In any kind of adaptive algorithm, a performance index has to be optimised. Such an index should meet at least two essential requirements. First, when the index is optimised, the objective of the controller, which is to increase the insertion loss of the partition, should also be achieved. Secondly, the index should be a function of the coefficients of the adaptive filters so that the index can be optimised by means of changing those coefficients. In the LMS-type algorithm used, the sum of the mean square values of the sound-pressure time series, measured at the control points by the so-called 'error microphones', is chosen to be the index. This index is found to be a quadratic function of the coefficients of the adaptive filters. Thus, a very simple gradient-descent method is employed to derive iteratively the optimal coefficient set.

The controller described above was implemented on a PC-based commercially available digital signal processing board equipped with a powerful TMS320C30 floating-point signal processor. The control algorithm was programmed in C code.

*Experimental results*

To verify the conclusions drawn from the theoretical analysis and the numerical calculations of the alternative approach, and more important to demonstrate the potential of the proposed approach, experiments were conducted with the ASANCA box. The incident noise was provided by means of a cube of six loudspeakers, placed inside the test box. The testing partition was mounted at the upper opening of the box. To avoid airborne and/or structure-borne flanking transmission as far as possible, the partition was resiliently supported and carefully sealed as described before. For most experiments a two-speaker/two-microphone configuration was used. A schematic of the test setup with the ANC system is shown in Figure 1.37.

The performance of the active noise control system was evaluated by using stepped sinusoidal excitation in the frequency range 55–220 Hz. Frequency response functions (FRF) of the primary excitation signal and the sound pressures at six microphone positions 20 cm above the testing partition were measured. The FRFs were then averaged and compared with and without the

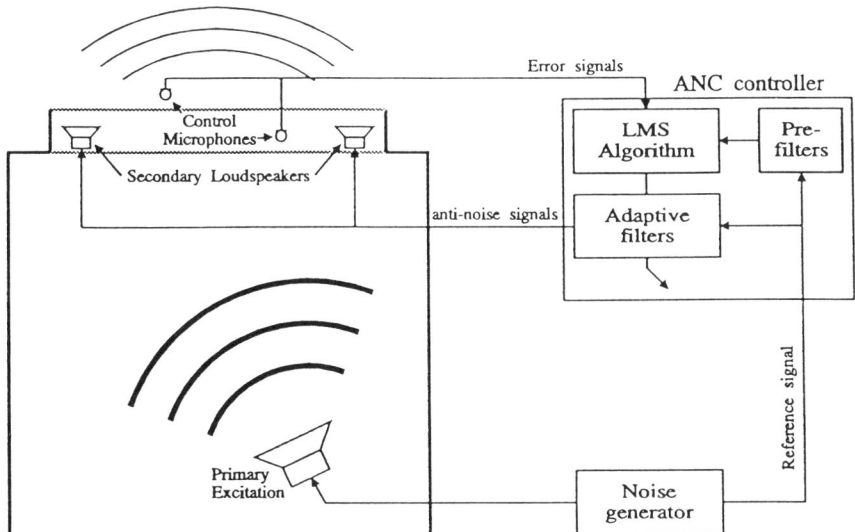

**Figure 1.37**  Scheme of ASANCA box setup with the active noise control system

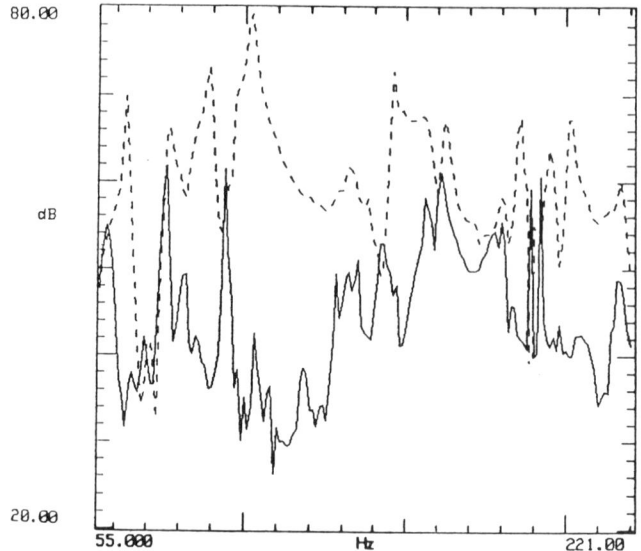

**Figure 1.38** Measured transmitted sound pressures of the partition with (——) and without (– – –) active noise control

active noise control system. Figure 1.38 shows one of the results. It is evident that a substantial noise reduction was achieved at and around all the resonance frequencies. At the resonance frequency of 100 Hz, a reduction as high as 48 dB was realised. At the resonance frequencies of 60–90 Hz, the noise reductions were up to 38 dB. At the other resonance frequencies, at least 10 dB reduction can be seen. The noise reductions were smaller (typically a few dB) at the antiresonance frequencies and in the frequency bands between resonances, and at some points a noise *increase* of several dB can even be observed. However, this normally will not raise any serious problems in practice since the noise levels at those frequencies are already low.

The experimental results also confirmed the predictions from the theoretical analysis and numerical calculations with respect to the effects of the number of secondary sources and their positions, the relation between the performances of the active noise control system and the compositions of the coupled modes, and the effects of the positions of the error microphones inside or above the partition.

*Conclusions*

The investigations showed that the insertion loss of a double-panel partition can be considerably increased by using small loudspeakers as control sources inside the air gap between the two plates.

It was shown analytically and verified experimentally that the general requirement for good performance of an active control system (i.e. having at least as many control sources as dominant modes of the system to be controlled) can be generalised for vibroacoustic systems, too. In this context, 'vibroacoustic' means that the behaviour of the system investigated is noticeably influenced by fluid–structure interaction phenomena between the acoustic and mechanical elements of the system.

The work on this subject was documented in four reports and the results met fully the planned objectives. Further work based on the initial research would be of great interest.

### 1.3.6  Advanced actuators

One of the biggest problems in implementing an active noise control system in an aircraft arises from the size and weight of transducers needed and their integration in the aircraft.

Among these transducers, the sensors for active noise control applications require less development effort than do the actuators. Available and reliable technology makes possible small, light and cheap microphones fulfilling most of the typical requirements. However, on the actuator side (shakers or loudspeakers) the weight of conventional units and the required performances in efficiency and frequency range are still two main obstacles.

Consequently, two partners started initial studies of new lightweight actuators, more or less integrated into the trim panels to save space and weight. Reson System focused its work on the development of loudspeakers using piezoelectric ceramic or electrodynamic driving units. Metravib worked primarily on the development of loudspeakers using a piezoelectric film as the driving unit, improved electrodynamic loudspeakers, and small electromagnetic shakers using a special concept. The results of these studies, which were also part of *Task 4*, are now presented.

#### Development of loudspeakers with piezoelectric actuators

The advantages of using the piezoelectric principle are simplicity and robustness, the high forces possible, and low weight. The major disadvantage is the small displacement. This disadvantage may, however, be overcome by utilising the bimorph principle illustrated in Figure 1.39. Here the piezoelectric material is combined with an elastic material, which amplifies the displacement. Based on this principle, two methods of integrating loudspeakers into the trim panels have been investigated:

- Using a number of small piezoelectric bimorph drive units based on a ceramic material.

- Applying a thin piezoelectric polymer film (PVDF) to the rear face of a trim panel, thus creating a simple and robust loudspeaker with practically no added weight.

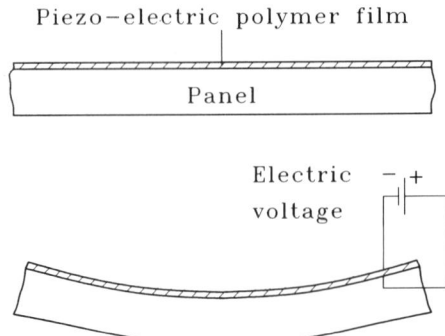

**Figure 1.39**   Explanation of the bimorph principle

In the first case, two different drive units have been tested—a small one with weight 2.4 g (shown in Figure 1.40) and a larger one with weight 9.0 g. The unit excited in parallel a large, stiff, sound-radiating area, which could be the trim or part of it.

The most successful piezoelectric loudspeaker studied used an array of 24 of the smaller piezoelectric drive units. The loudspeaker was integrated into a Dornier 228 ceiling trim panel to be tested. For the loudspeaker, a radiation plate with a sandwich design was used. The weight of the radiating plate was 165 g and the area 0.2 m². The increased depth and weight of the trim were less than 25 mm and 0.5 kg, respectively. The general design of the loud-speaker is shown in Figure 1.41.

For the acoustic tests, the trim panel with built-in loudspeaker was mounted over a 10 cm-deep cavity filled with lightweight mineral wool. This was done in order to simulate the mounting in the aircraft. The measured performance of the loudspeaker was 60–75 dB at 1 m for frequencies between 100 and 300 Hz.

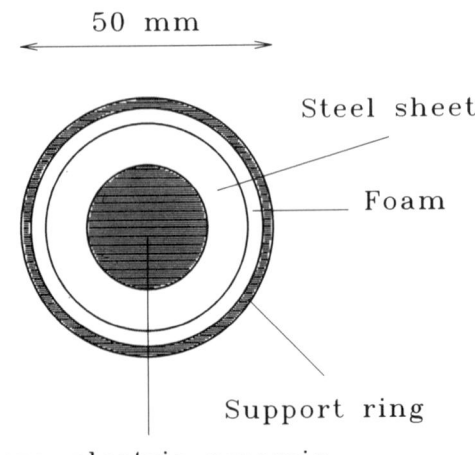

**Figure 1.40**   Design of the miniature piezoelectric loudspeaker drive unit (weight 2.4 g)

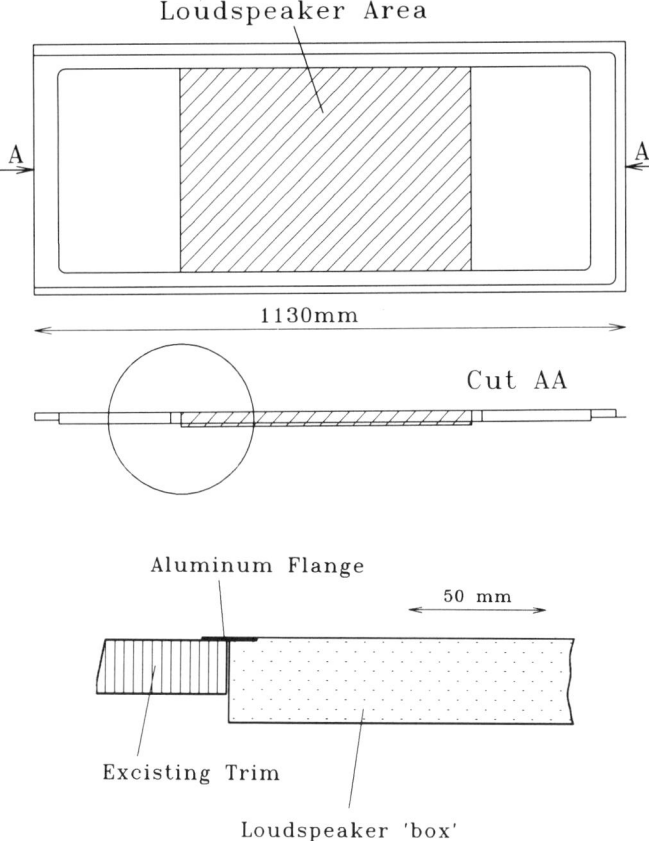

**Figure 1.41** The piezoelectric loudspeaker mounted in a Dornier 228 ceiling trim panel

The second development based on the piezoelectric principle involved piezo-electric polymer film (PVDF) as the driving unit. The related work program was splitted in three main steps:

- Actuator design.
- Choice and dynamic characterisation of a trim panel.
- Integration of multiple actuators in the trim panel.

Figure 1.42 shows the design of an elementary PVDF actuator. The main characteristics of a PVDF film are: length 5.5 cm, thickness 40 $\mu$m and coefficient $d_{31} = 9.5 \times 10^{12}$ pC/N.

Since the force developed by an elementary film is relatively low, a sandwich structure of five PVDF layers was built (the normal load obtained with this layered structure is about 15 N). After selecting a Dornier 228 ceiling trim panel and measuring its dynamic response, locations of actuators to be implemented

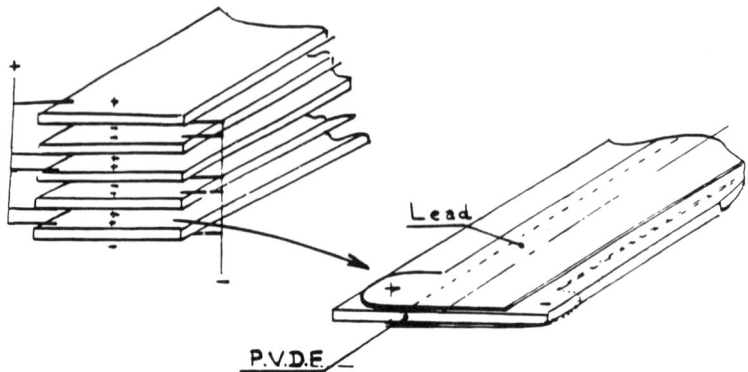

**Figure 1.42**   An elementary PVDF actuator

were determined. Three actuators were then installed on the trim panel with specific fixations to ensure mechanical amplification. Details of the installation are provided in Figure 1.43.

The best performance obtained with such a trim panel when loaded by the acoustic enclosure previously mentioned was around 70 dB at 1 m for 100 Hz and an input voltage of 80 V. As the modal response of the trim panel is used to reinforce the radiated pressure, the pressure levels obtained are very much

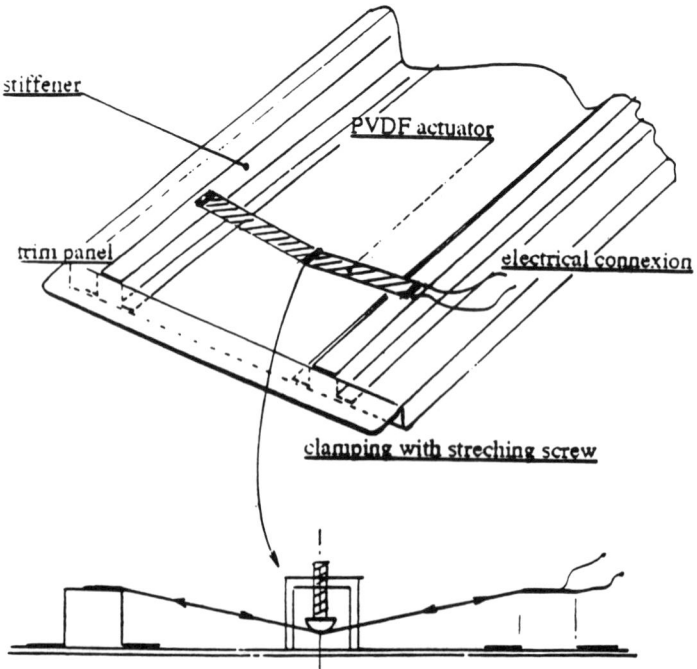

**Figure 1.43**   Installation details of PVDF actuators integrated in a Dornier 228 trim panel

affected by the frequency. On the other hand, the harmonic distortion is thus very small.

These results demonstrate that piezoelectric and PVDF actuators may be used in aircraft trim to give it the functionality of a loudspeaker without a large weight penalty and space requirements. However, the efficiency obtained is still very low and needs further improvement. In addition, the efficiency is connected slightly to the characteristics of the trim panel modal behaviour and to the boundary conditions applied to the trim, as may be seen for example in Figure 1.44.

### Development of loudspeakers with electrodynamic actuators

Conventional electrodynamic loudspeakers able to radiate high pressure levels in the low frequency range ($\leq 500$ Hz) have sizes and weights unacceptable for use in aircraft. However, owing to their sturdiness and our wide knowledge of the basic principle, and to updated manufacturing technology, two possible developments were investigated. The first requires small electrodynamic loudspeaker units acting as motors on a stiff flat plate integrated in the trim panel. The second consists in the improved design of a classic electrodynamic loudspeaker in order to increase its low-frequency efficiency with a minimisation of the diaphragm displacement.

**An integrated system** Instead of having a conventional loudspeaker with one coil and magnet, the new design used a number of smaller coil/magnet

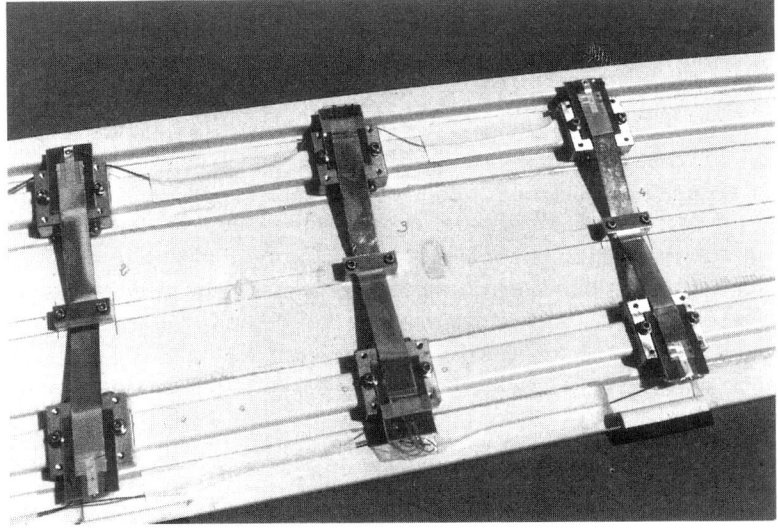

Figure 1.44  Dornier 228 trim panel with integrated PVDF actuators

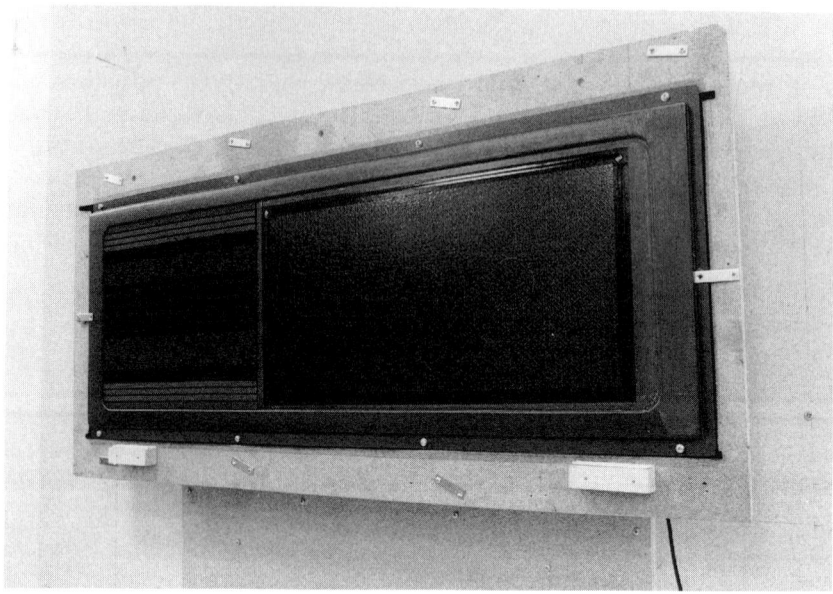

**Figure 1.45**  Loudspeaker using eight electrodynamic drive units mounted in a Dornier 228 ceiling trim panel

systems. The use of many small drive units decreases the radiation frequency of the loudspeaker because the traditional cardboard cone is replaced by a stiff, flat plate.

Several electrodynamic loudspeaker designs were tested using between five and eight small drive units of various designs. The most successful one used eight units. The design was implemented in a Dornier 228 ceiling trim panel (Figure 1.45). The radiating plate had an area of 0.16 m². The increased depth and weight of the trim panel were 35 mm and 570 g, respectively.

The frequency response of this electrodynamic loudspeaker is shown in Figure 1.46. It has a quite flat response with a resonant frequency very close to 104 Hz, the blade passage frequency of the Dornier 228.

Figure 1.47 shows the measured sound intensity level at 1 m for an input of 8 V (r.m.s.) at 107 Hz. The low distortion (around 1%) is noted. The level at around 450 Hz is due to the cooling fan in the computer used during the test.

Figure 1.48 shows the responses of both the piezoelectric loudspeaker described before and the electrodynamic loudspeaker, for four different frequencies relevant for the Dornier 228. The piezoelectric loudspeaker had a better response at all frequencies except the lowest (104 Hz).

However, frequency response is not the only measure of a loudspeaker's performance. Its ability to produce a high acoustic output with low distortion is very important in this context. Especially at the blade passage frequency — which is the lowest frequency for the loudspeaker to reproduce — this proves to be a problem. Figure 1.49 shows the distortion versus produced sound power

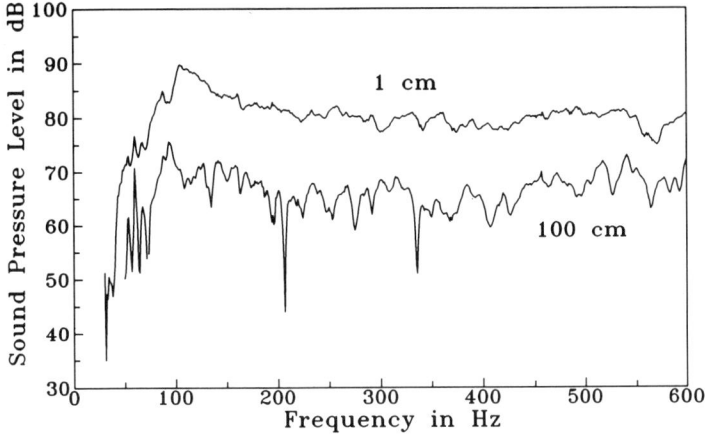

**Figure 1.46** Frequency response of the loudspeaker shown in Figure 1.45

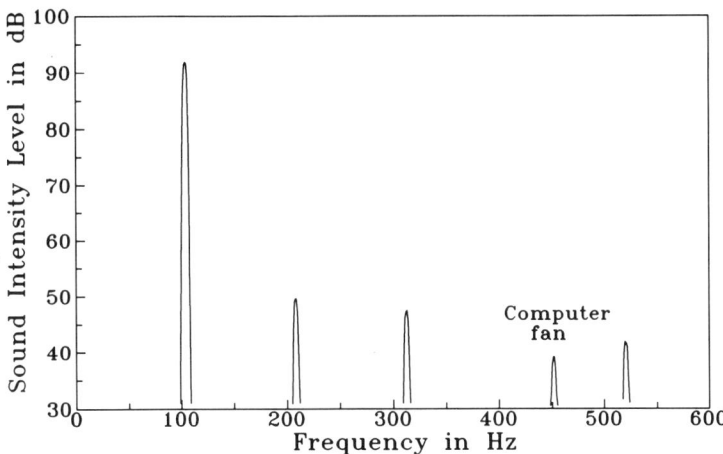

**Figure 1.47** Measured sound intensity at 1 m distance for the loudspeaker shown in Figure 1.45

level at the blade passage frequency for the two loudspeakers considered. Note that the loudspeaker with eight electrodynamic drive units is able to reproduce a higher sound power level at a lower distortion than the loudspeaker using piezoelectric drive units.

**An improved electrodynamic loudspeaker** Results derived from the other tasks of the project indicate that the generation of a high level of acoustic pressure might be necessary in only some locations of the cabin. Therefore, taking into account the performances of the developed trim panels, Metravib was also engaged in the project in a specific development to build an improved

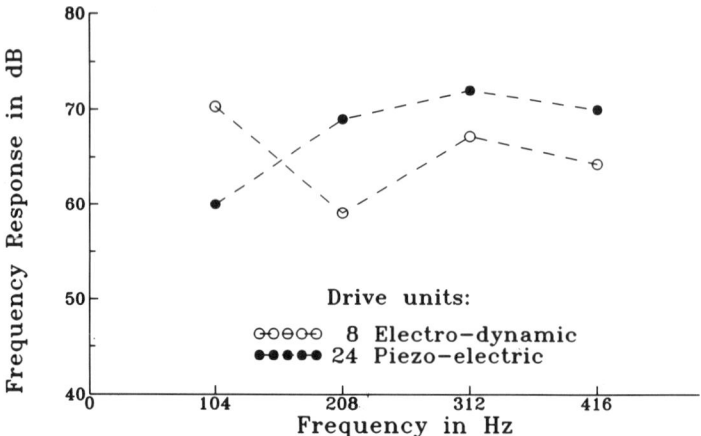

**Figure 1.48**  Frequency response of Dornier 228 ceiling trim loudspeakers at 1 m distance, with an input voltage of 1 V (r.m.s.)

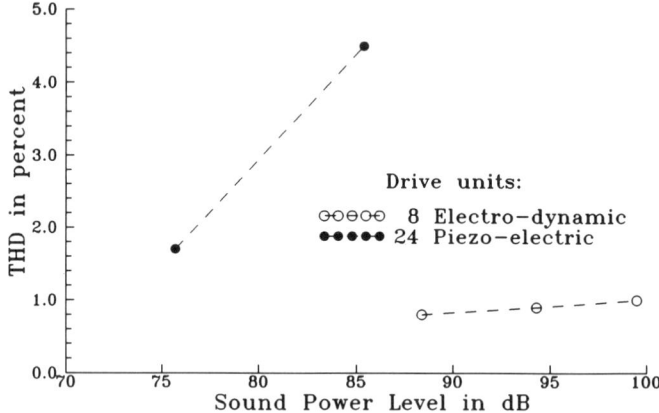

**Figure 1.49**  Total harmonic distortion of Dornier 228 ceiling trim loudspeakers for 104 Hz

electrodynamic loudspeaker with a very high efficiency in the low-frequency range.

The main concept to be dealt with in this work was minimisation of the large displacement of the diaphragm at low frequencies, associated with an increase in efficiency. Based on published knowledge [7], the work was oriented towards treatment of the moving parts of the loudspeaker (coil, cone and spider). The result was a 10-inch diameter 3.5-inch depth loudspeaker able to radiate 105 dB at 1 m for 1 W input at 100 Hz. This performance was maintained in the range 70–500 Hz. The harmonic distortion was slightly higher than for a nonoptimised loudspeaker (Figure 1.50) but remained compatible with the requirements of active noise control applications.

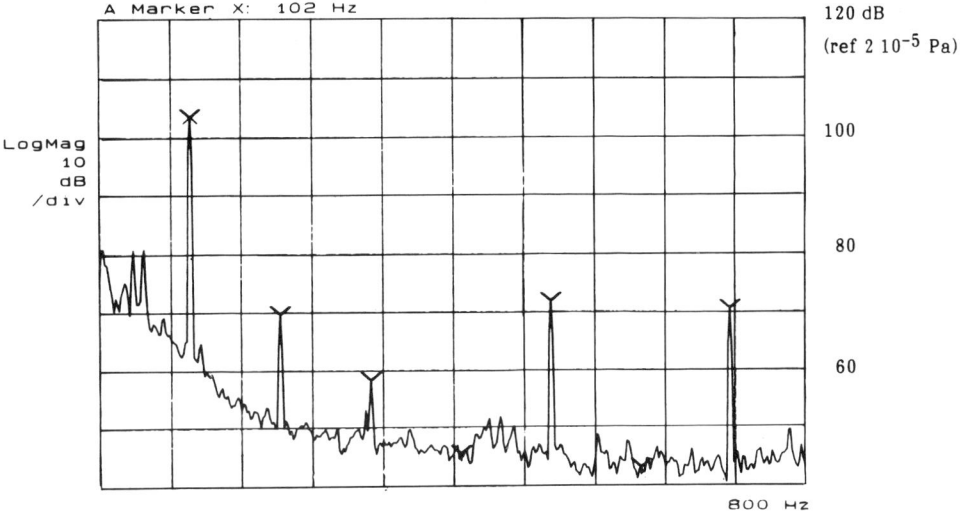

MEASUREMENT PAUSED

Figure 1.50   Sound pressure level in a normal room for sinewave excitation of the improved electrodynamic loudspeaker

## Development of the shakers

The first way to achieve noise reduction in an aircraft cabin is by the attenuation of original frame vibrations. To do this in active control, it is necessary to use small and light shakers and to match them to the frame at the desired locations. Moreover, such small shakers can be used as loudspeaker motors if attached to a trim panel.

Several transaction principles were investigated before an electromagnetic one was chosen for investigation. The actuator design was determined on the basis of the best ratio between its mass and mechanical efficiency. Figure 1.51 shows an example of electromagnetic shaker integration in a trim panel. The advantages of the configuration are mainly the use of a small and powerful magnet (rare earth), combined with an adjustable air gap. This adjustable gap (between 1 and 2 mm) permitted optimisation of the mechanical efficiency.

The shaker prototype issuing from this development work presents very interesting characteristics: 45 g total mass (core, coil, magnet and mechanical frame included); 25 N maximum load; 0.25 A maximum current in the coil; and 0.6 W maximum electrical power.

Besides showing these stand-alone parameters, this shaker was tested for acoustic performance when integrated in a Dornier 228 ceiling trim panel previously used for loudspeaker development. The frequency response reached a sound pressure level of 75 dB at 1 m and 105 Hz for 0.6 W electrical power. As before, it must be noted that the performances are obviously correlated to the dynamic structural behaviour of the trim panel and to the fixing conditions on the frame.

**Figure 1.51**   Electromagnetic shaker integrated into a Dornier 228 trim panel

*Conclusions*

Most of the main ambitious goals of this work were met. For example:

- The weight added by integration in a trim panel of any of the developed prototypes is 570 g maximum (eight electrodynamic drive units). This value has to be compared with the weight of conventional speakers of about 1.5 kg, or with shakers.

- The increased depth of the trim panels is only 30 mm, which means that they may be installed practically everywhere in the cabin.

- The integrated panel may be manufactured using materials approved for use inside passenger aircraft.

With regard to noise generation, the various prototypes exhibit quite similar performances although higher acoustic output combined with lower distortion is obtained with a set of electrodynamic drive units. Piezoelectric loudspeakers tightened to the trim panel itself showed performances more correlated with the structure panel dynamic behaviour and the fixing boundary conditions. Therefore, it is believed that these piezoelectric designs could have, after proper optimisation, performances close to that of the electrodynamic ones but with a lower weight.

With regard to electromagnetic shaker development, and the good related performances, it appears that size reduction is still possible. This has to be studied further, since smaller shakers seem to be today the only realistic active vibration control solution for real flight conditions.

The work on this topic was documented in 12 reports. Although the results achieved were very promising, the sound levels generated by the configurations are still too low for practical applications. Thus more detailed research on this topic is urgently needed.

### 1.3.7 Laboratory and flight testing

A very important part of this project was integration and testing of both complete active noise control systems, including flight tests in the Dornier 228. This work was part of *Task 8*. Selected controller-related results are reported in Section 1.3.4. Here, more details are given of laboratory and flight testing.

*Laboratory testing*

Real flight testing is very expensive, and several components developed by different partners all over Europe had to be integrated. Therefore, part of the plan was to do some laboratory tests before going into the aircraft, to learn how to handle and to set up the control units.

In order to have conditions for the initial laboratory tests as close as possible to reality, it was decided to perform these tests in the Dornier 328 'acoustic test cell' (ATC). The ATC had been built to perform advanced aircraft interior noise studies [8]. It consists of a fuselage test section with full cabin length, a realistic fuselage suspension and three exterior noise simulation rings equipped with 60 loudspeakers. To generate and control the noise excitation, a complex digital 60-channel computer/amplifier noise generation system is used. In addition, a multichannel digital data-acquisition system is available for measurement of the exterior and interior sound fields during the tests.

First, the system with the Matra control unit was tested in the ATC. Owing to the strong cooperation of all partners, it was possible to commission all the equipment in time and to integrate into the ATC the complete set of 48 microphones, microphone preamplifiers and power supplies as well as 32 power amplifiers and loudspeakers and certainly a lot of cabling. The 48 controller microphones were connected to the digital data-acquisition system for recording and evaluation of the results. Figure 1.52 is a photograph showing this test setup.

The first ATC tests with the complete integrated active noise control system showed that it was working. Several adjustments to the software of the controller were made, leading to better understanding of its behaviour and improved performance of the system. As these tests were planned only as preliminary checks it was not possible to test all ideas for improvements, but nevertheless an impressive global control of the sound field in the ATC was achieved. As can be seen in Figure 1.53, the average noise level of all 48 controller microphones of the first simulated propeller tone ($1 \times$ BPF) was reduced by about 14 dB, that of the second ($2 \times$ BPF) by about 16 dB and that of the third ($3 \times$ BPF) by about 10 dB.

**Figure 1.52**   Laboratory (ATC) active noise control test setup in the Dornier 328

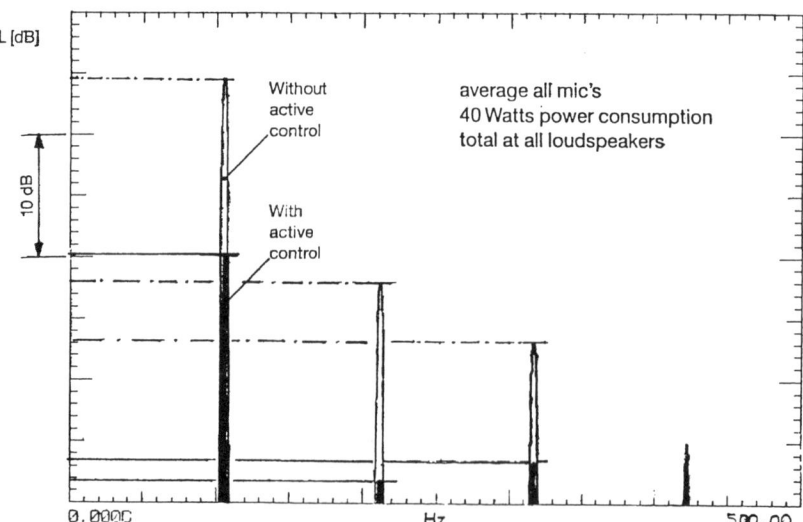

**Figure 1.53**   Result of ATC testing in the Dornier 328 using the Matra control unit

Owing to the flight schedule it was not possible to test the TNO control unit in a similar manner in the ATC. Therefore, this control unit was pre-tested at TNO, in a hardwalled rectangular chamber with dimensions $7.0 \times 5.0 \times 3.0$ m$^3$, were some absorption panels were added to reduce the strong standing-wave pattern and the reverberation time. The primary noise was generated by a dual loudspeaker with a spectrum comparable to that of the Dornier 228. An impressive result was obtained also in this case. The average noise level of all 48 controller microphones of the first simulated propeller tone was reduced by about 21 dB, that of the second by about 15 dB and that of the third by about 14 dB.

*Flight testing*

Flight tests were performed after the tests in the laboratories. For budgetary reasons only flight tests in the Dornier 228 could be conducted. During the tests, both control units were connected to the same setup of 48 microphones, microphone preamplifiers and power supplies as well as 32 power amplifiers and loudspeakers. All equipment was installed in the aircraft and the loudspeakers were integrated into the trim panels. In addition, two computer-controlled 24-channel tape-recorders were installed to record the microphone signals; these recordings were analysed later by using a multichannel digital data-acquisition system. Figure 1.54 shows the interior of the aircraft with integrated loudspeakers and microphones. Figures 1.55 and 1.56 show the Matra and TNO control units mounted in the aircraft.

For each control unit the flight tests were conducted in three stages. First, during ground tests, an artificial sound source was used to generate four sine tones at the frequencies of the first four propeller tones of the aircraft in the cruise condition, to check the components of the system and to identify possible problems. During these tests the 'sync' signal for the control unit was derived from a signal generator. In the second stage, ground tests with running engines were performed to check the behaviour under vibration, the

**Figure 1.54**   Test setup in the Dornier 228 for initial control-system flight testing

**Figure 1.55**    Matra control unit installed in the Dornier 228

**Figure 1.56**    TNO control unit installed in the Dornier 228

power supply required from the aircraft and the 'sync' signal of the left propeller. Finally, a flight of about two hours was conducted for each control unit.

During both flights, impressive results of global noise control were obtained. The average reduction of all controller microphones for the flight test with the Matra unit (Figure 1.57) shows about 10 dB at 1 × BPF, about 8 dB at 2 × BPF and 4 dB at 3 × BPF. At 4 × BPF, there is a slight *increase* of about 2 dB, which however can be ignored because of the very low levels at this frequency.

With the TNO control unit, the average results were reductions of about 15 dB at 1 × BPF and about 8 dB at 2 × BPF. At 3 × BPF, the noise was *increased* by about 8 dB. The noise of the forth tone was not controlled, as explained in Section 1.3.4.

A very important fact is that both control units harmonised the sound field to more constant levels by reducing it much more in the front of the aircraft, where the levels are higher, than in the rear part. At some locations in the rear a small increase was obtained; but since the propeller noise in this region is about 20 dB below the louder front part of the aircraft, this effect can be ignored. The reduction of the sound field for 1 × BPF is given as a 'colour' map for the Matra unit in Figure 1.58 and for the TNO unit in Figure 1.59.

*Conclusions*

It may be concluded that both controllers gave impressive results, and thus the main objective of this project was very successfully met.

For a better understanding of the mechanisms, for a more detailed adjustment of the control units, and for further improvements, much more testing time is needed. Future research on this topic is highly recommended.

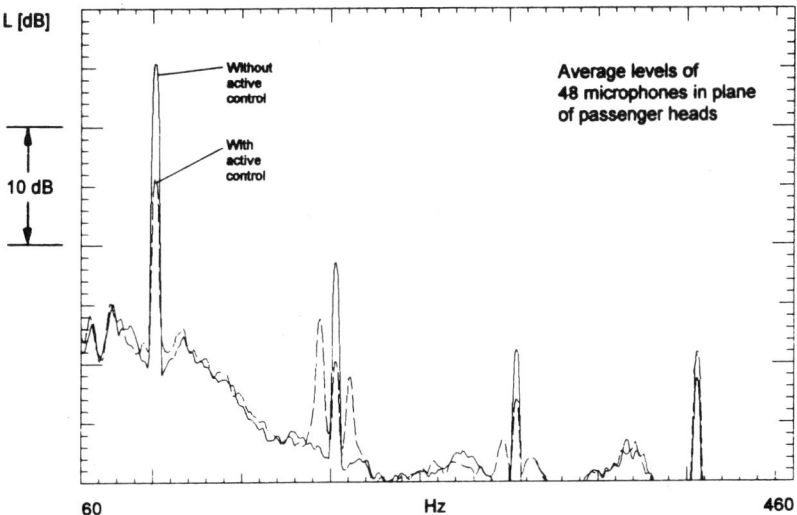

**Figure 1.57** Result of flight testing in the Dornier 228 using the Matra control unit

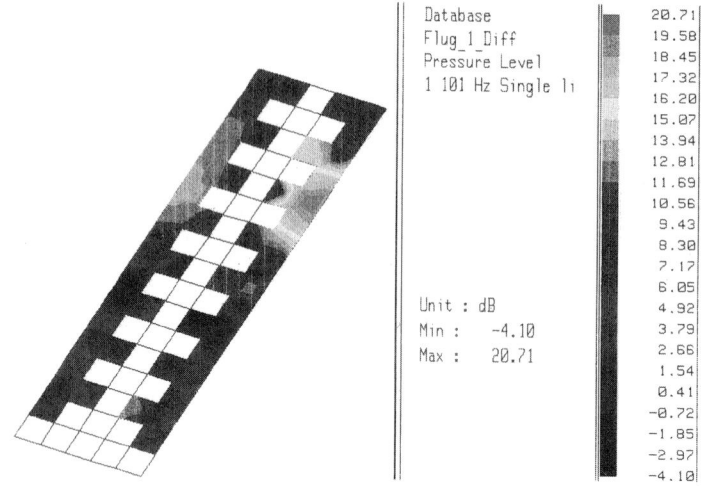

**Figure 1.58** Measured noise reduction in the Dornier 228 with the Matra control unit (fundamental tone 1 × BPF)

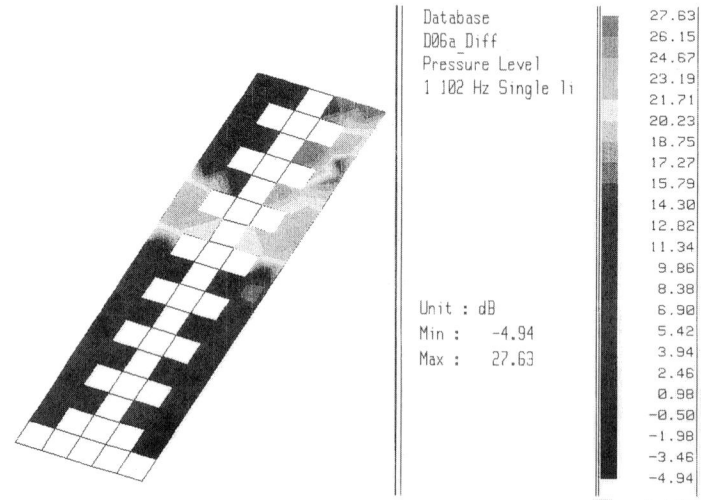

**Figure 1.59** Measured noise reduction in the Dornier 228 with the TNO control unit (fundamental tone 1 × BPF)

## 1.4 Helicopter related work

The objective for the interior of a helicopter cabin is to increase or maintain comfort. To this end, two matters must be addressed: the noise level and the allocation of weight of soundproofing and trim.

The only real competitor to the rotorcraft is fixed-wing aircraft, so long-term objective for reduced helicopter noise can be evaluated by considering trends

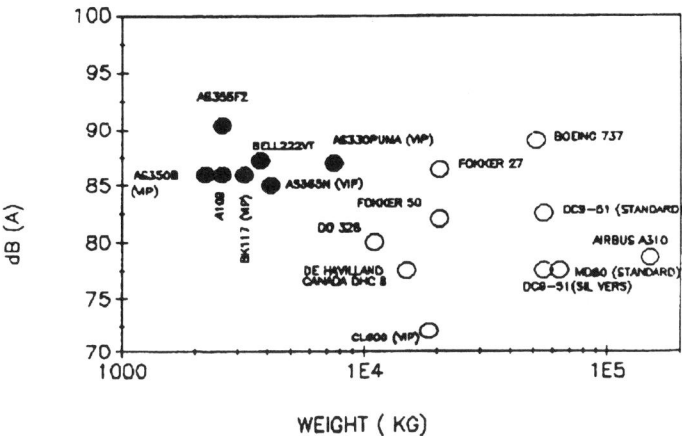

**Figure 1.60**  Helicopter (●) and aircraft (○) interior noise

and levels in fixed-wing aircraft. Figure 1.60 shows for comparison the noise levels in actual helicopters and aircraft.

As we can expect the noise levels in fixed-wing aircraft to be reduced, in order to satisfy customers' demands, a suitable target would appear to be around 75 dB(A). That can be achieved by changes in the craft's structure, propulsion, transmission and rotor systems etc., by the choice of materials which represent a good compromise between weight and capability to reduce noise, and by using the new technique of active control.

### 1.4.1  Critical review of existing knowledge

The three helicopter manufacturing participants, Agusta, MBB and WHL, organised a review of documentation about helicopter internal noise—its levels, sources and existing measures to combat it.

*Helicopter noise sources*

Identification of noise sources can generally be achieved by analysing a narrowband spectrum of noise recorded in flight and comparing this with the characteristic frequencies of the dynamic components—typically the main and tail rotors, transmission and engines. In this way it is possible to identify the sources of the major discrete tones, which are fixed for the maximum rotor r.p.m. (except for the compressor and gas producer sections of the engine, whose frequencies vary with the power setting).

A typical noise spectrum is presented in Figure 1.61. The main and tail rotor frequencies dominate the spectrum up to around 200–300 Hz. Above this,

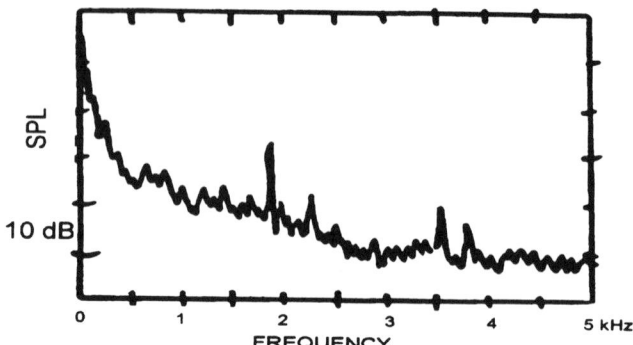

**Figure 1.61**   A typical helicopter noise spectrum

gear-mesh frequencies predominate in particular the frequencies of the planetary, bevel, tail rotor drive and transmission input gears. At around 2–3 kHz, frequencies of accessories such as the oil cooler are dominant. Engine noise is outside the frequency range considered in Figure 1.61.

Helicopter interior noise is typically dominated by structure-borne noise from the many gear meshes in the transmission. Sound travels to the cabin through the transmission mounting points.

For a rational approach to interior noise control it is necessary to identify and quantify the noise sources. However, the noise field is dominated by discrete tones which are more annoying than the same level of broadband noise.

*Noise-reduction treatments*

The techniques used to achieve cabin noise reduction have to include fuselage skin damping and panelling of the cabin to impede both the radiation and the propagation of noise. The following are some particular construction matters:

- Space and installation problems often limit the use of acoustic blankets, in particular on the ceiling.

- Insulation kits must be a wise balance between noise performance and weight.

- In the helicopter cabin there are many sound ingress points, like doors, windows and perimeter joints.

- The noise-reduction capability of acoustic treatments is limited by many additional requirements, such as interior design, customer requirements, etc.

Cabin noise control can be divided into source, path and interior treatments. Source treatment is the most correct approach but is the least easy to achieve on existing helicopters. For a new helicopter still at the design phase, it is possible to demonstrate that path treatment costs more than source treatment and increases with helicopter gross weight.

The aim of path treatments is to impede the transmission of energy between the source and the radiating surface, which can be done by isolation of the vibrating machinery or by damping the structure-borne transmission path. This is a compromise between source and interior treatments. It is realised typically by an impedance discontinuity in the transmission path.

Path treatments for structure-borne and airborne noise propagation depend on absorption or blocking of the energy. Blocking needs an impedance discontinuity in the transmission path, typically a mass layer. The barrier for a structure-borne path can be a low impedance, such as a soft rubber isolation mount, or a high impedance such as a solid and rigid machine foundation. In path treatments, absorption includes the damping material applied to the structure and acoustic absorptive material used to dissipate acoustic energy before it reaches the cabin.

Interior treatment reduces the sound level radiated into the cabin by blocking or absorbing the sound from the cabin panels. This can be done by adding mass, damping, stiffness or secondary walls. This is the easiest approach for an existing helicopter when structural modifications are not feasible.

### 1.4.2  Design, manufacture and testing of a model helicopter fuselage

The aim of this task was to design, manufacture and test a structure like a very simple model helicopter fuselage. This structure was used as a test case for the evaluation of different prediction approaches. The partners in this task were Agusta for design and manufacture of the ASANCA box (different, of course, from the box used for research on fixed-wing aircraft), B&K for the acoustic measurements, MBB for the modal analysis, and WHL for the assessment of panel damping levels and transfer of data to the Royal Aircraft Establishment.

*Design and manufacture*

The ASANCA box (Figure 1.62) had a removable wall (door) to allow instrumentation insertion and internal operation. Walls were connected together by means of angular brackets and rivets.

*Damping measurements*

To support the development of SEA prediction code, and boundary-element and finite-element methods, WHL performed out-of-situ damping measurements, a noise and vibration response survey and reverberation time measurements.

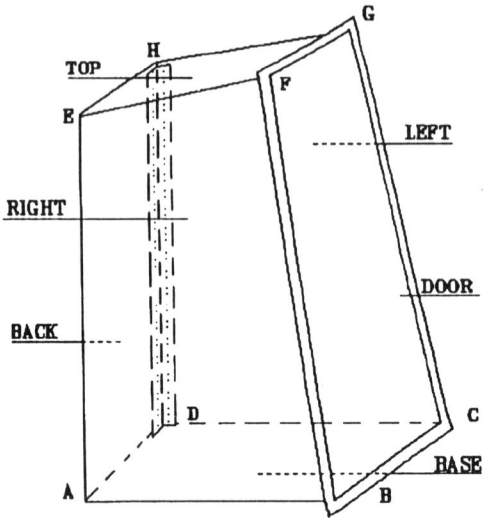

| MATERIAL | ALLUMINIUM ALLOY | | | | | |
|---|---|---|---|---|---|---|
| PANEL | BASE (ABCD) | TOP (EFGH) | DOOR (BCGF) | BACK (ADHE) | LEFT (DCGH) | RIGTH (ABFE) |
| THICKNESS mm | 4 | 4 | 3 | 2 | 3 | 2 |
| PANEL's AREA m² | 0.588 | 0.384 | 0.941 | 0.715 | 1.455 | 0.979 |

**Figure 1.62** Geometric and material characteristics of the ASANCA box for helicopter research

### Modal analysis

For the ASANCA box modal analysis, the measurement points on the structure were excited by a hammer, and a built-in force transducer yielded the input signal. The responses were measured at a fixed point with a one-dimensional acceleration transducer, to compute the transfer function. A modal test was performed on the left panel, on the 'frame' and on the whole box.

**Left panel**  One hundred and eighty-seven measurement points were distributed on the entire area of the panel. Points 1 to 50 were located close to the 'frame' (rivets) and points 51 to 187 were in the field of the plate. The distance between the points was about 80–100 mm.

**'Frame'**  In order to obtain the mode shapes of the frame, an analysis was carried out along the longer sides. One hundred and twenty-nine measurement points in three directions were distributed on this part of the structure. The maximum frequency investigated was 100 Hz. Higher modes were not investigated because of the relatively large separation of measurement points on the 'frame'.

**Whole box** One hundred and fifty-seven measurement points were distributed on the whole box. The analysis was limited to the *x* direction. A transfer function from point 123 (on the right panel) with calculated synthesis function, frequencies and damping table is shown in Figure 1.63.

To separate the modes and to assign them to the right structural element is difficult by modal analysis alone. Hence the analysis was performed step by step—at first an analysis of the 'frame' alone, and then with riveted plates because of the change of the boundary conditions. For the global mode determination, it is helpful to have more measurement points, especially when higher modes are to be analysed.

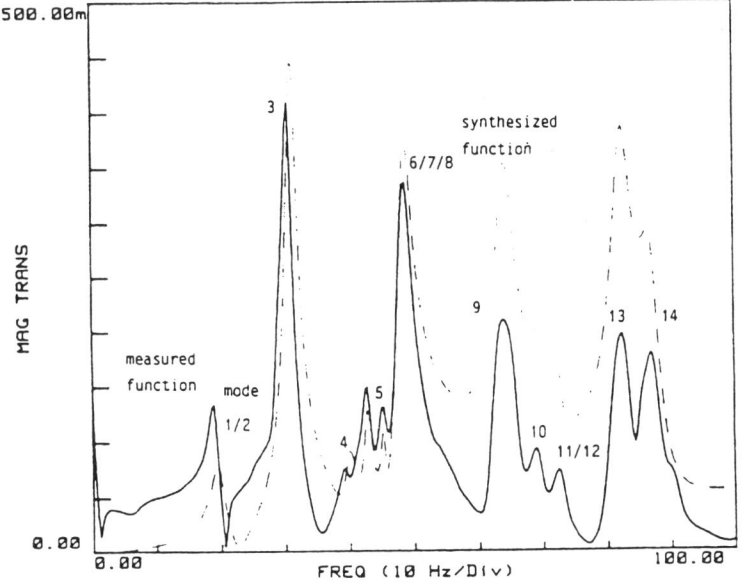

Frequency & Damping Table

| Mode | Freq (Hz) | Damp (%) | Damp (Hz) |
|------|-----------|----------|-----------|
| 1  | 19.7 | 3.72 | 0.73 |
| 2  | 21.2 | 3.16 | 0.67 |
| 3  | 31.2 | 2.89 | 0.90 |
| 4  | 40.0 | 1.98 | 0.79 |
| 5  | 45.5 | 1.99 | 0.91 |
| 6  | 48.7 | 0.76 | 0.37 |
| 7  | 50.4 | 1.80 | 0.91 |
| 8  | 53.7 | 2.36 | 1.27 |
| 9  | 63.4 | 2.77 | 1.76 |
| 10 | 69.5 | 1.56 | 1.08 |
| 11 | 73.4 | 2.95 | 2.16 |
| 12 | 76.5 | 1.29 | 0.98 |
| 13 | 85.5 | 1.17 | .99 |
| 14 | 88.3 | 2.46 | 2.18 |

**Figure 1.63**  Transfer function and calculated synthesis function of the ASANCA box

*Acoustic and vibration measurements on the ASANCA box*

The box was excited by four vibration exciters (type B&K 4810) mounted on the left panel, using white noise. The following measurements were performed.

**Acoustic measurements** The *sound pressure distribution* was mapped in a plane parallel to the left panel by moving a microphone around with an automated positioning system. The pressure was measured in a $10 \times 14$ grid, 50 mm away from the left panel.

The *sound intensity flow* (between 125 Hz and 6 kHz) inside the box was measured by moving around a vector intensity probe in a three-dimensional grid.

The *reverberation time*, of the box and laboratory, at high frequencies was calculated.

**Vibration measurements** *Input power* was measured with only one exciter active and then with all four exciters active from 50 Hz to 10 kHz.

The mechanical energy transmitted from the vibration exciters flows through the panel, to be distributed on the whole box and eventually absorbed by the damping material. The *vibration intensity* was measured on the left panel with all four exciters active. The measurement was performed up to a frequency of 1.6 kHz.

With regard to *acceleration measurements*, the vibration levels were measured in three ways: at the vibration exciter positions, as mappings on the door, left and right panels and on all six panels, in five randomly selected positions.

*Mobility* was measured using a force transducer and an accelerometer.

*Input power* was evaluated, at each excitation point, both with a single source and with a multiple source, by using force transducers and accelerometers.

The results of all these experiments were combined with the results of the numerical analysis described below.

### 1.4.3 Prediction codes assessment and comparison with experimental results

The aim of this task was to produce a numerical analysis of the ASANCA box and to compare experimental and numerical results.

*Finite-element analysis and comparison with experimental data*

Both structural and coupled structural/acoustical FEM analysis was performed by CIRA. Structural analysis was performed on the left panel and on the whole box using the mesh size shown in Figure 1.64. This mesh, used also in BEM analysis, was chosen taking into account CPU time and RAM space.

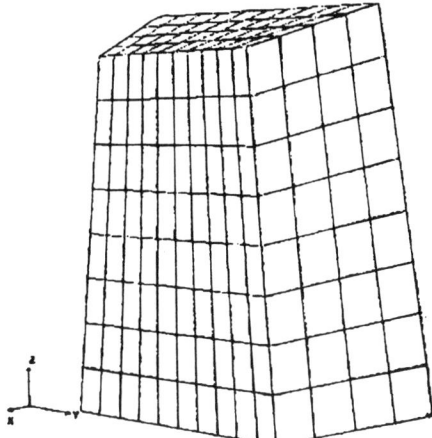

**Figure 1.64**   FEM and BEM mesh for the ASANCA box

**Modal analysis**   As described above, the experiments separated the different dynamic behaviours of the test-article elements: the plates, the frame, and the total model. Using the mode shapes and the eigenvalues found experimentally, it is possible to summarise the results for the left plate (Table 1.5).

Analysis of the frame element is a little more difficult. In fact, the numerical models are really insulated from the global structure, while in the experimental test they were extrapolated. Two finite-element models were assembled with the same degrees of freedom, differing only in their modelling of the mass distribution of the longerons of the removable (door) plate. The results are given in Table 1.6.

Identification of the mode shapes, and comparison of the related eigenvalues, was relatively simple using model A, while the effect of refining the mass distribution was to alter the geometry of the previous ones. The only one identified with reasonable certainty was that itemised by MBB as number 8. The numerical mode shapes are typical of a frame behaviour; in fact it is possible to find in the first 20 eigenvectors all the possible relative rotations among the pairs of faces associated with the bending mode of the beams. Clearly this means that the analytical methods (including the FEM and the BEM), the experimental modal analysis, and the correlations between the results are far from being ideal. Nevertheless, the analytical methods have to

**Table 1.5**   Natural frequencies of the left plate, simply supported

| Mode number item | | Natural frequencies (Hz) | |
| MBB | CIRA | Experimental | FEM |
| --- | --- | --- | --- |
| 1 | 1 | 19.72 | 20.00 |
| 4 | 2 | 31.11 | 30.15 |
| 8 | 3 | 49.25 | 47.35 |
| 13 | 4 | 85.63 | 87.00 |

Table 1.6    Natural frequencies of the frame, completely free

| Mode number item | | Natural frequencies (Hz), FEM | | |
| MBB | CIRA (Model A) | Experimental | Model A | Model B |
| --- | --- | --- | --- | --- |
| 1 | — | 20.20 | — | — |
| 2 | 1 | 31.65 | 28.60 | — |
| 5 | 2 | 48.07 | 47.04 | — |
| 8 | 3 | 69.68 | 65.50 | 65.68 |
| 10 | 4 | 85.49 | 86.00 | — |

be used, not forgetting that all the numerical and experimental tests have to be read in this light.

The modal analysis of the complete box will not be reported here. In fact, the four global mode shapes found by MBB are included in the set of numerical ones. The problem is that they are one part of a general set in which there are also others of difficult correlation. The presence of these modes could be a numerical effect of the routines for the eigenanalysis. Clearly with such modal densities it is quite impossible to separate the real and spurious mode shapes. As before, however, where it is possible to compare the mode shapes, the numerical results are in good correlation with the experimental ones.

The Agusta structural dynamics team also developed a finite-element model of the box, and they found the same difficulties of comparison with the experimental results. The Agusta model was a little more refined than CIRA's for the stiffness evaluation of the longerons, but there were some discrepancies in terms of the first three natural modes at 10 Hz, which is primarily a bending mode of the right panel. This was not found in the experimental test.

**Acoustic results**    The experimental excitation used in the FEM and BEM analyses is shown in Figure 1.65. The FEM acoustic numerical results, compared with the experimental, are shown in Figure 1.66 (only for microphone 1). The following points can be noted:

1. The finite-element method can give good predictive levels in the 200 Hz frequency range. Confidence in the results will depend on the level of modelling details. It is not possible to establish *a priori* the amount of error in reading the numerical response.

2. It is not practical to use the FEM for the highest frequencies because the computational costs are unacceptable. Also, a modal representation of the model will not be totally correct owing to truncation errors.

3. Owing to uncertainties related to the boundary condition simulations, internal connections, and modelling of damping effects, it is useful to use the FEM at the simplest level; with, say, a low number of degrees of freedom, and with only a general fit of the geometric and dynamic properties.

4.  The nature of the coupling between the fluid and the structure can be defined as weak, so that structural dynamics can be solved without fluid simulation, neglecting the fluid loading. The FEM or the BEM can solve the fluid equations forced by the structural dynamics.

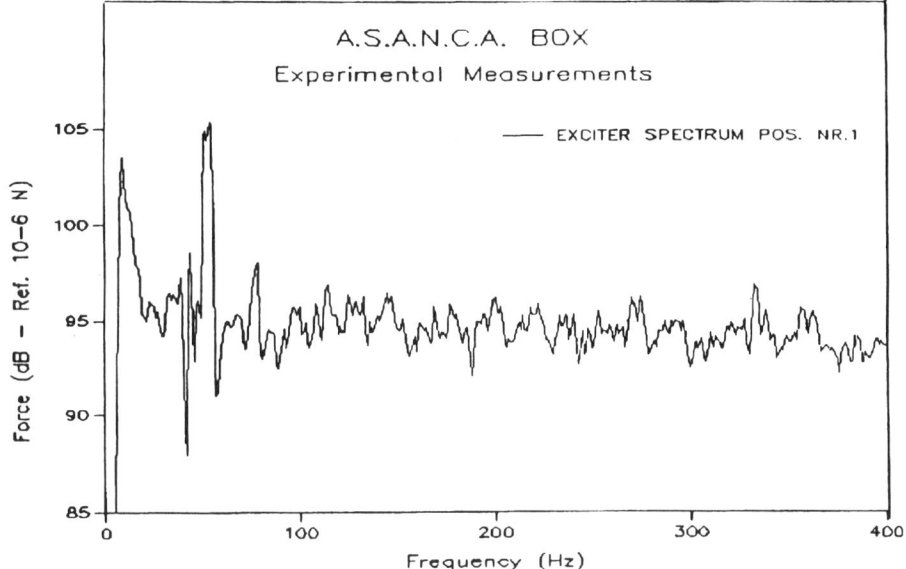

**Figure 1.65**   Experimental excitation used in the FEM and BEM analyses

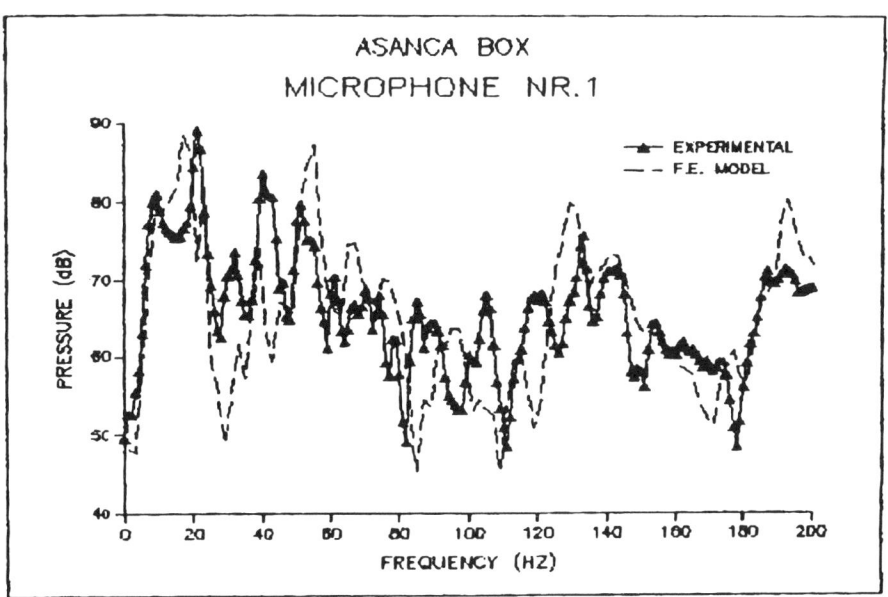

**Figure 1.66**   FEM experimental/numerical comparison

*Boundary-element analysis and comparison with experimental results*

Agusta performed a numerical analysis of the ASANCA box using a boundary-element approach. From an industrial point of view the objectives were to assess the potential of the tool, keeping in mind cost, time (CPU and man time), available personnel, etc. An evaluation of the tool cannot be divorced from this type of consideration.

The elasto/acoustic behaviour of the box (i.e. consideration of both structural and acoustic aspects) was investigated by trying to simulate the experimental 'setup' used to test the box in the laboratory. The mesh used is shown in Figure 1.64.

The structural aspect was evaluated using an FEM approach, coupled with acoustic analysis of the responses at the microphones (located at the experimental positions). Comparison between predicted and measured sound pressure levels was achieved at six microphone positions. The excitation was imposed on the left panel at a point whose coordinates were $x = 0$, $Y = 300$ mm, $z = 850$ mm.

The input spectrum is shown in Figure 1.65. A 0.04 modal damping was assumed. A BEM computer code for elasto/acoustic analysis was employed to analyse the ASANCA box. Structural modes, for the structural acoustic coupling, were evaluated with numerical finite-element code.

Table 1.7 compares the experimental (MBB) and numerical (Agusta) results for the three lowest mode frequencies.

Next, the coupled elasto/acoustic analysis was performed, and the relationship between acoustic pressure and frequency was evaluated in the frequency range 1–200 Hz with a step of 1 Hz. CPU time, in this analysis, using an HP9000/375, was 900 seconds per frequency.

An example of the comparison between experimental structural results and Agusta numerical analysis is shown in Figure 1.67 for microphone number 2. With regard to the acoustic results for all six microphones, the best fit is in the low-frequency range (under 100 Hz), for frequency matching but not amplitude matching (where the difference was 6–10 dB). Above 100 Hz there is also a shift in frequency.

*SEA assessment and comparison with experimental results*

In collaboration with WHL, the DRA has utilised its existing background theory on 'statistical energy analysis' (SEA). The theory was first enhanced to

**Table 1.7** Experimental (MBB) and numerical (Agusta) comparison of lowest three mode frequencies (Hz)

| MBB | AGUSTA | EXPERIMENTAL | FEM |
|-----|--------|--------------|-----|
| 1 | 1 | 19.72 | 15.24 |
| 2 | 2 | 21.16 | 20.20 |
| 3 | 3 | 31.221 | 28.58 |

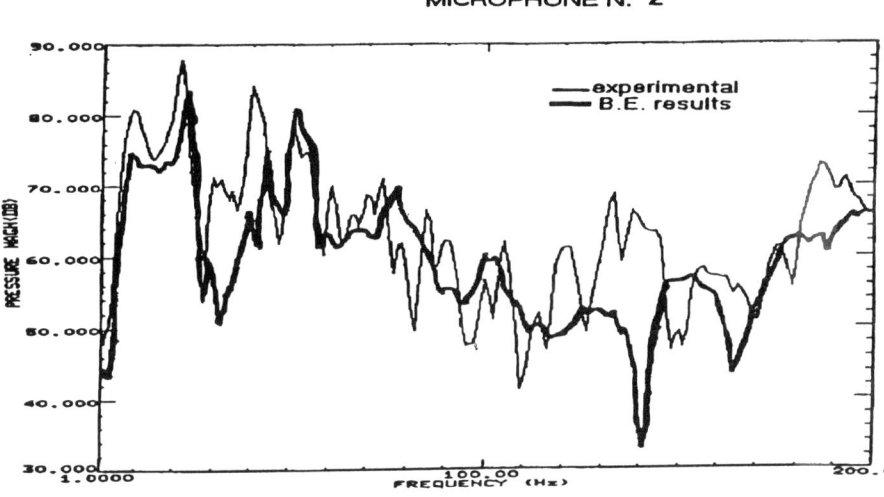

Figure 1.67    FEM experimental/numerical comparison

cope with the ASANCA box design, and the resultant theory used to predict the response of this box structure.

SEA was originally developed using a modal summation and averaging approach, and this has led to the many well-known and restrictive SEA assumptions. The SEA version used here involves little less than a complete redevelopment of the theory using a wave approach coupled with the inclusion of all wavetypes in each plate (out-of-plane bending, in-plane longitudinal, and in-plane shear). The theory depends on the ability to predict all the transmission coefficients at each line junction when the line is assumed infinite and the plates semi-infinite. By a novel approach using 'linewave' impedances, it is possible to predict these transmission coefficients accurately for any junction geometry, provided only that the plates are isotropic.

For this project it was first necessary to extend the method to include a beam at the line junction, and this was done successfully even for the most general of beams. This new theory was then checked against known results in the literature [9, 10] and thus validated at least in a few relatively simple cases.

The next theoretical enhancement of the existing DRA code was to include the acoustic couplings, and although in principle this is straightforward it was decided to use the most up-to-date theory available [11–14].

Finally, the enhanced SEA code was used for the ASANCA box. The input data included all the design details, the WHL measured plate dampings, and the B&K measured cavity and laboratory reverberation times. The DRA took the basic response measurements and reduced them by taking spatial and frequency averages to produce 90% confidence intervals for comparison with the SEA theory.

These final comparisons showed excellent agreement between theory and experiment. The internal cavity noise was predicted with an accuracy of better than 2 dB across the whole frequency range of 200 Hz up to 8 kHz (roughly equivalent to a full-size helicopter frequency range of 70 Hz up to 3 kHz). Of course it must be remembered that the ASANCA box is a very simple structure when compared with a helicopter, and one should be aware that much needs to be done both theoretically and experimentally before one can achieve such accuracy in predicting helicopter internal noise levels. However, these results are very encouraging and point the way forward to the future development and use of SEA.

### 1.4.4   Laboratory validation of the path-identification method

The DRA had developed a unique method [15] for identification of the structure-borne path in the transmission of vibrational energy from a helicopter gearbox to the cabin, where it is then radiated as noise. The work under the present contract was aimed at a full experimental validation of the method prior to future flight applications.

The method involves taking detailed measurements of the gearbox/cabin interface vibration levels as well as of the internal noise levels. Twenty-four accelerometers (six at each gearbox foot) were used, together with 12 internal microphones. The method requires both the magnitude and inter-transducer phase to be measured, and thus data acquisition and analysis is necessarily complicated and time-consuming. A 40-channel digital data-acquisition system was employed throughout. The DRA path-identification method also involves calculating the statistical accuracy of the various predictions and displaying the results as confidence intervals; in many ways this statistical approach is vital since it is necessary to know the quality of any result as well as its numerical value. The heart of the DRA method is in the way these statistical results are calculated as well as the way statistical confidence tests are used to drive and steer the necessary data-fitting processes.

The validation trial was conducted using a Lynx helicopter on the ground. It involved creating a series of known flight conditions, and then applying the path-identification method to reproduce the known structure-borne noise levels.

In the course of the work it was found to be possible to use the data both to refine the method and to predict the airborne as well as the structure-borne noise levels. The main objective of validation was achieved not only with respect to the basic predictions but also with respect to the statistical confidence intervals, in the sense that a 90% confidence interval prediction was indeed wrong about 10% of the time! This work was completed successfully ahead of schedule and summarised in a report. The report concluded:

1.   A new method for the identification of a structure-borne noise path in the overall transmission and generation of internal noise has been fully validated.

2.  The method has been revised and improved during this validation process, and the predicted statistical confidence intervals for the results have been shown to be accurate.

3.  The method has been extended to include airborne noise prediction, and prediction of the spatially averaged noise levels.

4.  The method can now be considered to be an engineering tool of proven quality, able to diagnose accurately and identify quantitatively the different paths.

### 1.4.5  Review and appraisal of active noise control methods

Many of the most important techniques for active control have been reviewed in relation to controlling the noise level in helicopter cabins. Some of the approaches considered are now reasonably well understood. Other techniques remain more speculative and draw upon the latest developments in material science. These techniques therefore require further research at the most fundamental level before their potential for practical control can be assessed. Helicopter acousticians should therefore be aiming at both short-term and long-term objectives.

Global noise control in the helicopter cabin using loudspeakers to minimise the 'sum-of-squares' pressure at a number of microphones distributed around the cabin can be implemented. The technology has progressed and the underlying physics is reasonably well understood and proven. According to computer simulations, using just seven loudspeakers is predicted to afford substantial reductions in the low-frequency sound pressure level throughout the helicopter cabin, 15 dB being typical below 200 Hz. This frequency range includes all the harmonics up to the fifth of the main rotor and the first few harmonics of the tail rotor. It must be emphasised that sound pressure level reductions are generally less by about 5 dB for the WG30, whose cabin volume is approximately three times greater than that of the smaller craft (namely the BK117 and A109), giving rise to a corresponding increase in the number of significantly contributing acoustic modes. Whilst being very effective at low frequencies, this control strategy has been shown by computer simulation to be inadequate for dealing with the noise field due to the transmission system, which begins to be important at about 500 Hz.

Another technique which is sufficiently well developed and understood to be implemented in the short term is zonal control around the head of the seated passenger. This technique seeks to drive a single loudspeaker in order to cancel the acoustic pressure at a closely spaced microphone to produce a small but significant zone of quiet in the vicinity of the control microphone. This type of approach offers several advantages over the global control philosophy.

First, the region of active control can be aimed more precisely at the passenger. Second, since the acoustic pressures at the microphones are dominated by the direct field from their respective loudspeakers, each unit (one loudspeaker and one microphone) can probably be 'tuned' independently of its neighbour, which allows for a more flexible, modular approach to active control since each unit can be fully integrated into the seat. The disadvantage is the potentially large number of units required. Under the assumption that two units per seat are required, to fully equip the WG30 a maximum of 34 units are needed. Another difficulty is the positioning of the microphone, since theory has shown that the size of the quiet zone increases with the microphone-loud-speaker separation increases. Microphone far from the loudspeaker may interfere with the free movement of the passenger's head.

For completeness, the effectiveness of active headsets has also been investigated. Judged solely on performance and weight, the active headset is far superior to all the other techniques considered here. However, the principal objection to headsets is that they cause discomfort and are intrusive. Whilst being acceptable for military applications, the compulsory wearing of headsets means that helicopters will never be able to compete on equal terms with fixed-wing aircraft, which is the objective set by many helicopter manufacturers. One approach to make the headset more acceptable to fee-paying passengers is to use it in addition to a control of the helicopter structure and to cover a sufficiently wide frequency range to be able to deal with noise generated by the gearbox as well as noise from the rotors. For example, the working frequency range of the Westland's system is currently limited by actuator performance to 100 Hz. It is anticipated that the new generation of mechanical actuators, such as those formed from magnetostrictive, piezoceramic and piezoelectric materials, will be increasingly used. The second concerns further research into the complicated relationship between sound radiated into the cabin and the principal sources of vibration. Once this is established the noise control engineer will be better informed as to where active control may be most effectively applied.

Passive treatments for noise reduction are limited by properties inherent in mechanical structures, such as gear teeth elasticity. The development of sophisticated actuators, together with increasing computer power, offer the potential for considerably reducing noise levels in the helicopter cabin by active techniques. There is little doubt, therefore, that active control can play an important role in reducing the notoriously high noise and vibration levels in helicopters, although a significant amount of research and development is still needed to bring this potential to fruition.

### 1.4.6   Discussion of results and conclusions

The main aim of this phase of the project was to identify the most promising research methods and ideas for reducing helicopter internal noise.

*BEM, FEM, SEA*

The BEM and FEM methodologies both require, when taking into account the structural and acoustic interactions, a previous structural numerical evaluation.

This raises fundamental problems that, in a last analysis, do not lead to very good agreement.

Inaccuracies present in analytical models are due mainly to boundary conditions, unknown characteristics of the damping materials, inadequate modelling of the joints and couplings among the structural parts, the hypothesis of linearity, model size, and lumped parameters.

Modal identification algorithms identify a limited number of modes at a limited number of coordinates, and comparison of mode shapes is difficult owing to the different sizes of the models (numerical-solution grid points and experimental acquisition points).

The differences between the BEM and FEM results are not so great as to justify using both methodologies in future investigations.

From a research point of view the quality of the results can be said to be good. The deterministic approaches are very useful when interest is in *trends* of the unknown under observation—the structural and the acoustic levels. However, from an industrial point of view the results are not good enough. The deterministic approaches cannot be used as predictive tools because the link with experimental testing is fundamental, and because the hypotheses with which it is common to work are too approximate. Use of the deterministic techniques will be justified only when more refined models of the various effects are available.

Conventional methods such as the FEM can become impracticable owing to the large number of degrees of freedom needed to model structural deformations. Indeed, at these frequencies, different design realisations produce different internal noise levels, and deterministic methods cannot predict this variability. Statistical energy analysis (SEA) was developed in an attempt to overcome these difficulties.

The results presented using SEA are only a beginning, because many problems associated with nonisotropic plates, curvature, tunnelling, in-situ beams etc. have to be solved before we can make predictions for real structures. However, the results here are very encouraging.

Excellent agreement has been obtained between SEA predictions and experimental results on the ASANCA box. This modern approach to SEA theory, without recourse to the usual SEA factors, points the way to SEA becoming a true predictive tool to stand alongside FE methods.

### Noise-path identification

A new method for identification of a structure-borne noise path in the overall transmission and generation of aircraft internal noise has been fully validated. The method has been revised and improved during this validation process and the predicted statistical confidence intervals for the results have been shown to be accurate. It has been extended to include airborne noise prediction, and the prediction of spatially averaged noise levels. The method can now be considered to be an engineering tool of proven quality, able to diagnose accurately and identify quantitatively the different paths.

On the other hand, attempts to predict airborne noise essentially by subtract-ing the predicted structure-borne levels from the flight total have been less successful. Perhaps this is trying to push the method too far, or maybe we have not yet discovered the correct procedure. Either way, at the moment any flight application of the method should concentrate on the structure-borne path and view any deduced airborne path results with caution, particularly when the structure-borne levels are predicted close to the flight total noise levels .

It should be remembered that at the outset of this validation work no airborne prediction was envisaged; it was only as the work proceeded that it became clear that such an attempt could be made. The fact that one has only been partially successful in this attempt at airborne noise identification is still a bonus over the original work program.

The introduction of mean or r.m.s. predictions is also new and an important improvement. Not only does this reduce the predicted confidence intervals, but for helicopter applications it also allows us to answer the question of whether the structure-borne path is dominant.

### Active control

In our survey of the most promising potential techniques for actively control-ling helicopter cabin noise, particular emphasis has been given to the cabins of the Westland W30, the Agusta A109 and MBB's BK117. These techniques may be classified broadly into two distinct categories: acoustic and structural con-trol.

Acoustic control, which uses an array of loudspeakers to minimise the 'sum-of-squares' pressures at a number of microphones within an enclosure, has already been applied successfully to fixed-wing aircraft and cars to provide good global reductions in the sound pressure level at the low blade-passing and engine harmonic frequencies. This technology has progressed and the underlying physics is now reasonably well understood. The results of com-puter simulations indicate substantial sound pressure level reductions using this method at frequencies below about 200 Hz. This makes it suitable for controlling the tones due to the low-order harmonics of the main and tail rotors, but unsuitable for dealing with the tonal noise due to the transmission system (which begins to be important at around 500 Hz). Another example of controlling the sound field in the cabin directly is zonal control, which entails cancelling the pressure at a point to produce a small but significant size of quiet zone around the head of the seated passenger.

The second type of approach, which is far less developed and understood than acoustic control, involves applying mechanical actuators to the helicopter structure itself with a view to reducing the sound radiated into the cabin. Two very distinct control philosophies have been discussed. The first uses mechani-cal actuators in the important load paths, such as the engine and the gearbox support struts, to reduce the vibrational energy transmitted from the engine and rotors. The effectiveness of this approach is difficult to assess in the absence of a detailed path identification analysis. The other technique uses mechanical actuators to shake the cabin walls of the helicopter with the objective of minimising the radiated acoustic energy. Although laboratory experiments have confirmed some of the principles of both these approaches, their applicability to helicopter noise reduction remains to be established.

### Conclusions

Encouraging results have been obtained from numerical methods, but more work is needed to develop a methodology that can be a useful tool for noise control at the design stage. FEM and SEA in particular show promise as prediction codes.

The main noise sources are the rotor and the transmission system gearbox, and there have been recent advances in gear noise analysis that have identified the importance of parameters such as gear eccentricity and misalignments.

Validation of noise path identification methods has demonstrated that it is possible to produce an initial important analysis of the global noise. The relative importance of structure-borne and airborne noise can be determined, and consequently the most effective noise-reduction approach.

The promising results for active noise control in fixed-wing aircraft offer encouragement also for helicopters. The goal will be to reduce the internal noise level to no more than 75 dB(A).

A final word can be said about the need for cooperation in complex research such as this. The good results are a testimony to the cooperation that was achieved. It is very important to exchange information freely and to compare results at each stage in order to produce good technical results.

## 1.5   Overall summary and conclusions

This project work has shown that active noise control may be an effective means for solving the critical low-frequency interior noise problem for fixed-wing and rotary-wing aircraft. Noise reductions above 10 dB at the critical low frequencies were measured during the Dornier 328 ATC ground test and during the Dornier 228 flight test, as a result of using the ASANCA active noise control systems.

The predictive noise-control calculations were based primarily on finite-element analysis and advanced analytical methods. This provided detailed theoretical inputs for selecting noise-control approaches, including the optimum number and locations of actuators and sensors, during the preliminary phase of an aircraft design as well as during development of improved versions of an existing aircraft.

The numerical codes developed for the active noise-control system configuration (in terms of microphone and speaker number and locations) can be also used in the future for evaluation of detailed noise measurements, both on the ground as well as in flight.

The algorithms developed during the project (time-domain and frequency-domain algorithms), and their implementation on the control units, can be easily tailored for propeller and jet engine aircraft of different sizes and powers.

As the preferred noise-control approach, loudspeakers in the trim panels of the Dornier 228 were chosen. In order to achieve the highest feasible noise reductions, 32 output and 48 input channels were selected, and a very versatile system for future application on other aircraft was made available. The final demonstration performed on the Dornier 228 also underlined important items that need to be developed in future research projects. In fact, the size and weight of the developed system need to be drastically reduced for a commercial aircraft environment.

Furthermore, the system was tested primarily only during a selected cruise condition. No test was performed to prove the efficiency of the system during transient conditions (e.g. turbulence, fast change of speed, or engine torque) or during other cruise conditions (takeoff, climb, descent, and so on).

Very promising results have also been obtained during laboratory tests on trim panels with integral piezoelectric or electrodynamic drive units. These results have shown that advanced actuator designs have great potential for aircraft active noise and/or vibration control. However, in order to define and provide feasible configurations, additional detailed research is required.

In addition, a promising alternative active noise-control technique has been studied during specific laboratory tests. This technique is based on active control of the sound waves travelling in the cavity between the aircraft primary structure and the trim panels, by means of loudspeakers and/or shakers. It was demonstrated on a laboratory double-wall mock-up.

Further detailed research on this topic is required, for optimum control system definition, for system weight and size minimisation, and for the establishment of new alternative control strategies which seem to be promising.

The results achieved were made possible by excellent cooperation between all the partners, which is in line with the spirit of mutual cooperation suggested by the Commission for this kind of project.

# References

[1] Fuller, C.R. and Jones, J.D. Experiments on reduction of propeller-induced interior noise by active control of cylinder vibration. *J. Sound Vib.*, 1987, **11**(2), 389–395.

[2] Dowell, E.H., Gormann, G.F. and Smith, D.A. Acoustoelasticity: general theory, acoustic natural modes and forced response to sinusoidal excitation, including comparisons with experiment. *J. Sound Vib.*, 1977, **52**(4), 519–542.

[3] Fahy, F.J. *Sound and Structural Vibration: Radiation Transmission and Response.* Academic Press, London, 1985, pp. 249–252.

[4] Sas, P., Van de Peer, J. and Agusztinovic, F. Modelling the vibroacoustic behaviour of a double wall structure. *Proceedings of the 14th Aeroacoustics Conference, Aachen, Germany, 11–14 May 1992*, vol. 2, pp. 561–570.

[5] Sas, P., Bao, C., Augusztinovic, F. and Van de Peer, J. Active control of sound transmission through a lightweight double-wall partition. *Proceedings of the 17th ISMA, Leuven, Belgium, 21–25 Sept. 1992*, pp. 743–764.

[6] Bao, C., Sas, P. and Van Brussel, H. Adaptive active control of noise in 3-D reverberant enclosures. *J. Sound Vib.*, 1992, **159**(2), 535–548.

[7] Bruneau, M. and Venet, G. Electromechanical characterisation of a loudspeaker. *Acoustica*, 1980.

[8] Hackstein, H.J., Borchers, I.U., Renger, K. and Vogt K. The Dornier 328 acoustic test cell (ATC) for interior noise tests, and selected results. *Proceedings of the 14th Aeroacoustics Conference, Aachen, Germany, 11–14 May 1992*.

[9] Heron, K.H. The development of a wave approach to statistical energy analysis. *Proceedings of InterNoise 90, Gothenburg, Sweden*, pp. 973–976.

[10] Cremer, L., Heckl, M. and Ungar, E.E. *Structure-borne Sound*. Springer-Verlag, Berlin, 1973, pp. 406–414.

[11] Langley, R.S. and Heron, K.H. Elastic wave transmission through plate/beam junctions. *J. Sound Vib.*, 1990, **143**(2), 241–253.

[12] Leppington, R.S., Broadbent, E.G. and Heron, K.H. The acoustic radiation efficiency of rectangular panels. *Proc. Roy. Soc. London*, 1982, **382A**, 245–271.

[13] Leppinton, F.G., Broadbent, E.G. and Heron, K.H. Acoustic radiation from rectangular panels with constrained edges. *Proc. Roy. Soc. London*, 1984, **393A**, 67–84.

[14] Leppington, F.G., Broadbent, E.G., Heron, K.H. and Mead, S.M. Resonant and non-resonant acoustic properties of elastic panels. I: The radiation problem. *Proc. Roy. Soc. London*, 1986, **406A**, 139–171.

[15] Heron, K.H. and Davies, C. *A New Path Identification Method in Structural Acoustics*. Technical report TR89060, Royal Aircraft Establishment, 1989.

# 2 *Acoustic fatigue and related damage tolerance of advanced composite and metallic structures (ACOUFAT)*

## D. Tougard*

This report, for the period September 1990 to February 1993, covers the activities carried out under the BRITE/EURAM Area 5 'Aeronautics' Research Contract No. AERO-CT90-0025 (Project AERO-P1079) between the Commission of the European Communities and the following:

Dassault Aviation* (coordinator)
Instituto Superior Tecnico, Lisboa
Office National d'Etudes et de Recherches Aerospatiales, ONERA
British Aerospace Defence Limited
Institute of Sound and Vibration Research, ISVR-Southampton
SAAB-Scania AB.
Dornier Luftfahrt GmbH
Deutsche Aerospace Airbus GmbH
IABG GmbH
Per Udsen Co.
Fokker Aircraft B.V.
Nationaal Lucht-en ruimtevaart Laboratorium, NLR
Société Anonyme Belge de Constructions Aéronautiques, SABCA
Katholieke Universiteit Leuven, KUL

Contact: Mr. D. Tougard, Dassault Aviation-DGT/DEC/CS, 78, Quai Marcel Dassault-Cédex 300, F-92552 Saint-Cloud Cédex, France

*Advances in Acoustics Technology* Edited by J.M. Martin Hernandez. © ECSC-EEC-EAEC, Brussels–Luxembourg, 1994. Published in 1995 by John Wiley & Sons Ltd.

## Abstract

The objectives were to enhance our knowledge and understanding of acoustic fatigue strength data of selected advanced composite and metallic materials, and to develop analytical/computational and experimental methodologies to predict the fatigue life of modern civil aircraft structures subjected to acoustic excitations. The following three main areas have been investigated by means of experimental and theoretical tasks:

- dynamic loads applied to structures through acoustic excitation

- structural dynamic responses

- acoustic fatigue strength data.

*Acoustic loads*. On the basis of wind-tunnel (WT) calibration tests, a semi-empirical model of the spatiotemporal characteristics of the aero/acoustic loads exerted on a flat panel by the turbulent field created by a flap (simple configuration of a typical turbulence) has been developed and utilised as 'load data input' for finite-element (FE) calculations. Using the same panels, investigations were conducted in a progressive wave tube (PWT) in an attempt to match the strain spectral densities obtained in the WT.

Even for a simple aerodynamic configuration, the excitation of structures by aero/acoustic loads may not be simulated fully in a PWT, by simply modifying and correctly shaping the spectral content. The effect of the spatial distribution of the loading is clearly different, and the tested specimen endurance may be significantly different. It is clear that a theoretical approach based on correctly predicting responses to both types of environment is required. The development of a semi-empirical model has produced encouraging results. The WT tests have been reasonably well represented and the initial good results of modelling suggest further work related to PWTs.

*Structural dynamic response*. The main available computational methods is dynamic FE analysis taking into account aero/elastic or acoustic coupling with the fluid. The method needed to be validated by comparison with structural dynamic test results.

Several computer codes were available for this study. Their ability to deal with the problem has been recognised by comparison with each other and with test results. These assessments improve predictive capability, define the analysis assumptions and provide rules of use. They are a necessary support to the testing activity.

*Acoustic fatigue strength data*. Standard $S/N$ endurance curves have been elaborated for five advanced materials (two CFRPs + GLARE + aluminium–lithium + SPF/DB titanium) with different designs representative of aeronautical structures. The commonly used methodology for this type of test has been critically analysed. On the basis of these tests, analytical work on damage initiation and damage propagation/accumulation was performed for CFRP materials.

The 'frequency degradation' criterion which is commonly applied to standard endurance data of classical metallic materials has been evaluated for CFRP materials. This criterion was not considered suitable as the sole parameter for determination of specimen 'failures'. Further work is required to investigate the 'settling phase' observed in CFRP materials. It is suggested that a suitable criterion should be based on the degradation of the mechanical properties of the specimen.

An assessment of coupon failures compared with complex structural failures is required, to validate the use of current coupon designs and endurance data.

## Selected abbreviations and definitions

| ESDU | Engineering Sciences Data Unit |
|------|--------------------------------|
| GLARE | GLass Aluminium REinforced |
| JAR 25 | Joint Airworthiness Regulation, part 25 (for the transport aircraft) |
| NDT | Non Destructive Testing |
| OASPL | Over-All Sound Pressure Level |
| PSD | Power Spectral Density |
| PWT | Progressive Wave Tube (noise tunnel) |
| RT | Room Temperature |
| SPF/DB | Super Plastic Forming/Diffusion Bonding (titanium) |
| SPL | Sound Pressure Level |
| TWGX | Task Working Group X (Partners of Task X) |
| WT | Wind Tunnel |

| Specimens | Items to be tested as coupons, panels or components |
|-----------|------------------------------------------------------|
| Coupons | Flat items of simple geometry |
| Panels | Representative of skin panels of aircraft |
| Components | Built-up structures representative of aircraft parts (e.g. boxes) |

## 2.1 Introduction

### 2.1.1 Definition and general objectives

The term 'acoustic fatigue' implies the structural fatigue produced by the predominantly resonant response of structural components subjected to radiated sound fields (fluctuating pressures). Acoustic/aerodynamic excitation aspects are coupled with structural dynamic behaviour and endurance aspects.

All the European civil aircraft manufacturers are concerned with comprehensive studies of acoustic fatigue, with the following general objectives:

- To qualify their structures according to the requirements of the Joint Airworthiness Regulation (JAR) 25 on transport category airplanes.

- To reduce maintenance problems which are known from experience to be very costly (repairs, modifications, down-times, etc.), when they are due to acoustic sources of excitation.

- To understand and define rational solutions when acoustic problems appear.

The fundamental technical objectives of the ACOUFAT program were:

1. To enhance our knowledge and understanding of the acoustic fatigue strength data of selected advanced composite and metallic materials.

2. To develop the analytical/computational and experimental methodologies needed (for dimensioning and certification) to predict the fatigue life of modern civil aircraft structures subjected to acoustic excitations.

The safety and efficiency of aircraft operations can then be improved, and thus the maintenance costs of future aeronautical structures can be reduced.

## 2.1.2 Fields of study

To predict the acoustic fatigue life of structures, three main issues have to be addressed:

- *Loads* applied to the structures (dynamic characteristics of the acoustic excitation).

- *Structural dynamic response* evaluations which provide stresses $S_{rms}$ and frequencies.

- *Acoustic fatigue strength data* for the materials and the selected designs (rivets, etc.).

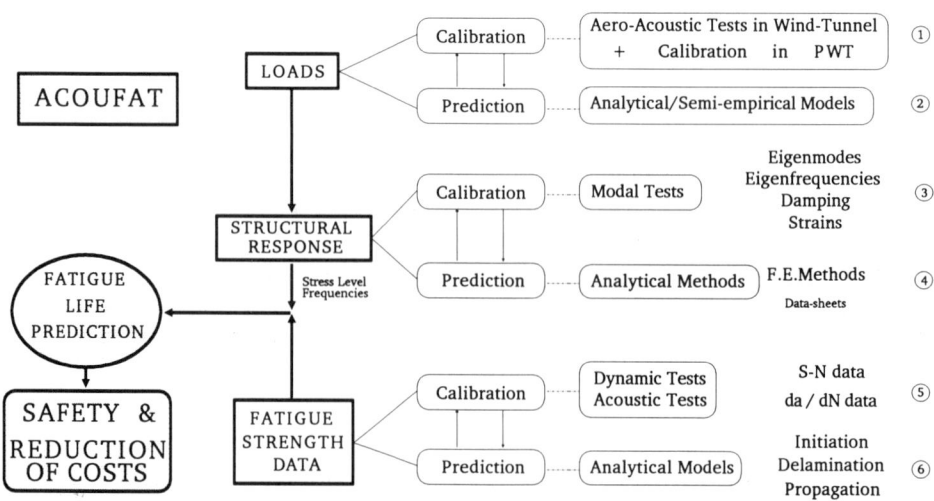

**Figure 2.1** The objectives of the ACOUFAT project

For each of these three main fields of study, objectives have been defined in this project, with experimental tasks (calibration) and theoretical tasks (prediction), as summarised in Figure 2.1.

### 2.1.3   Main tasks of the ACOUFAT study

The following are the four main tasks of the study:

1.   Preparatory investigation, design and manufacture of the specimens
     Subtask 1.1: Selection of the proper design for test specimens
     Subtask 1.2: Manufacture and control of the test specimens

2.   Dynamic tests by shaker excitation
     Subtask 2.1: Dynamic tests of simple specimens
     Subtask 2.2: Dynamic tests of specimens with initial damage

3.   Acoustic fatigue tests,
     Subtask 3.1: Aero/acoustic loads data
     Subtask 3.2: Acoustic tests on panels

4.   Theoretical analysis of acoustic fatigue and damage tolerance
     Subtask 4.1: Loads
     Subtask 4.2: Structural response analysis
     Subtask 4.3: Damage initiation and crack propagation analysis.

Some subtasks are linked in the investigation as follows:

* For the acoustic-fatigue strength data study: subtasks 2.1, 2.2, 3.2 (experimental work) and 4.3 (analytical work).

* For the aero/acoustic loads study: subtasks 3.1 (experimental work) and 4.1 (analytical work).

* For the structural dynamic response evaluation: subtask 4.2.

Also, *Task 1* is related to the definition and manufacture of all the ACOUFAT specimens ($\sim$ 550 coupons and 7 panels) to be tested in *Tasks* 2 and 3.

## 2.2   Research objectives and activities

### 2.2.1   Acoustic fatigue strength data

*State of the art at the start of the program*

'Working tools' concerning acoustic fatigue are currently available in the Engineering Sciences Data Unit (ESDU) Series, collected by a qualified staff with the assistance of many establishments throughout the world in research, industry and teaching, to give general guidance on the design of structures to avoid acoustically induced fatigue failures.

ESDU datasheets provide authoritative information on aluminium and titanium alloys, and limited data on carbon/epoxy materials. For the classical metallic materials (aluminium and titanium), the datasheets provide endurance data (acoustic fatigue strength diagrams: $S/N$ curves) which are widely used by all the aeronautical companies for their civil applications. These endurance diagrams of acoustic fatigue present (for metallic materials) $S_{rms}$, which is the root-mean-squares value of stress at a reference position on the failure line, away from the effects of stress concentration, or (for composite materials) $\varepsilon_{rms}$, the root-mean-squares value of strain etc., plotted against $N_r$, which is the equivalent endurance (half the number of zero-crossings to failure of the stress–time function). The diagrams are for specimens typical of aircraft structural elements excited by narrowband loading of random amplitude with zero mean load.

The data are, however, very limited or not available for advanced metallic and composite materials, yet they are needed to meet design requirements for advanced civil aircraft structures.

*ACOUFAT objectives*

Aeronautical companies use advanced metallic and composite materials widely, and they need stress endurance data ($S/N$ curves) and crack or damage growth-rate data ($da/dN$ or $dA/dN$ curves) as working tools to satisfy the JAR25 requirements for acoustic fatigue.

The first objective of this study was elaboration of these diagrams for the widely used advanced materials on the basis of selected specimens, similar to the ESDU specimens and representative of the civil aircraft sensitive configurations.

Another objective was to evaluate a 'frequency degradation' criterion, as usually applied to classical metallic materials and early carbon fibre reinforced plastic (CFRP) materials.

*Summary of the work performed*

Two families of endurance tests were performed. First there were tests of coupons by shaker excitation. About 550 coupons were tested in this way (random vibration). Standard $S/N$ endurance curves were elaborated for five selected advanced materials (two CFRP, GLARE, aluminium–lithium, SPF-DB titanium) with different designs representative of aeronautical structures. This work was performed with *Task 2*. Examples of tested coupons are shown in Figure 2.2.

The second family of tests was on panels in PWT. Six large panels ($\sim 1\,m^2$), representative of aircraft panels, were tested to confirm the failure modes and

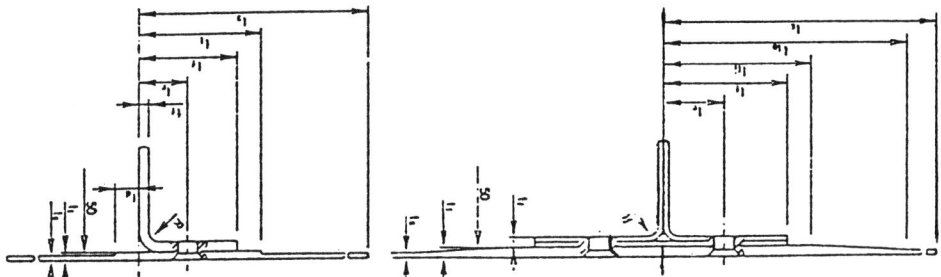

**Figure 2.2**  Examples of coupons tested by shaker excitation

Example of tested panel :

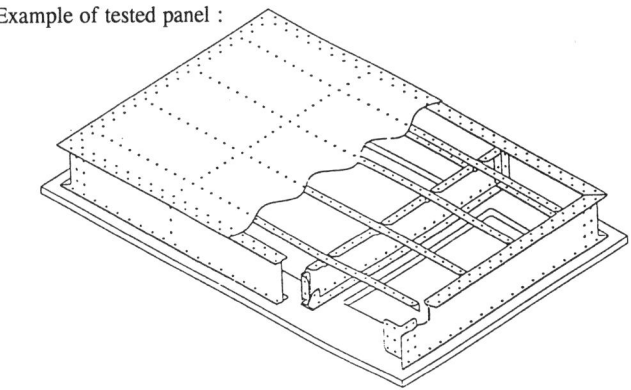

**Figure 2.3**  Example of a panel tested in PWT

$S/N$ data gathered during the shaker tests, and to assess the contributory effects of modes other than the fundamental. This work was performed within *subtask 3.2*. An example of a panel thus tested is shown in Figure 2.3.

The methodology commonly used for this type of test has been critically analysed.

The elaborated $S/N$ diagrams, and a critical analysis of the applied criteria, are important results of this study. The results of this part of the ACOUFAT program are presented in Section 2.3.

On the basis of these tests, analytical work concerning damage initiation and damage propagation/accumulation has been performed for CFRP materials (*subtask 4.3*).

### 2.2.2  Aero/acoustic loads

*State of the art at the start of the program*

Acoustic fatigue failures have occurred in structures lying close to, or in the path of, the jet efflux. Similar failures have occurred in other regions of pressure fluctuation, such as within the intake duct of fan engines, close to propeller tips and in regions of separated flow near control surfaces, such as elevators, flaps, rudders, or near items such as spoilers which are used on some aircraft during manoeuvres. Another example of particular interest when considering acoustic fatigue is the blown flap which may be used on STOL vehicles during both take-off and landing as a lift augmentation device.

At present, aero-acoustic load data are very sparse, and there is no general database. The available data concern mainly the time fluctuation, rather than the space distribution—which, nevertheless, is strictly essential to calculate the structural response.

In addition, it is not obvious that we can ever represent aero/elastic coupling by tests in a reverberant chamber or in PWT: the response of the structure depends on the spatial coherence of the excitation, which can be different in flight from that in the classical acoustic tests in reverberant chambers or PWT noise tunnels, which are commonly used to perform acoustic studies. Owing to this shortage of knowledge, the testing strategy using common acoustic noise facilities (reverberant chambers, PWT noise tunnels, sirens, etc.) may not be representative of real flight conditions. Improvements are needed in this field of study.

*ACOUFAT objectives*

The objectives of the ACOUFAT program in this field of study were:

- To develop and validate a semiempirical model of acoustic loads, with space distribution and time fluctuation, on the basis of information collected from the open literature and pressure measurements obtained during experiments.

- To study experimentally, with regard to the structural response of a representative aircraft panel, the variation with flight conditions (simulated by wind-tunnel tests), and with loads applied during acoustic fatigue tests with the commonly used acoustic testing facilities. A specific aim was to define a PWT testing strategy to get the same structural response with aero/acoustic loads and in a PWT (if possible).

*Summary of the work performed*

Within the framework of the ACOUFAT aero/acoustic loads study, four experimental studies and one analytical study were performed.

1. Identification of the pressure-field loads (time and *space* distribution) in a wind-tunnel (ONERA, S1 Modane) behind a full-scale deflected surface (representative of spoilers, airbrakes or other surfaces providing separated flows) by a series of measurements with sensors behind this surface.

2.  A specimen (box/panel, $\sim 1$ m$^2$, representative of aeronautical structures) was set up in the wind-tunnel behind the same deflected surface. The structural response of this specimen was identified by strain-gauges (and FE calculations).

3.  This specimen (with its instrumentation) was then set up in a first PWT (at IABG) to calibrate the acoustic excitation and to try to define a testing strategy to get the structural response which was identified in the wind-tunnel by the strain-gauges. To check this calibration in another PWT, the panel was also tested in a second PWT (by BAe).

4.  Study of the scale-effect in the wind-tunnel with a smaller deflector at a reduced distance.

This experimental work was performed within *subtask 3.1.*

5.  On the basis of wind-tunnel (WT) calibration tests, a semiempirical model of the spatiotemporal characteristics of the aero/acoustic loads exerted on a flat panel by the turbulent field created by a flap (simple configuration of a typical turbulence) was developed and utilised as load data input for finite-element calculations.

This analytical work was performed within *subtask 4.1.*
    The results of this significant part of the ACOUFAT program are presented in Section 2.4.

### 2.2.3 Analytical support (structural dynamic response evaluation, and damage initiation and crack propagation analysis)

*State of the art at the start of the program*

To prepare a computer prediction of the acoustic fatigue life of an aeronautical component, it is necessary to have the stress history of the sensitive parts of the component (calculated by computational methods) and cumulative failure criteria obtained by simulation models and corroborated with test results.
    The main available computational methods for calculation of the structural response are dynamic finite-element analysis, taking into account the aero/elastic or the acoustic coupling with the fluid. These methods already existed for this type of calculation, but they needed to be applied and validated by comparison with structural dynamic test results.
    In addition, mathematical modelling work concerning damage tolerance, linked with the previous acoustic fatigue response, was beginning in Europe, but the various proposed models needed extensive validation and possible improvements in accuracy.

*ACOUFAT objectives*

One of the main objectives was to develop analytical methods for dimension-ing and for qualification of structures subjected to acoustic loads. The experi-mental test results of *Task 2* (shaker tests) and *Task 3* (WT and PWT tests) were to be analysed with the aim of extending existing prediction techniques in the fields of aero/acoustic loads, calculation of the structural response, and damage initiation and crack propagation analysis.

*Summary of the work performed*

**Work related to the aero/acoustic loads study**   The structural behaviour of an aluminium panel, tested in the WT and the PWT tests related to the aero/ acoustic loads study, was evaluated by finite-element analysis under loads: (a) with the spatiotemporal characteristics measured during the WT tests; (b) with the spatiotemporal characteristics from the developed semiempirical model of the aero/acoustic loads; or (c) with different assumptions (parameter study) related to the spatiotemporal characteristics of the loads in the PWTs.

**Calculation of the structural responses of the ACOUFAT specimens**   Four computer codes were available for study of the dynamic structural behaviour of the specimens. No development of new modelling methods was included in this project, but only the application of available methods.

Modelling was done of the simple specimens for the dynamic tests by shaker excitation and the seven panels for the acoustic tests in WT and in PWT. Two of these panels were modelled by all the 'calculating' partners to compare predictions between them and test data. The ability of the computer codes to deal with the problem was then evaluated by comparison with each other and with test results.

**Work related to the damage-initiation and crack-propagation analysis**   For the prediction of crack initiation and growth rate, suitable models were applied and compared with each other and with the tests results. Analytical and finite-element methods were used. The dynamic test results obtained in *Task 2* were one set of inputs for this fracture mechanics analysis.

## 2.3   Research results: acoustic fatigue strength data

### 2.3.1   *Random vibration endurance investigations by shaker excitation*

*Introduction: Task 2 presentation*

This section summarises all the experimental shaker testing activity performed within *Task 2*. A range of materials and specimen designs were investigated for their endurance qualities in a pseudo-acoustic loading environment. The main purposes of this activity were to enhance our knowledge and understanding of the acoustic fatigue strength data of selected advanced composite and metallic materials, and to verify if accurate life predictions of aircraft panel/stringer constructions subjected to acoustic loading can be made, using simple coupon tests.

*Task 2* was subdivided into two parts. *Subtask 2.1* dealt with the majority of specimens which were manufactured in an undamaged condition, and *subtask 2.2* dealt with damaged CFRP specimens. With increasing awareness of damage tolerant design of structures, a comparison between damaged and undamaged specimens was required.

An assumption which dictates the design of a coupon is that panel behaviour can be adequately represented using a thin strip (beam behaviour). This approximation is valid for panels with large aspect ratios where edge effects in the long direction of the panel can be ignored (Figure 2.4).

In order to subject a test coupon to a pseudo-acoustic load, it is assumed that panel/stringer constructions exposed to an acoustic field have predominant modes of vibration associated with symmetrical bending of the panel skin around the substructure. The reason for this assumption is that the fluid–structural coupling effect is thought to be most efficient in vibration modes of this type.

The above assumptions will become invalid if panel aspect ratios are small, or if different structural modes are excited due to spatial effects of the acoustic field.

Owing to the number of materials being investigated, the failure modes associated with each material could be expected to be manifested in different ways. Based on preliminary work by the CFRP test partners, and on the ESDU series (see ESDU item number 72015 or 73010), it was decided to use a 2% reduction in the settled frequency of the specimen as 'failure'. The 2% criterion adopted by the CFRP partners is not strictly applicable to the failure of metallic

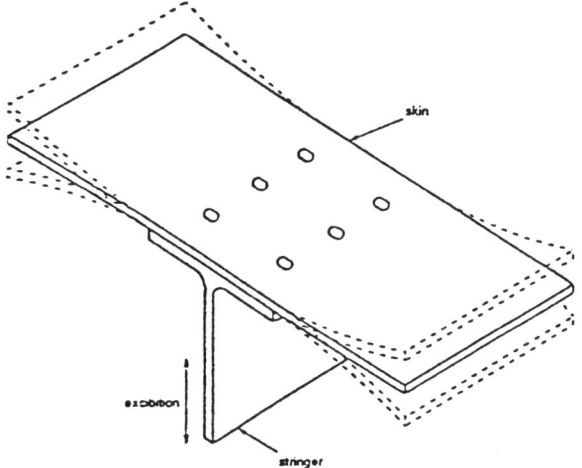

**Figure 2.4**   Coupon specimen type for acoustic fatigue testing

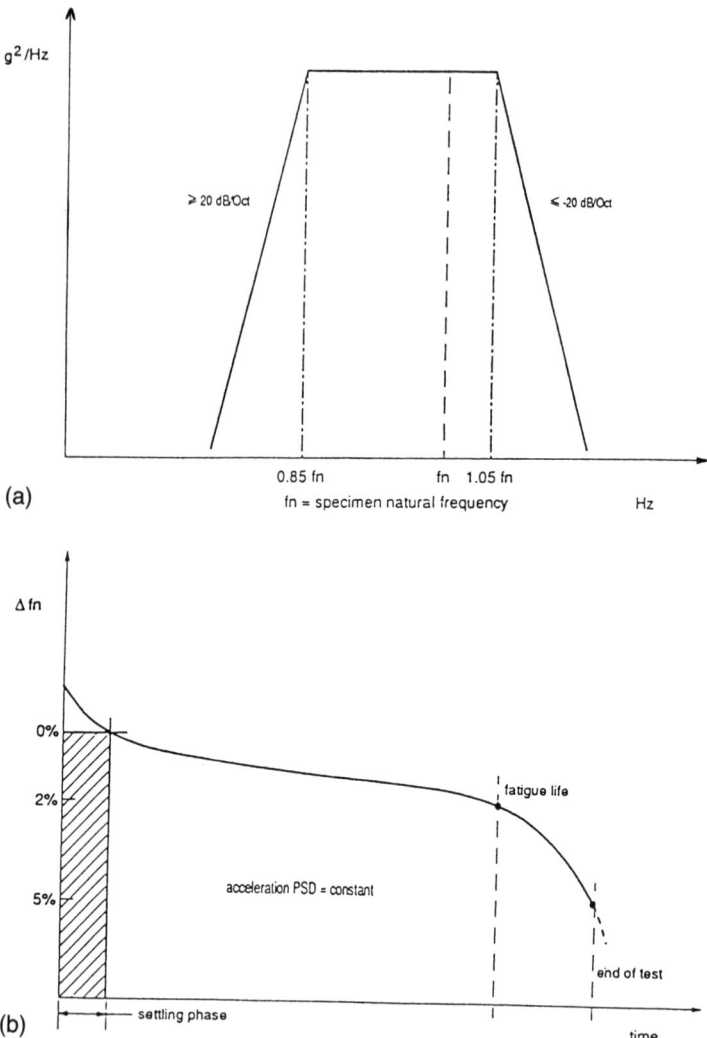

**Figure 2.5** Tests by shaker excitation: (a) driving spectrum, and (b) coupon natural frequency trend

specimens, but can be used as a guide to failure of hybrid materials. The following sections provide an overview of the results and endurance properties of the materials and constructions investigated.

### The materials investigated

The materials investigated in the ACOUFAT study were as follows:

- Two metallic materials: SPF/DB titanium (see Figures 2.7 and 2.8) and aluminium–lithium (Figures 2.9 and 2.10).

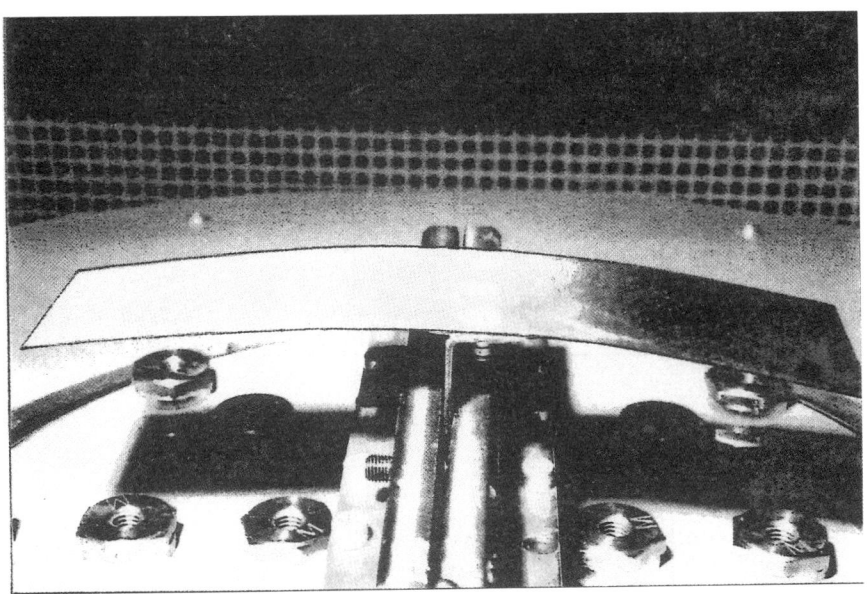

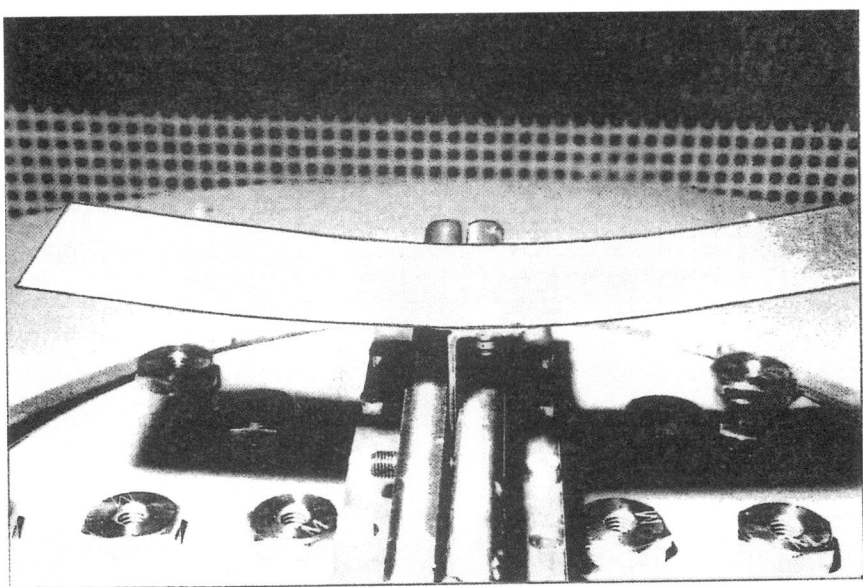

**Figure 2.6** Bending mode of coupons with sinewave excitation

- Three composite materials: GLARE (see Figures 2.11–2.16), CFRP T800/BSL924 (Figures 2.17–2.22) and CFRP HTA/6376 (Figures 2.23–2.36).

*Specimen designs*

Specimen designs used in the endurance testing performed by all Task Working Group 2 (TWG2) partners followed a common approach. From the

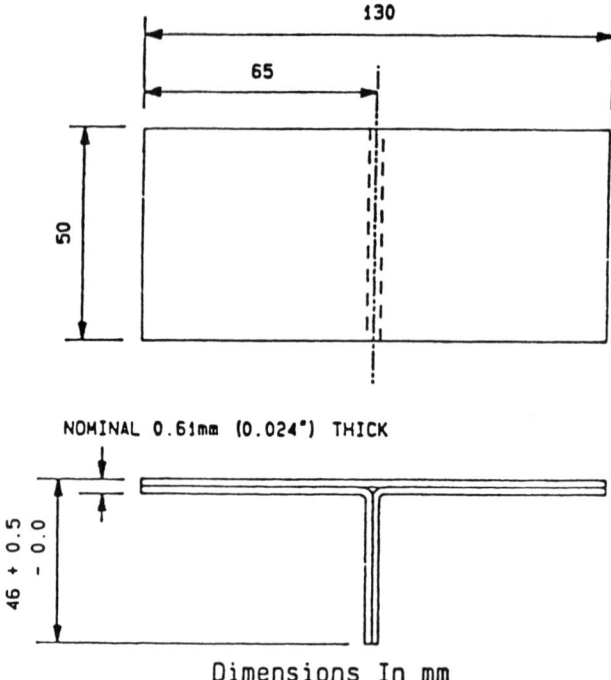

**Figure 2.7** Dimensions of SPF/DB titanium random vibration fatigue specimens

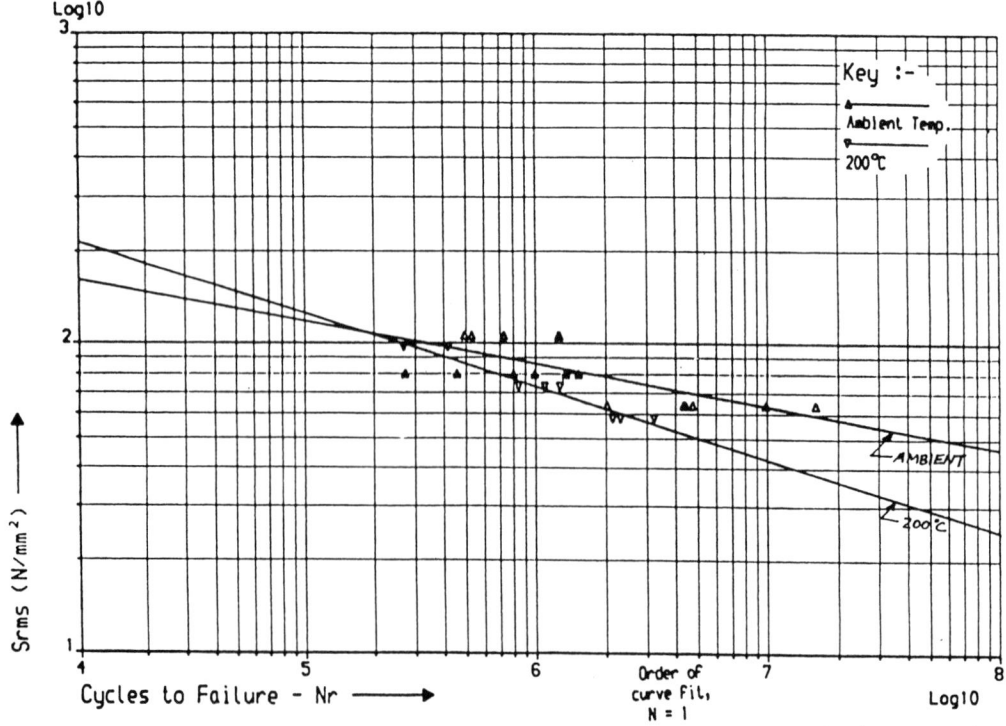

**Figure 2.8** *S/N* data for SPF/DB titanium coupon specimens

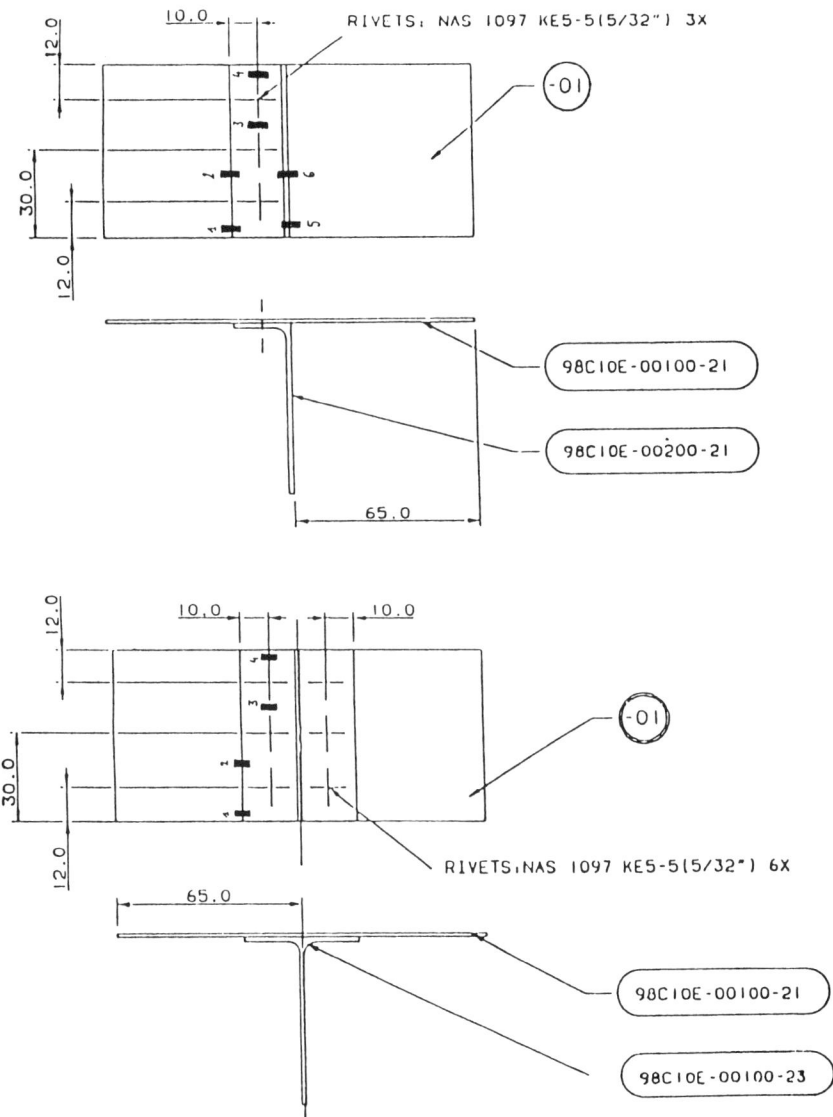

**Figure 2.9**  Dimensions of aluminium–lithium coupons

assumptions highlighted in the introduction, the design used was generally a section of skin material attached to a substructure. A flat skin coupon was also investigated. The test coupon was excited in the first bending mode, using a narrowband random load applied through the substructure or clamp (see Figures 2.5 and 2.6). The work performed by Fokker had a slightly different arrangement, whereby the coupon was turned on end and the base excited through the skin to cause first-mode bending of a section of stringer (see Figure 2.12).

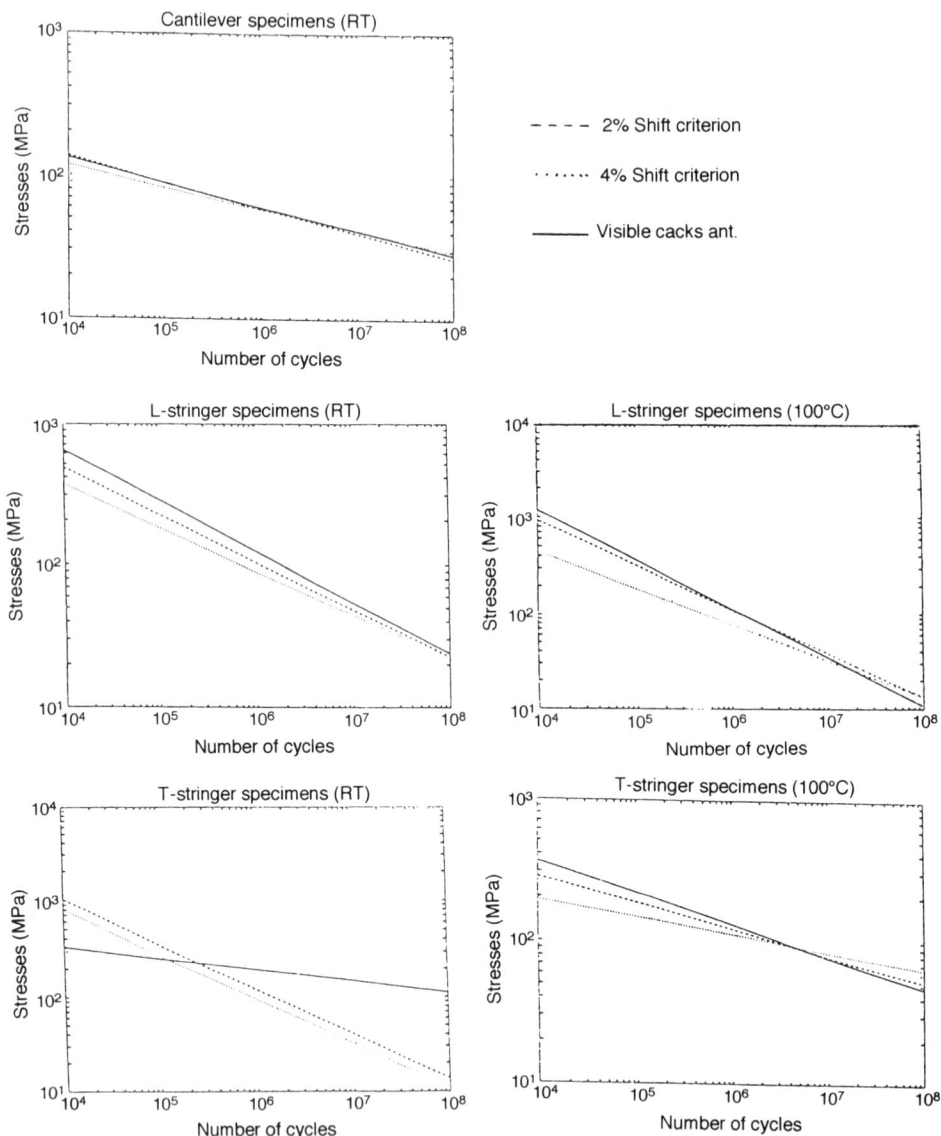

**Figure 2.10**   Endurance data for aluminium–lithium coupons

Two substructure profiles investigated were the L-shaped and T-shaped stringer sections. A number of attachment regimes were used to connect the substructure to the section of skin (see Figure 2.23).

The work performed by NLR investigated two attachment types, bonded or riveted. A flat cantilever specimen was also tested to investigate the effects of sudden sectional changes on fatigue life of the GLARE material. Testing performed by Fokker on GLARE materials investigated a riveted stringer to skin coupon which was excited in 'rib mode vibration' (see Figure 2.12).

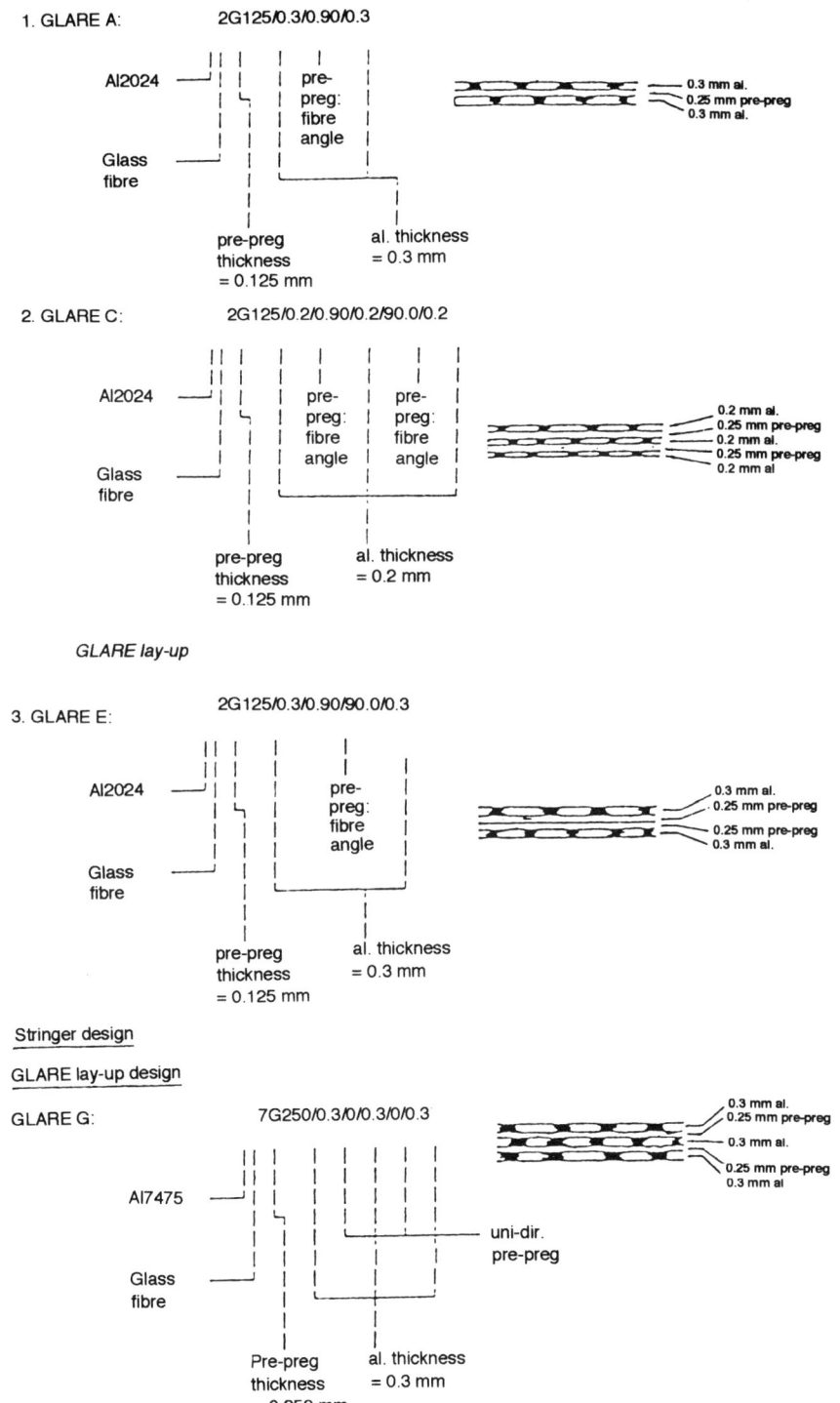

**Figure 2.11**  Definitions of GLARE A, C, E and G materials

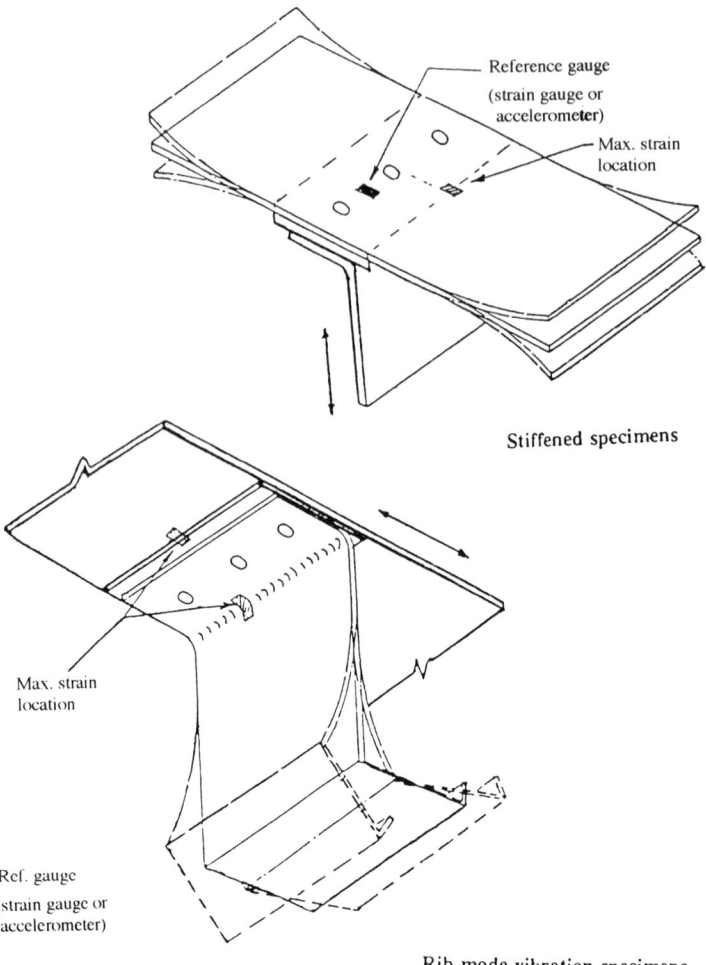

**Figure 2.12**   GLARE coupon types, and gauge locations for regular tests

SABCA tested L and T substructure designs which were both riveted to the skin, while BAe Airbus investigated a T-shaped titanium specimen that was manufactured using the SPF/DB technique—this specimen had a continuous substructure to skin attachment (see Figures 2.7 and 2.9).

The CFRP partners also investigated L and T substructure designs. The work performed by DASA investigated eight different specimen types (see Figure 2.23). Each specimen design varied in skin thickness from 1 mm to 2 mm, with or without a landing which was cocured to the skin. Three coupon designs were investigated by BAe. Two of these coupon designs were the L and T substructures with a tapered landing design (see Figures 2.18 and 2.19). A flat laminate coupon was used to determine material endurance (see Figure 2.11).

The effects of initial damage on CFRP coupons were investigated by Dornier.

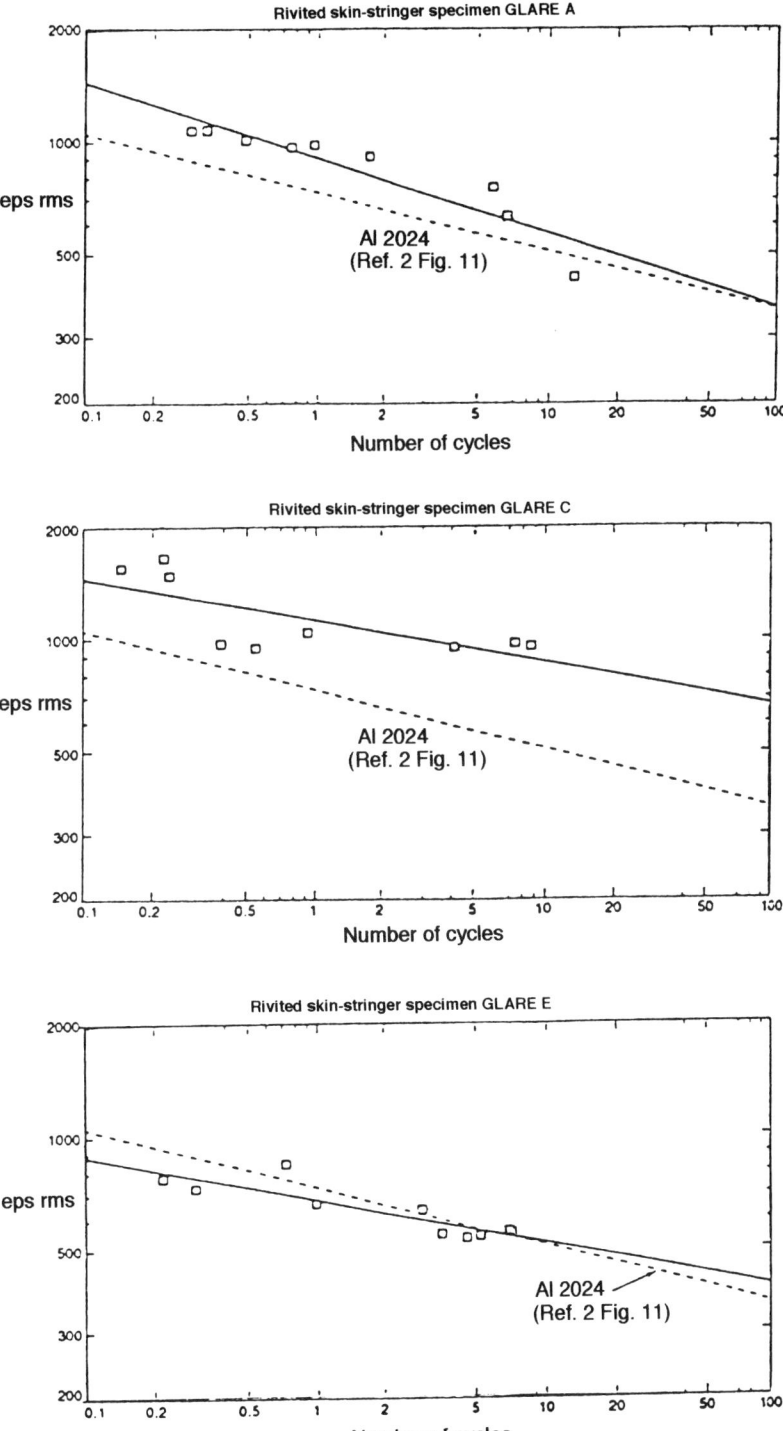

**Figure 2.13**   GLARE coupons endurance data (riveted skin–stringer coupons)

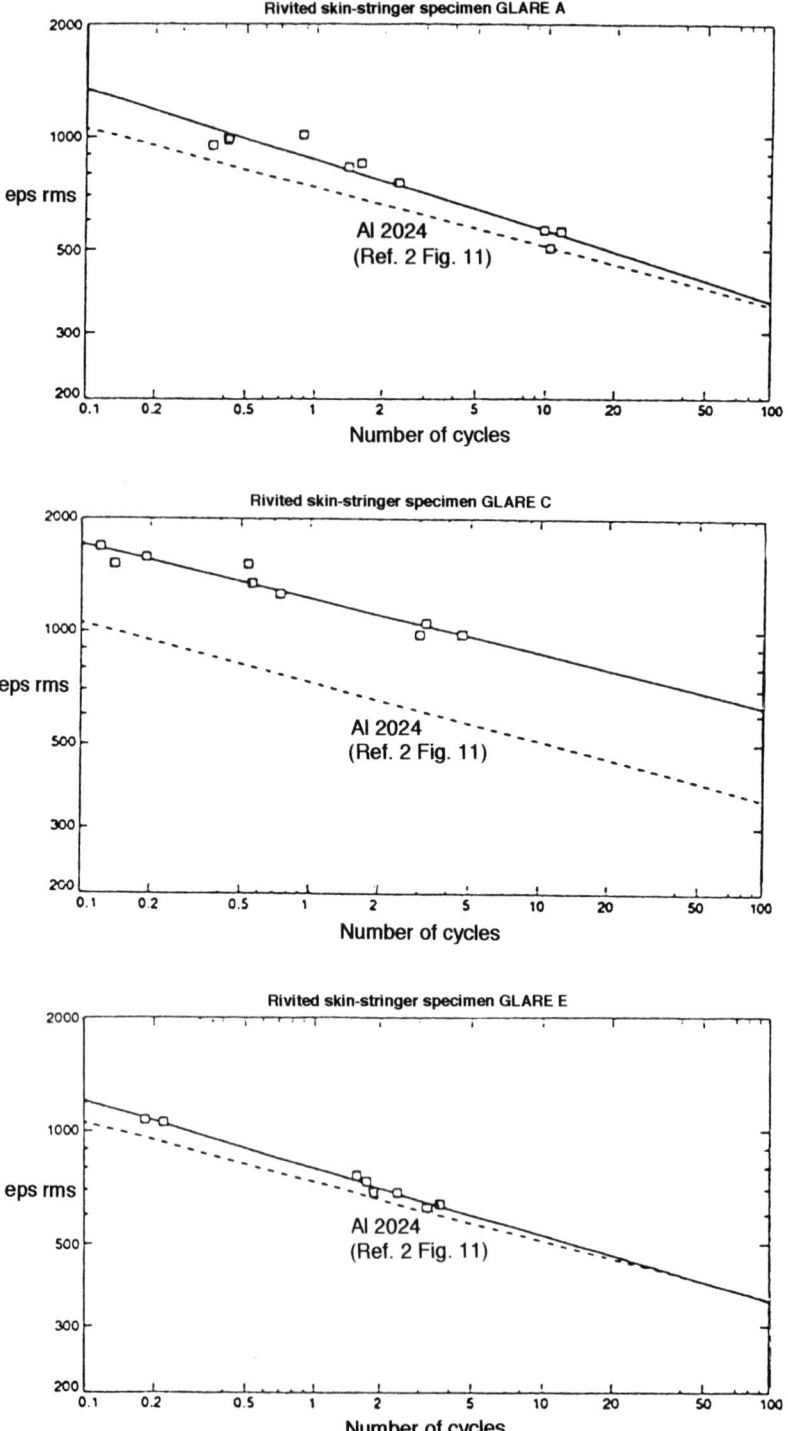

**Figure 2.14**  GLARE coupons endurance data (bonded skin–stringer coupons)

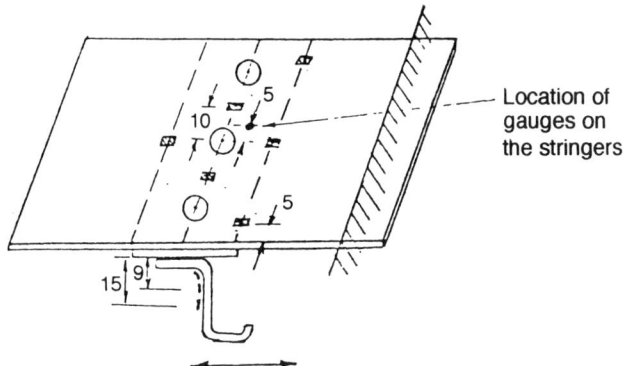

**Figure 2.15**  Details of GLARE coupons for rib-mode vibration tests, showing strain-gauge locations

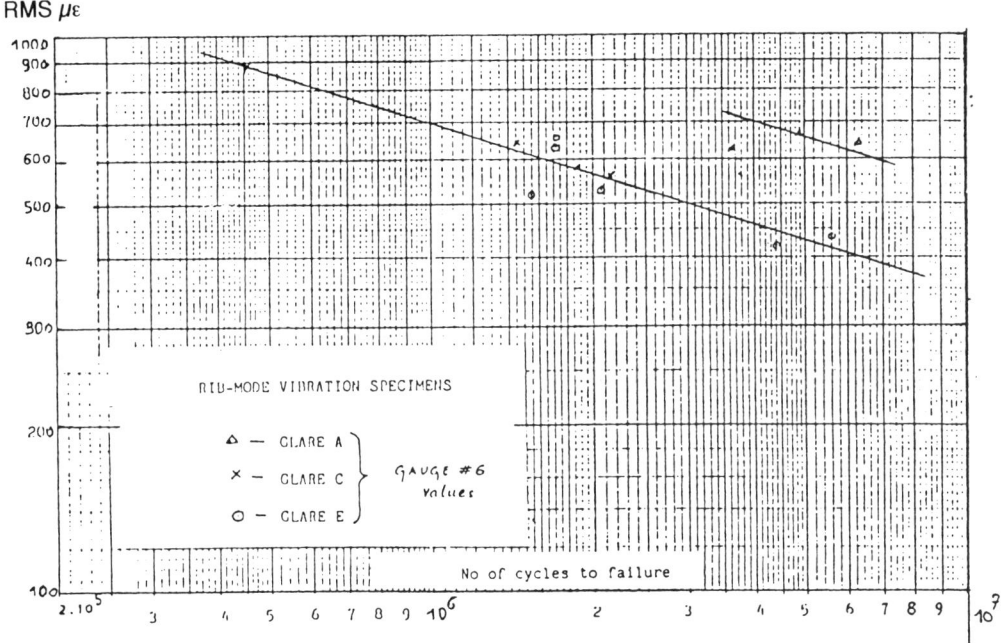

**Figure 2.16**  Fatigue lives of riveted GLARE specimens under rib-mode vibration

The two coupon designs investigated were similar in design to the types tested by DASA (DO2/MBB2 and DO6/MBB6 coupons—see Figures 2.30 and 2.32).

## Specimen settling characteristics

Several types of settling characteristics were observed in the shaker tests.

SPF/DB Ti coupons apparently had no settling phase, whereas the Al–Li specimens exhibited two characteristics. The coupons with substructure had a

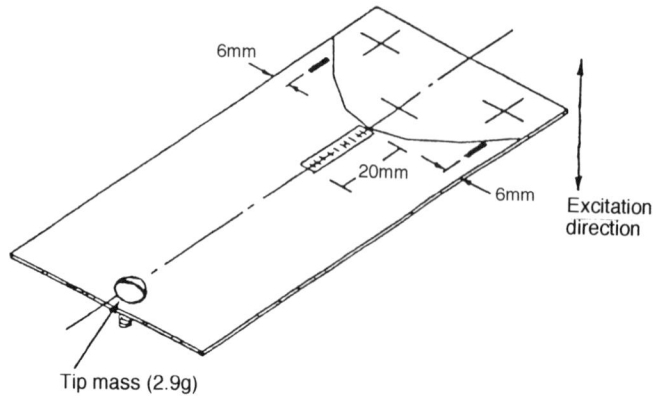

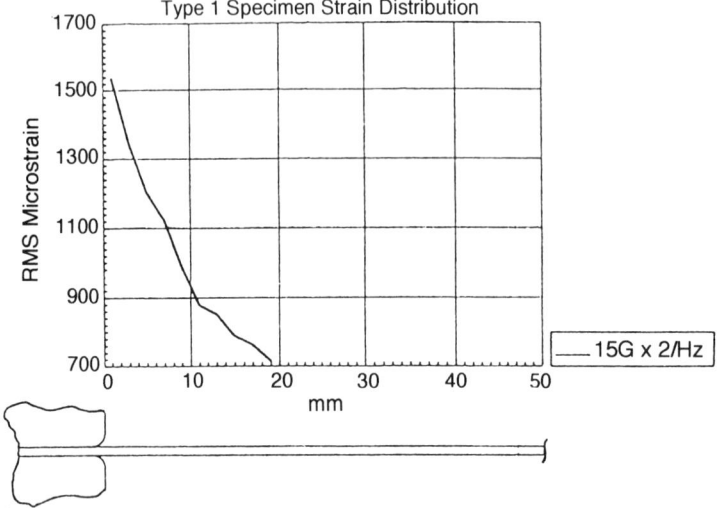

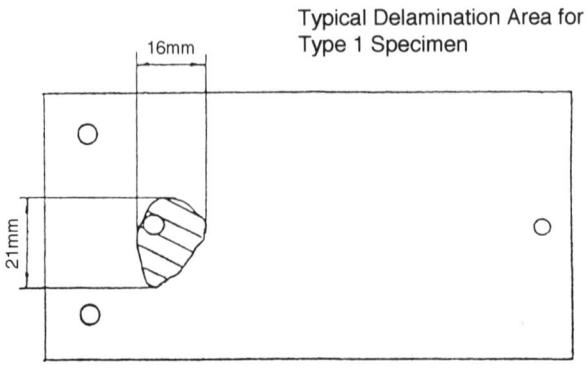

**Figure 2.17**   T800/BSL924 cantilever (type 1) coupons

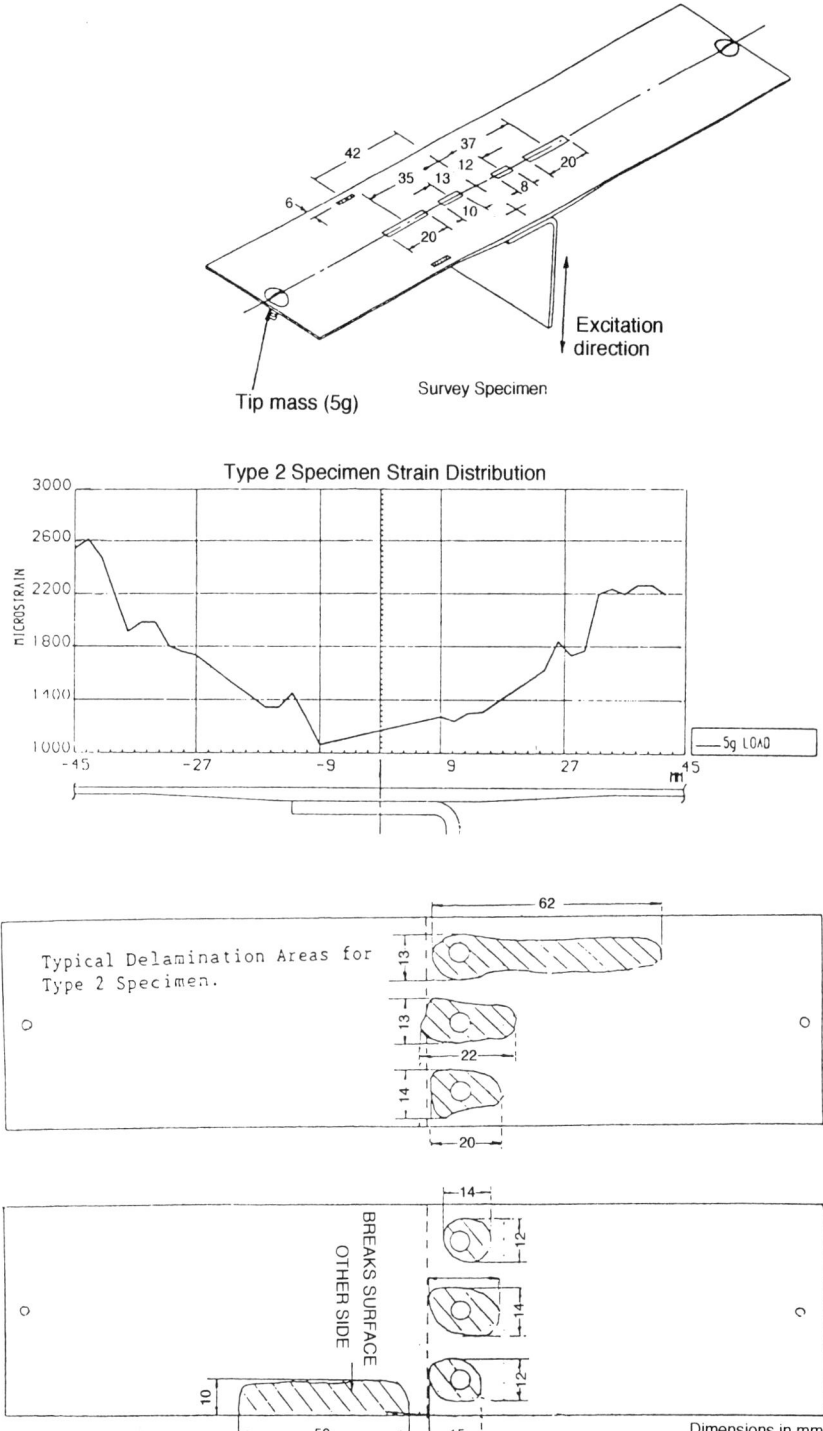

**Figure 2.18** T800/BSL924 L-stringer (type 2) coupons

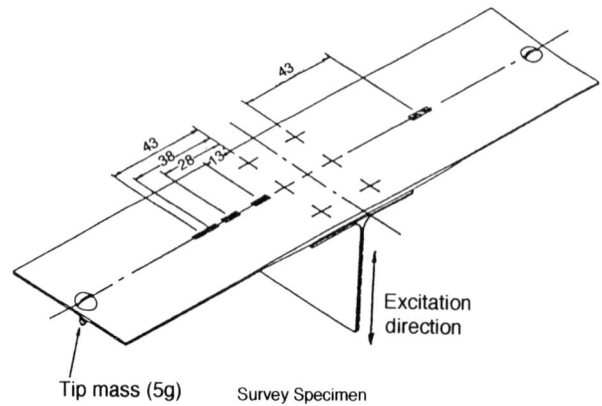

Tip mass (5g)          Survey Specimen

Dimensions in mm

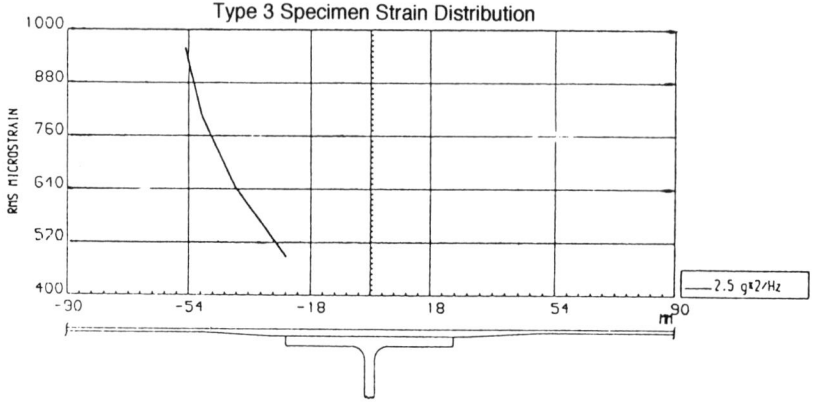

**Figure 2.19**   T800/BSL924 T-stringer (type 3) coupons

sudden drop in frequency at the start of the test which was due to the breakage of a lime film between the skin and substructure. The frequency then decreased linearly until failure. The frequency of the flat skin coupon remained constant for most of the test and then decreased when failure was approached.

The GLARE materials revealed no noticeable settling phase for all coupon designs tested. Instead, the coupon frequency remained constant for the major part of the test and dropped away sharply as cracks propagated in the material.

From the results of testing on CFRP HTA/6376, it was shown that there was a fast drop in coupon frequency due to the partwise separation of the skin from the substructure (see Figure 2.24(a)). Consequently, it was difficult to determine where the actual settled point occurred.

The CFRP T800/924 materials demonstrated a large settling period for the plain laminate coupons, which is thought to be due to the stiffness of the clamping arrangement. The coupons with substructure had a significantly

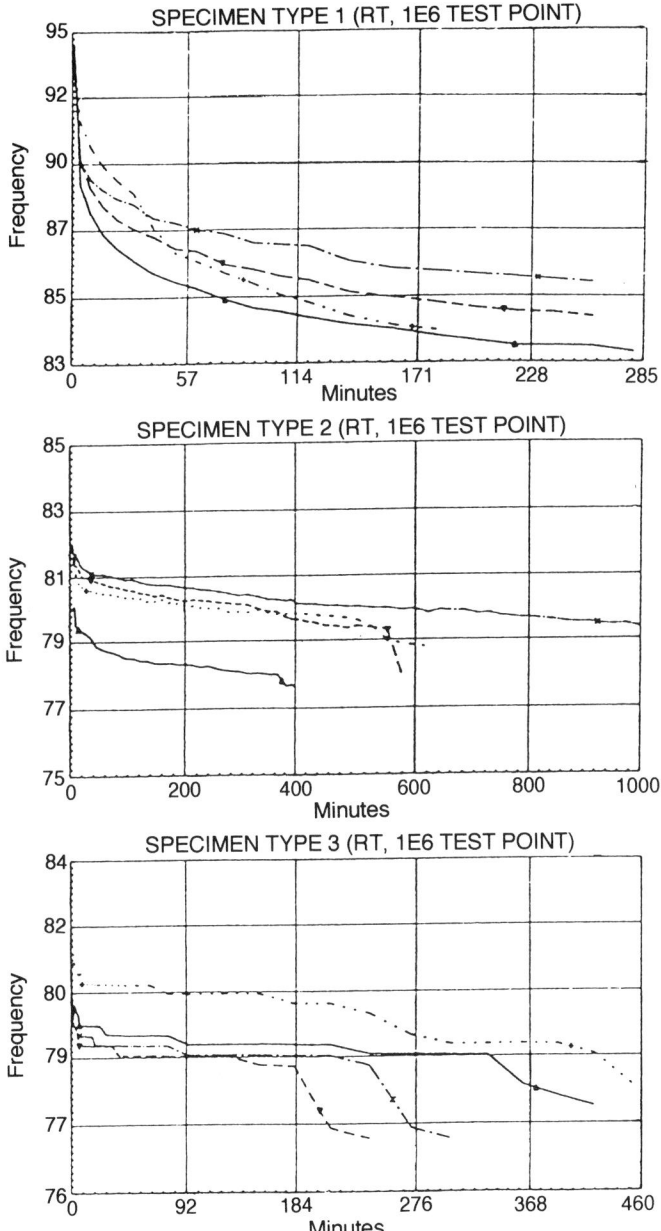

**Figure 2.20**   Typical settling characteristics for T800/BSL924 coupons

reduced settling phase which had a gradual change in frequency before the onset of linearity. There was no sharp frequency drop as observed in the work performed on HTA/6376. This is due to the taper of the bolted region instead of the step design which debonds.

Work on predamaged coupons made from CFRP HTA/6376 demonstrated that there was a significant settling phase which consisted of a sharp reduction in frequency followed by a more gradual decrease until the onset of linearity (see Figures 2.30–2.33).

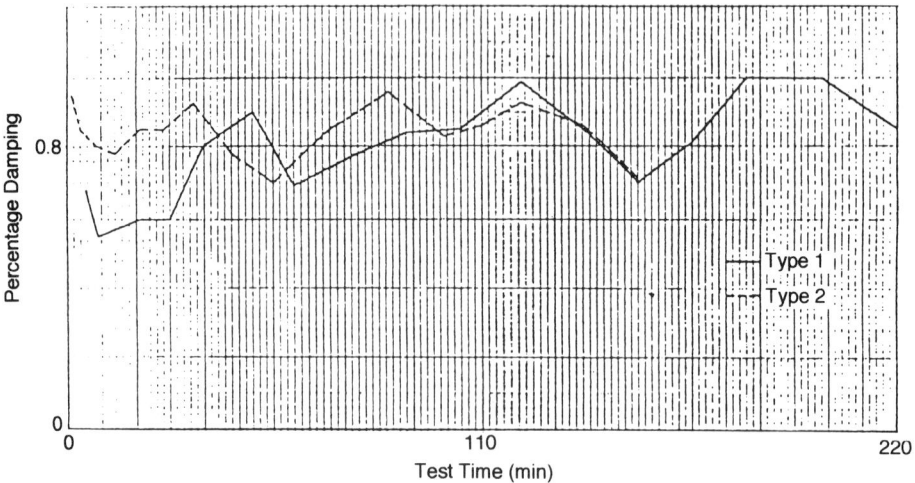

**Figure 2.21**  Typical percentage settling characteristics for T800/BSL924 coupons types 1 and 2

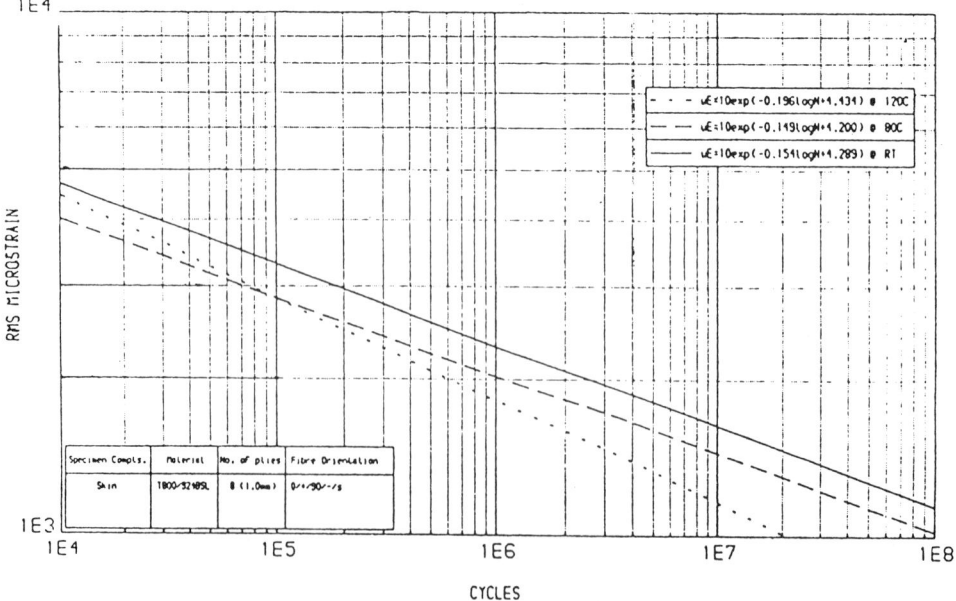

**Figure 2.22**  Specimen endurance curves for T800/BSL924 coupons types 1, 2 and 3

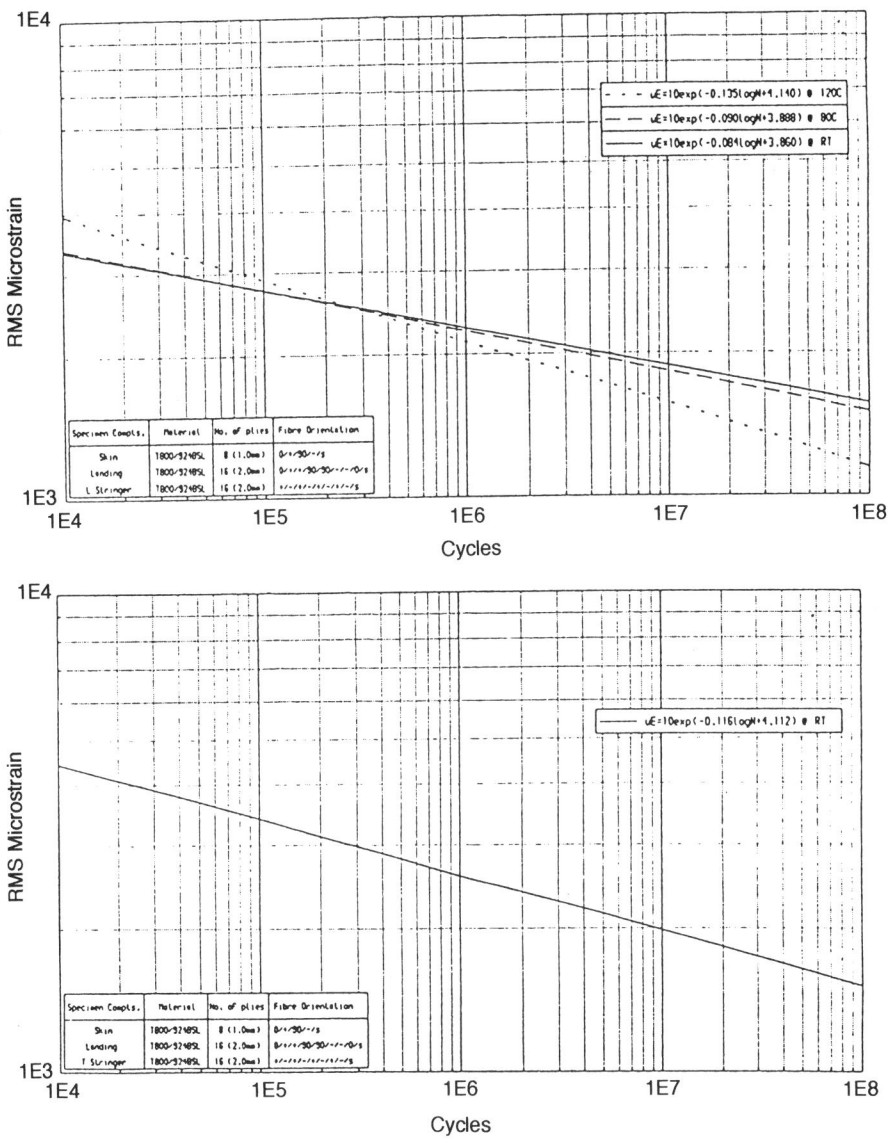

**Figure 2.22**   (*continued*)

*Effect of initial damage*

Dornier was the only TWG2 partner to investigate the effects of initial damage
on specimens. Damage regions were simulated by the incorporation of elliptical Teflon inserts in the layup of the HTA/6376 specimens. The two specimen
designs investigated by Dornier were of a similar design to those being
investigated by DASA (see Figures 2.30 and 2.32), so a comparison of the
endurance could be made for the damaged and undamaged specimens.

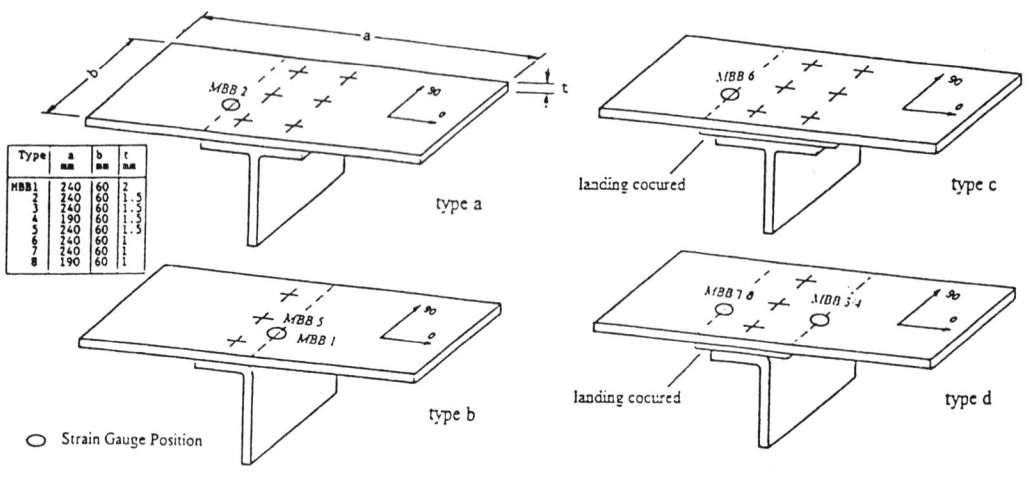

| type | | No | mat. | skin | | | landing | | stringer | | |
|------|---|-----|------|------|---|---|---------|---|----------|---|---|
| | | | | t [mm] | l [mm] | w [mm] | t [mm] | l [mm] | t [mm] | l [mm] | |
| 1 | | 9 | CFA | 2.0 | 120 | 60 | | | 2.5 | 20 | Hi-Lok-SENK |
| 2 | | 24 | CFA | 1.5 | 120 | 60 | | | 2.0 | 20 | " DAN 6-5-3(5/32") |
| 3 | | 48 | CFA | 1.5 | 120 | 60 | 2.125 | 25 | 2.0 | 20 | " + collar HL 94 |
| 4* | | 9 | CFA | 1.5 | 95 | 60 | 2.125 | 25 | 2.0 | 20 | " |
| 5 | | 12 | CFA | 1.5 | 120 | 60 | | | 2.0 | 20 | " |
| 6 | | 12 | CFA | 1.0 | 120 | 60 | 1.625 | 25 | 2.0 | 20 | " |
| 7 | | 39 | CFA | 1.0 | 120 | 60 | 1.625 | 25 | 2.0 | 20 | " |
| 8* | | 9 | CFA | 1.0 | 95 | 60 | 1.625 | 25 | 2.0 | 20 | " |
| | | | * other fibre direction | | | | | | | | |

**Figure 2.23**   Definitions of HTA/6376 coupons types MBB1 to MBB8

It was found that the endurance level of the damaged specimens appeared to
be better than the comparable undamaged specimens, which conflicted with
expectations (comparisons between coupons DO2 in Figure 2.31 with coupons
MBB2 in Figure 2.29; and coupons DO6 in Figure 2.33 with coupons MBB6 in
Figure 2.29). This result was due to the nature of control being adopted in the
tests. The DASA tests were load-controlled; therefore the strain level could

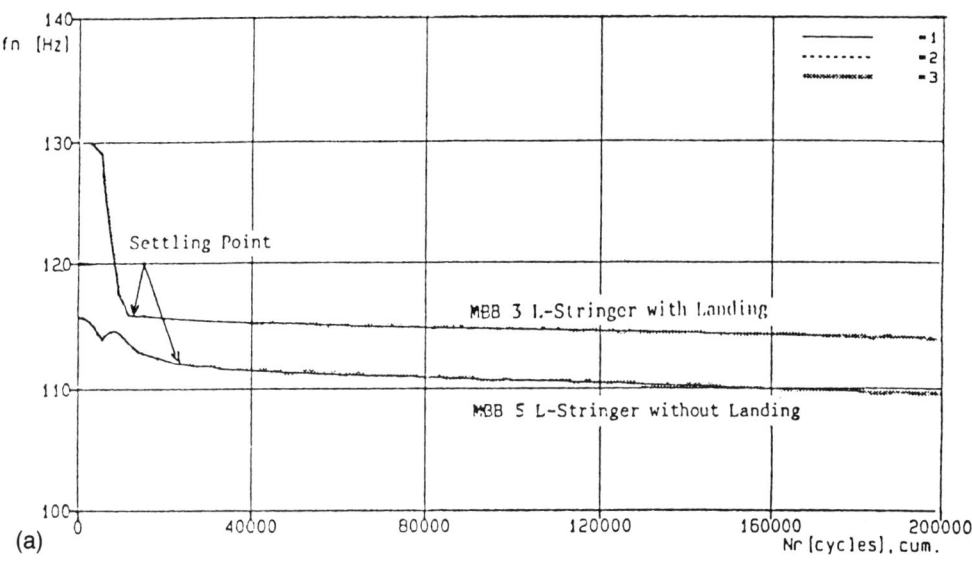

(a)

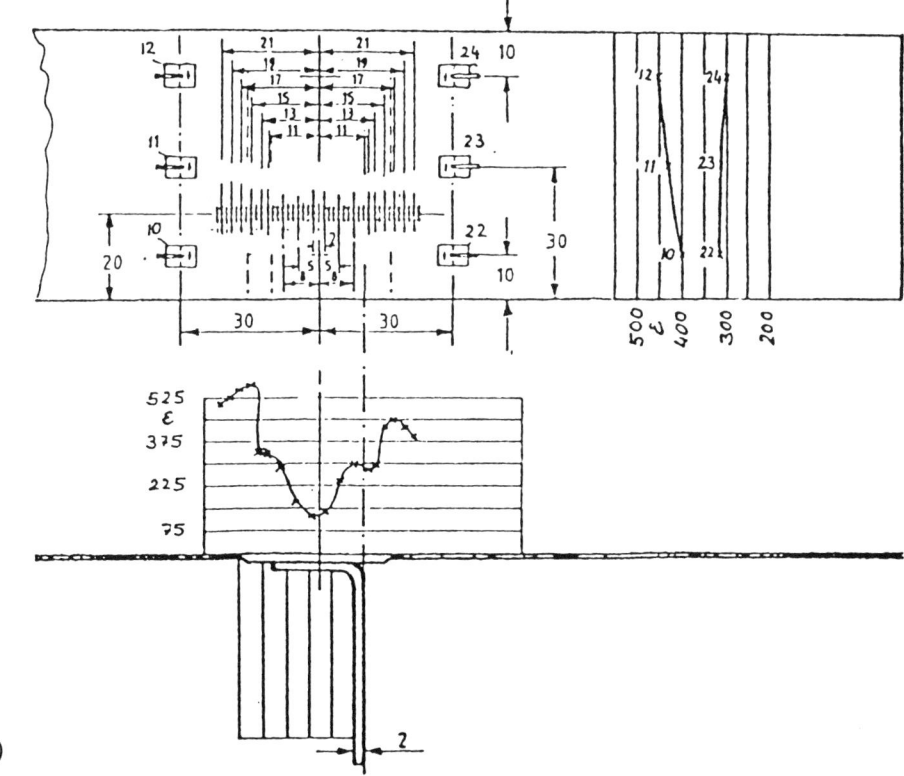

(b)

**Figure 2.24**  HTA/6376 coupons: (a) example of frequency curve (eigenfrequency) during settling/landing separation phase; (b) example of preliminary test, strain gauge locations and measured strain distribution of coupon type MBB7

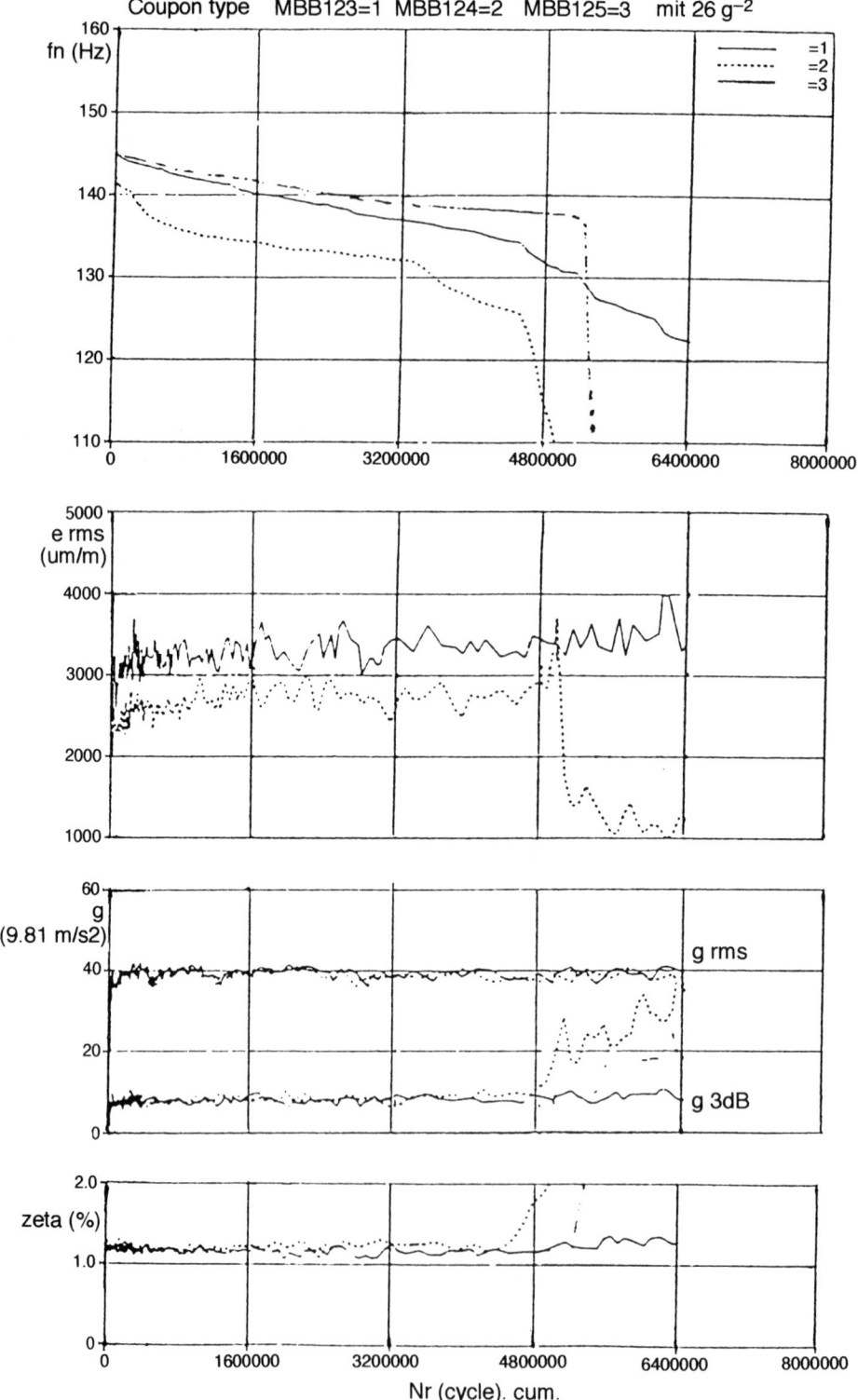

**Figure 2.25** HTA/6376 coupons: examples of measured data for MBB type 1, with an input load of 26 g$^2$/Hz

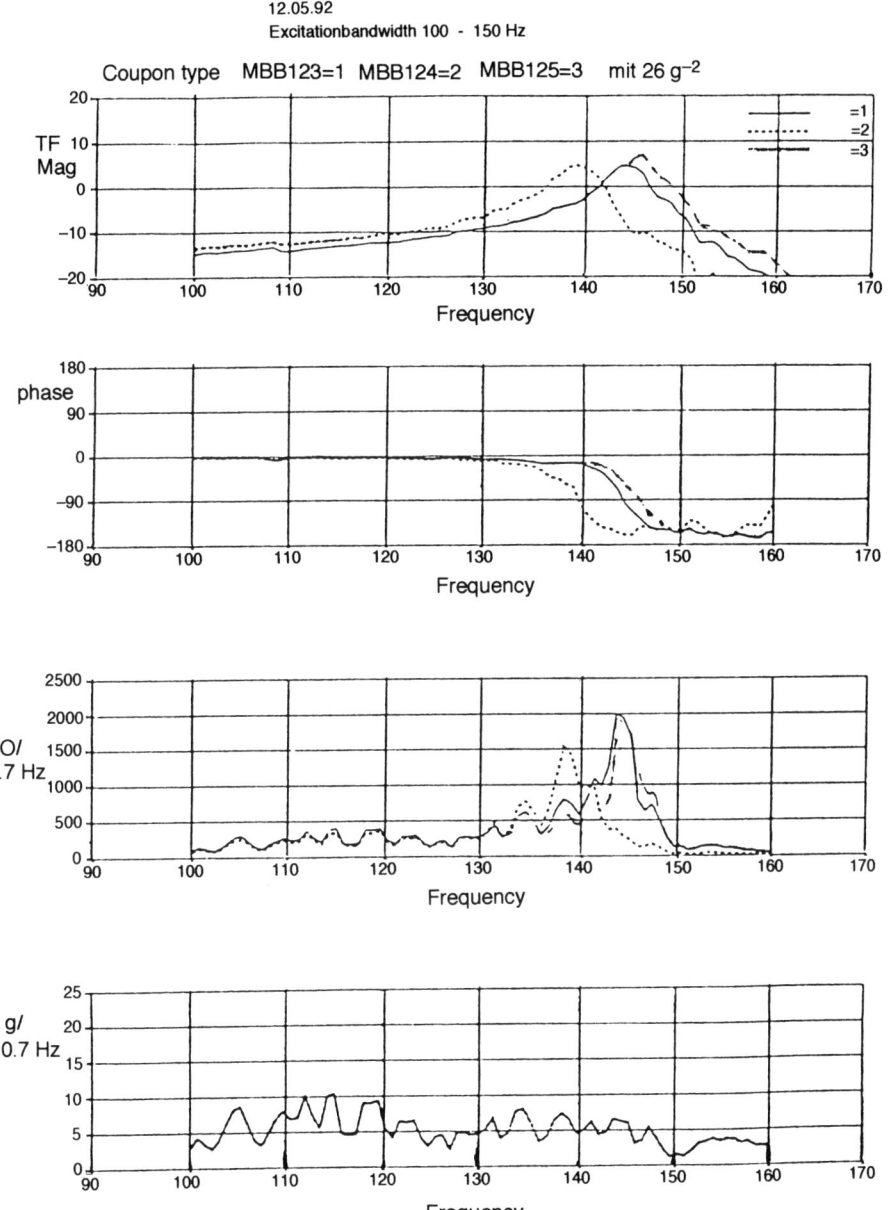

**Figure 2.26** HTA/6376 coupons: examples of analysis data for MBB type 1, with an input load of 26 g²/Hz

increase during the test. The Dornier tests were strain-controlled; therefore strain remained constant during the test. Owing to the nature of the DASA test technique, the endurance was marginally underestimated since the strain level used was taken from the initial stages (lowest strain level) of the test. A

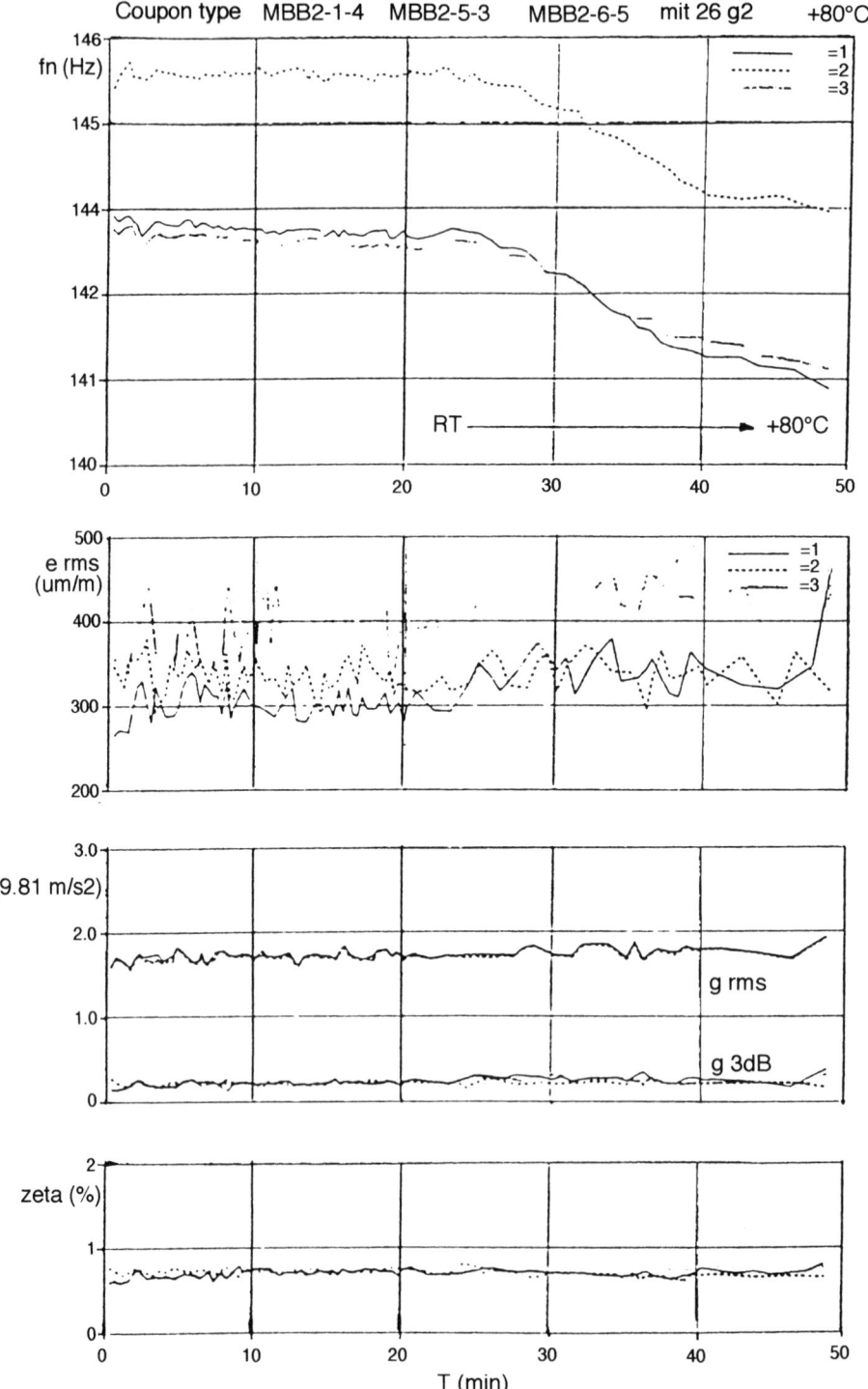

**Figure 2.27**    HTA/6376 coupons: measurement during heating phase, RT > 80 °C

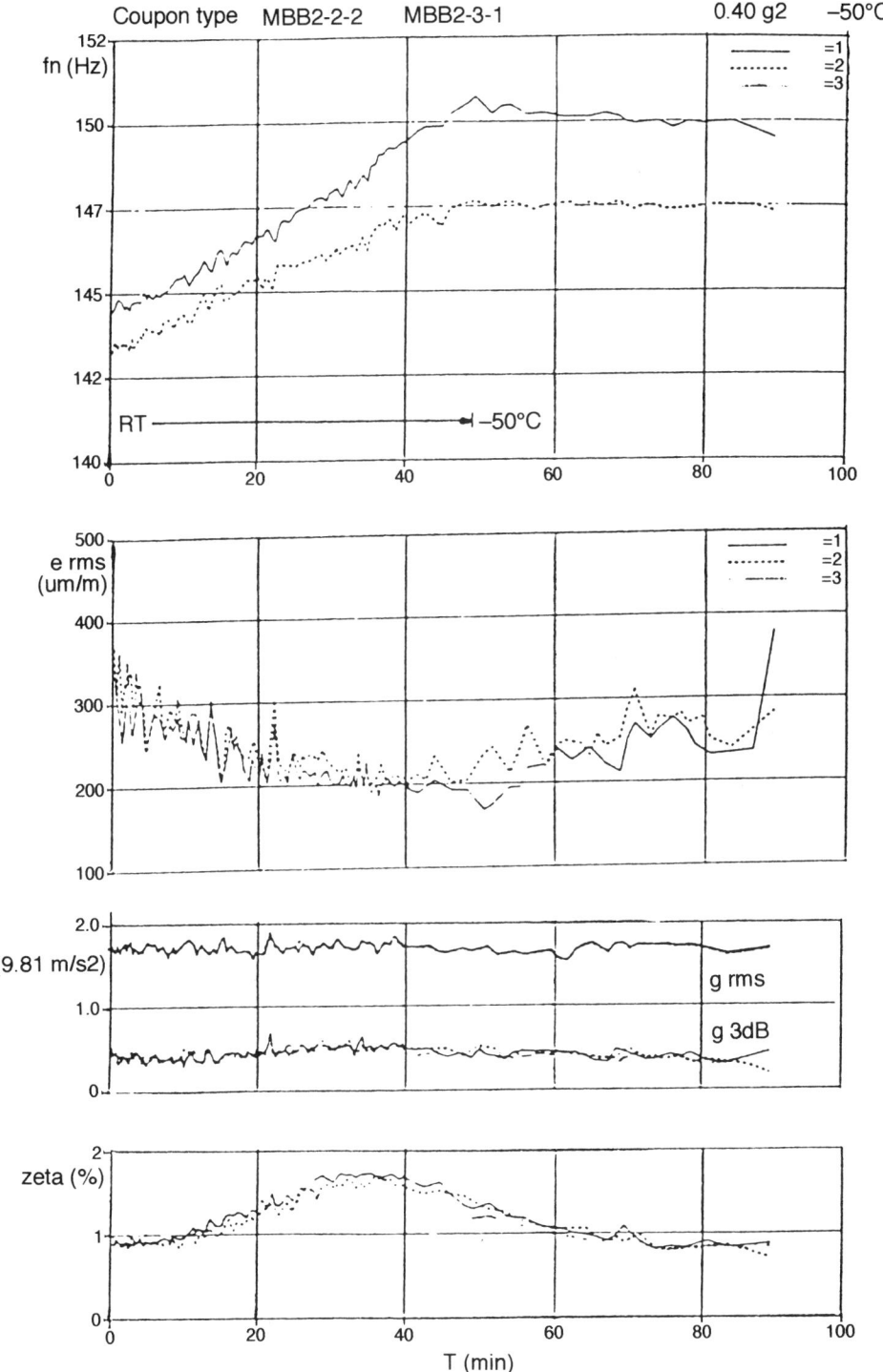

**Figure 2.28**   HTA/6376 coupons: measurement during cooling phase, RT > −50 °C

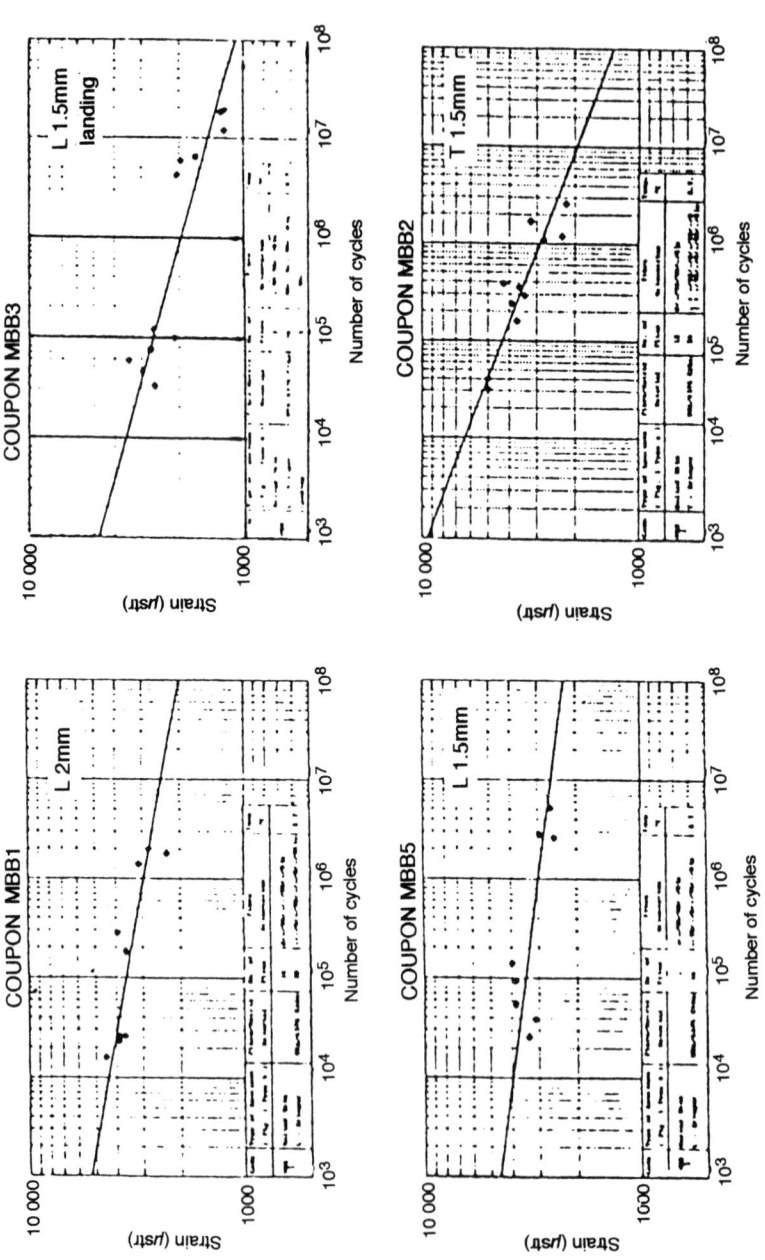

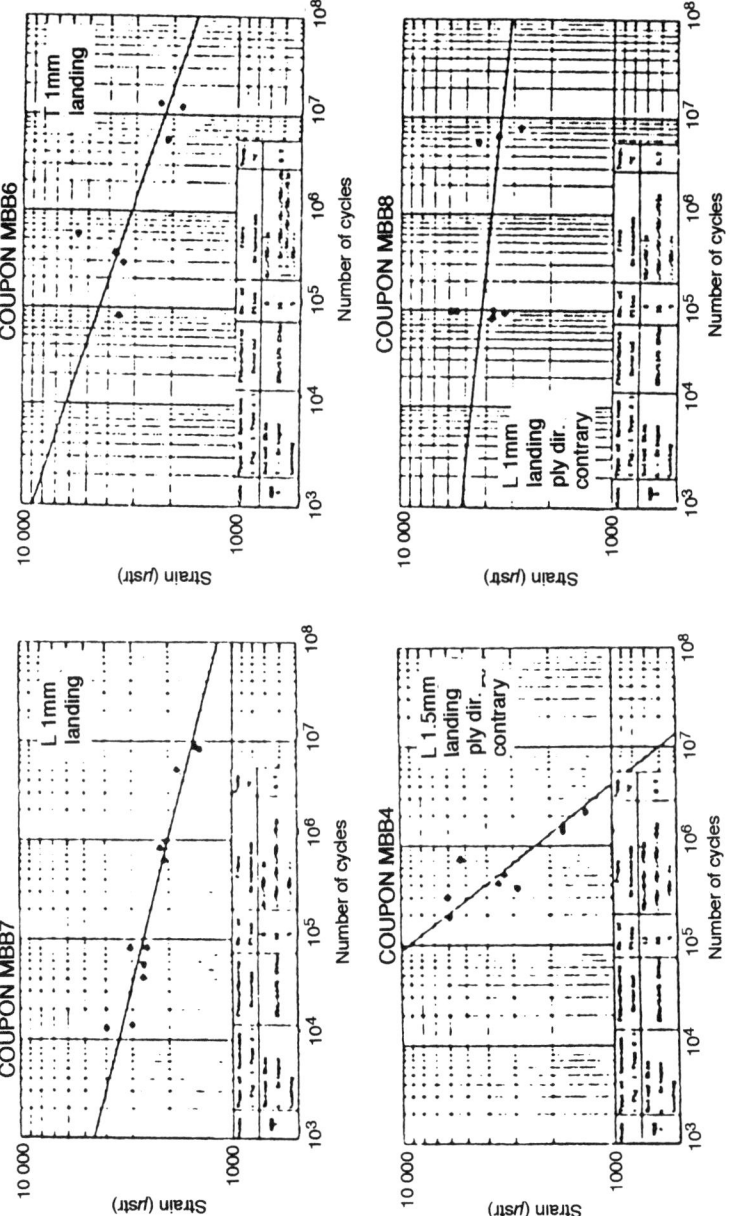

**Figure 2.29**   HTA/6376 coupons endurance data: comparison of $\varepsilon_{\mathrm{rms}}/N$ data for coupon types MBB1 to MBB8, RT

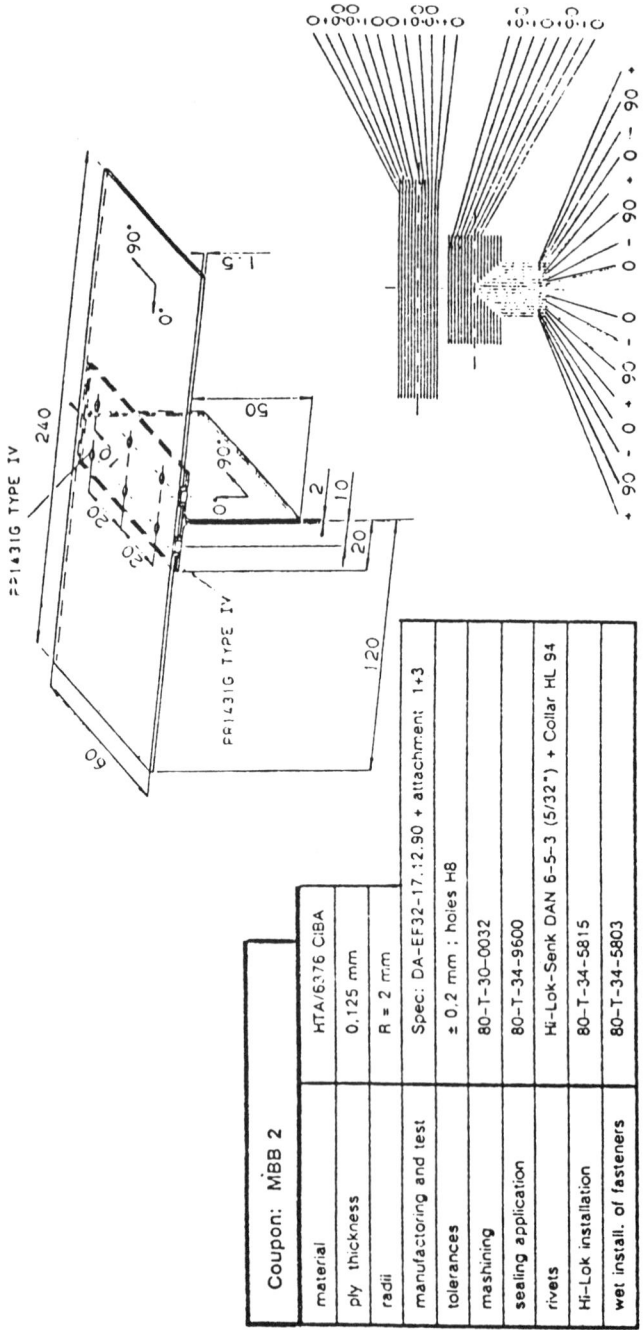

| Coupon: MBB 2 | |
|---|---|
| material | HTA/6376 CIBA |
| ply thickness | 0,125 mm |
| radii | R = 2 mm |
| manufactoring and test | Spec: DA–EF32–17.:2.90 + attachment 1+3 |
| tolerances | ± 0,2 mm ; holes H8 |
| mashining | 80–T–30–0032 |
| sealing application | 80–T–34–9600 |
| rivets | Hi–Lok–Senk DAN 6–5–3 (5/32") + Collar HL 94 |
| Hi–Lok installation | 80–T–34–5815 |
| wet install. of fasteners | 80–T–34–5803 |

**Figure 2.30**   Details of HTA/6376 coupon MBB2 with initial damage

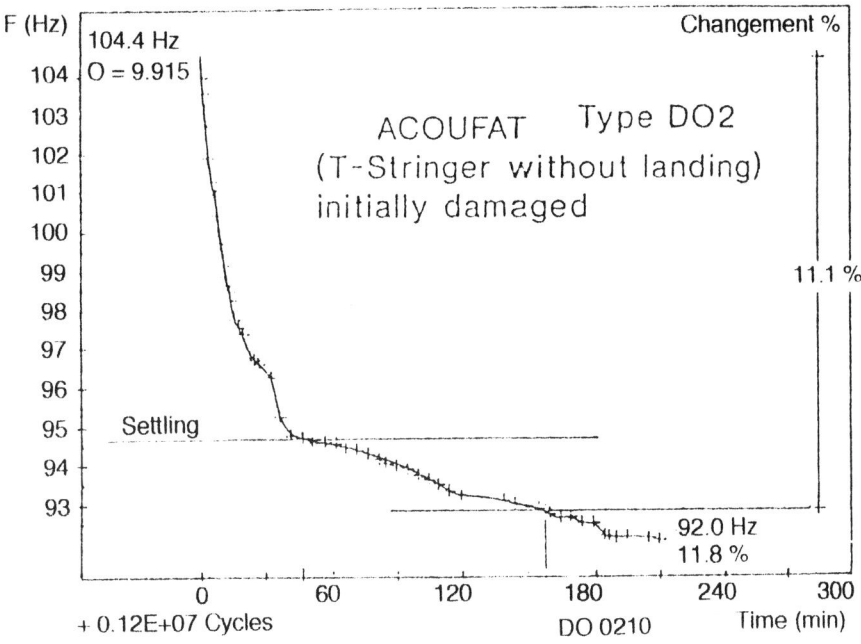

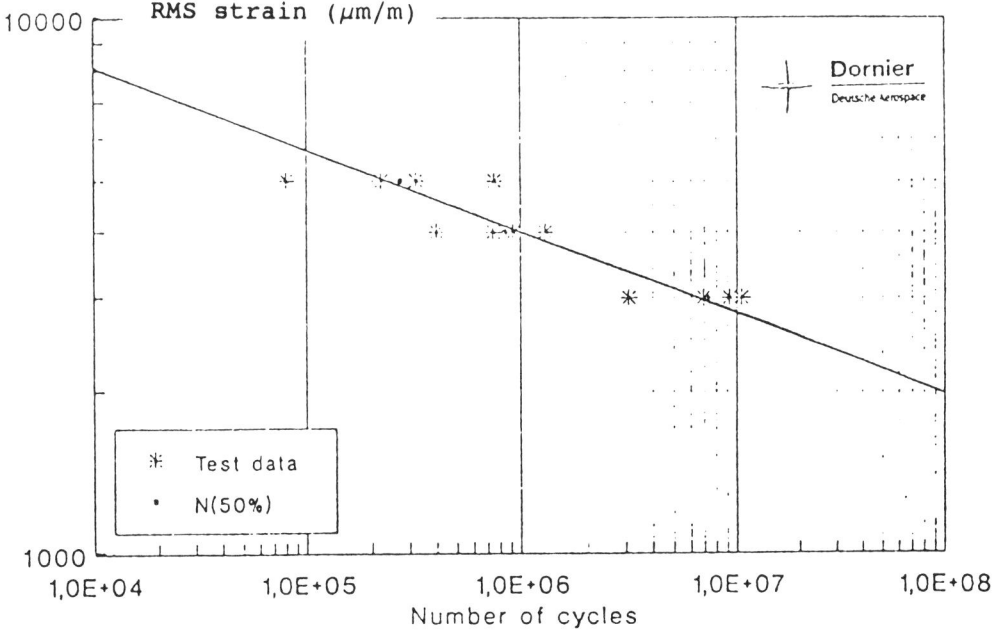

**Figure 2.31**    HTA/6376 coupon MBB2 with initial damage: measured data

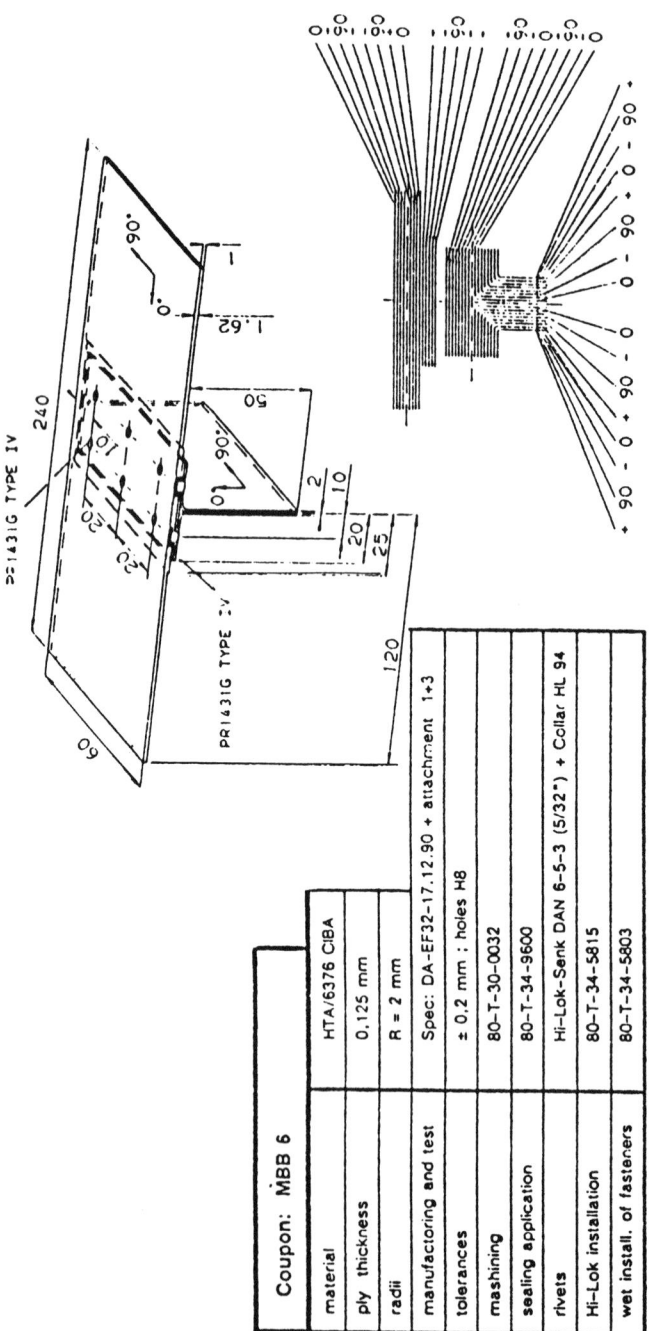

**Figure 2.32** Details of HTA/6376 coupon MBB6 with initial damage

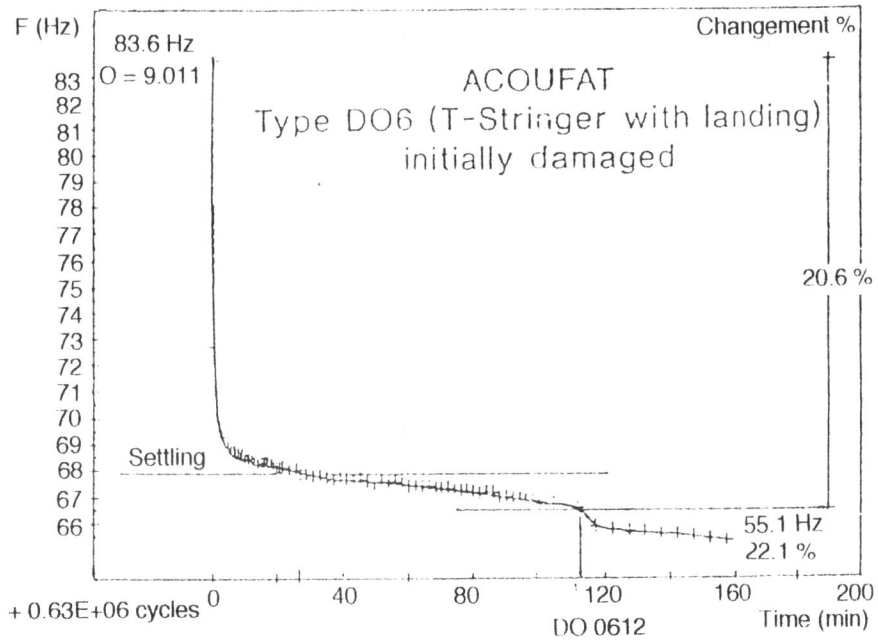

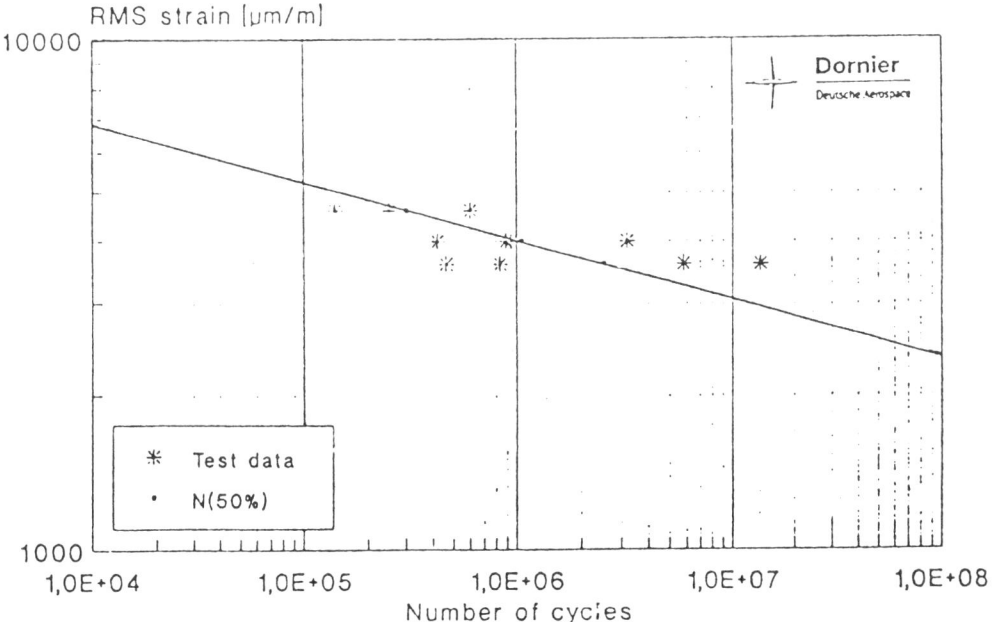

**Figure 2.33**  HTA/6376 coupon MBB6 with initial damage: measured data

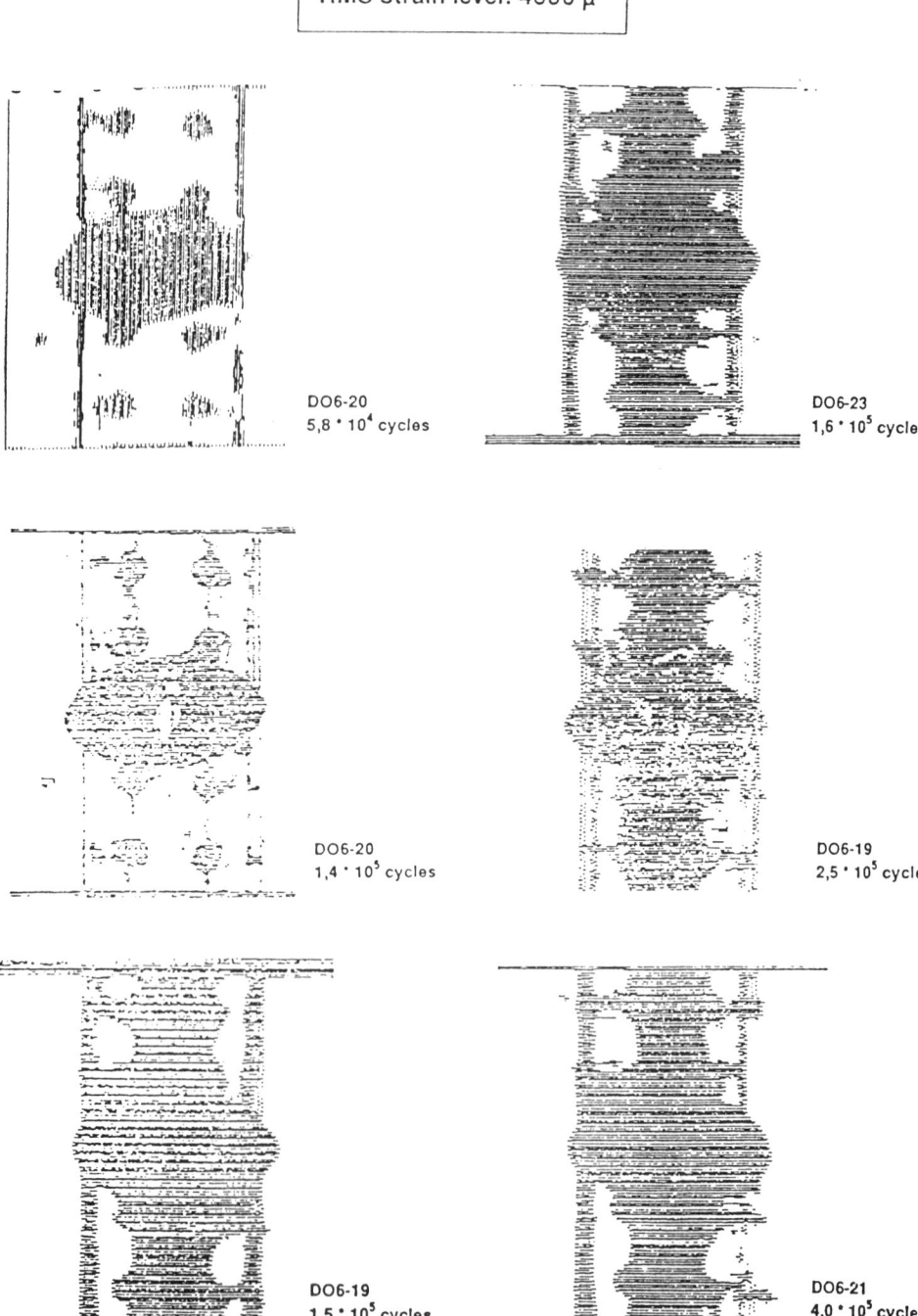

**Figure 2.34**   HTA/6376 coupons with initial damage: C-scan results

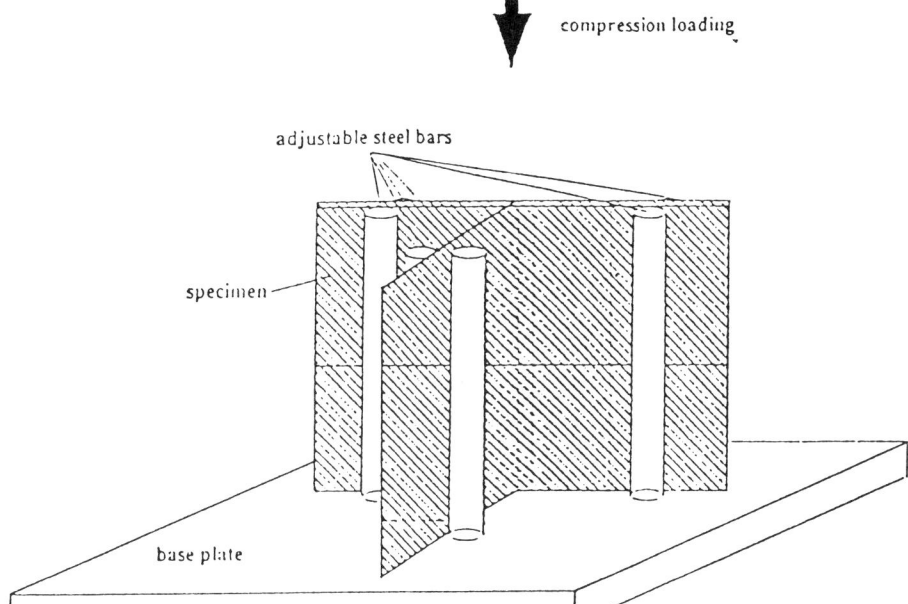

**Figure 2.35**  HTA/6376 coupons with initial damage: setup for residual strength tests in compression

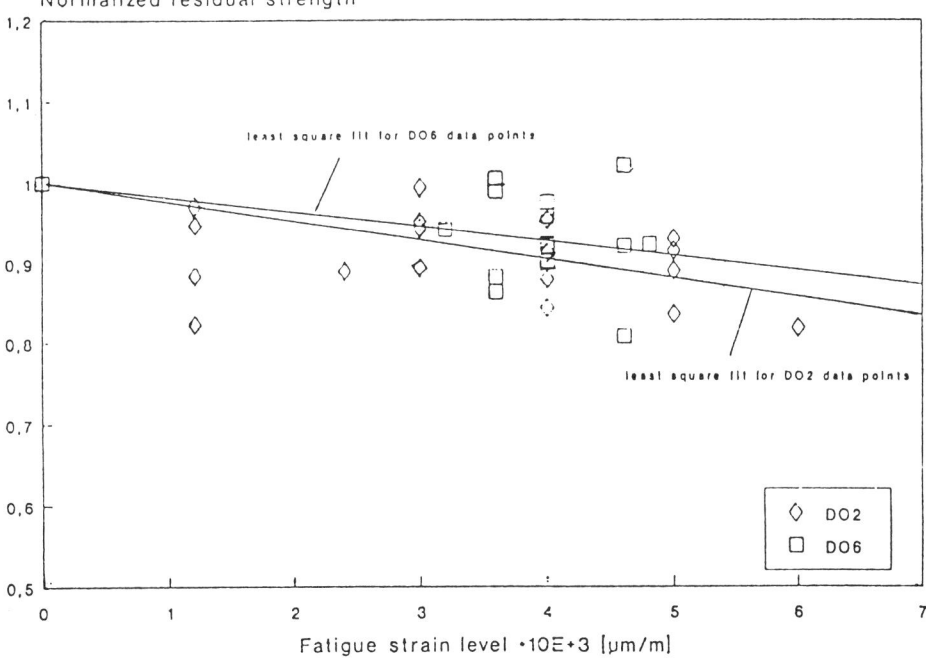

**Figure 2.36**  HTA/6376 coupons with initial damage: normalised residual strength

Dornier specimen was tested by DASA which demonstrated that the endurance of the predamaged specimen was similar to that of the undamaged type.

*Failure criteria*

Owing to the scope of materials and coupon designs being investigated, it was difficult to define an adequate failure criterion. Based on ESDU reports, it was agreed that a 2% reduction in specimen settled frequency would constitute 'failure'. As was apparent from the various tests performed, this criterion was difficult to apply, particularly for CFRP materials. It is generally recognised from all work performed by the partners that the 2% criterion was not sufficient to categorise failure.

Work performed on aluminium–lithium used a 4% criterion and a break criterion to establish failure of specimens (see Figure 2.10). The type of failures occurring in the specimens was questioned for their representativeness to true complex structural failures.

Problems arose during testing of SPF/DB titanium specimens owing to malfunctioning of the strain gauges; therefore specimen frequency was difficult to determine. Failure of the specimens was determined visually by the development of a crack across the width of the specimen.

Work performed by Fokker on GLARE materials demonstrated that, for a 2% drop in frequency, there was in some cases no apparent failure of the specimen. Further testing was carried out which demonstrated that there was considerable fatigue life in the specimen after the 2% criterion. Typical failures of the GLARE material consisted of cracking of the aluminium on the rivet line. However, this was dependent upon the type of GLARE material being investigated.

Tests performed by NLR on GLARE materials demonstrated that the 2% criterion was associated with cracking of the outer aluminium skin near the landing of the specimen.

Residual strength testing of HTA/6376 coupons demonstrated that a 30% reduction in strength occurred at 2% frequency drop. Despite that, no visible failures were evident in some cases.

Results of the testing performed on T800/924 coupons revealed that various regions of delamination were present at 2% drop in settled frequency (see Figures 2.17 and 2.18). Some of the failure modes encountered were thought to be unrepresentative of panel type failures (the edge effect—see Figure 2.18). Also, the 2% criterion was considered insufficient to categorise specimen failure.

As a result of the work performed on predamaged HTA/6376 coupons (see Figures 2.34–2.36) it was suggested that further investigations were required, to establish what the mechanisms were that contribute to settling of CFRP specimens. The use of a 'frequency degradation' criterion was not considered suitable for determination of specimen failures. It was suggested that a suitable failure criterion could be based on the degradation of the mechanical properties of a specimen.

*Endurance data*

The work performed on aluminium–lithium specimens (see Figure 2.10) demonstrated that the T-shaped substructure design had better endurance qualities than the L-shaped design. The cantilever specimen had the worst endurance. From the results of the temperature tests performed on the L coupons, it was inconclusive that temperature had an effect on endurance. The temperature tests performed on the T coupons demonstrated that fatigue life decreases with increasing temperature.

Results from temperature tests performed on the SPF/DB titanium coupons (see Figure 2.8) revealed that the endurance life decreased with increase in temperature.

From the work performed on GLARE materials (see Figures 2.15 and 2.16), it was shown that for the given coupon geometry tested and rib-mode vibration, GLARE A had four times the endurance (number of cycles for the same reference r.m.s. elongation) of GLARE C or E, which had similar endurance qualities.

Various geometries constructed from GLARE materials were investigated by NLR (see Figures 2.13 and 2.14). It was shown that the plain specimen design had the highest endurance when compared with the substructure coupons. It is thought that this is due to beneficial clamping effects. It was also shown for the plain specimen design that GLARE C had the highest endurance. For the other substructure designs, GLARE A and E had better endurance qualities than GLARE C. The effect of temperature was not noticeable for GLARE A and E endurance (see definition Figure 2.11). GLARE C endurance deteriorated with an increase or decrease in temperature.

The testing performed on CFRP HTA/6376 materials demonstrated that the endurance marginally decreased with increasing temperature. From testing performed at −50 °C, the endurance was shown to improve. From the numerous specimen designs investigated, it was shown that the endurance data for the given material and ply orientation at the outer regions of the skin depended mainly on the joint configuration.

The testing performed on the CFRP T800/924 materials demonstrated that the endurance was only marginally affected by increasing test temperature. The T substructure coupon was shown to have similar endurance characteristics to the L coupon (see Figure 2.22). The plain laminate coupon did not show any marked increase in endurance due to the nature of the failure mode.

The work on predamaged HTA/6376 specimens demonstrated (as noted above) that the endurance was not significantly affected in comparison with undamaged specimens.

*Conclusions*

1.  Generally, metallic and hybrid coupons do not exhibit a 'settling phase' except in some cases where a sudden frequency drop occurs, owing to

substructure behaviour. This 'settling' of the specimen is not an appreciable length of test time.

2.  The CFRP coupons exhibited a settling phase. However, it was difficult to establish specimen 'settling' in some cases, owing to substructure behaviour.

3.  It was generally agreed that a 'frequency degradation' criterion was not sufficient to categorise specimen failure.

4.  A possible failure criterion could be based upon residual-strength testing since, after specimen vibration, there can be a reduction of up to 30% in static compressive strength for CFRP materials.

5.  Increase in temperature, generally, has the effect of reducing fatigue life for all materials tested.

6.  The effect of single-impact damage on HTA/6376 material does not affect fatigue life.

7.  The failure modes observed in some test coupons were in some cases considered unrepresentative of complex structural failure (edge effects).

*Recommendations*

1.  Further work is required to investigate the reasons for the 'settling phase' observed in CFRP materials.

2.  A suitable failure criterion is required for all types of specimens tested, possibly based on residual strength characteristics of the specimen.

3.  An assessment of coupon failures compared with complex structural failures is required in order to validate the use of present coupon designs.

### 2.3.2   Acoustic fatigue tests of panels in progressive wave tubes (PWTs)

*Introduction: subtask 3.2 presentation*

This section summarises all the experimental endurance testing activity performed on panels within subtask 3.2.

From the shaker specimen endurance tests, a first assumption which dictates the design of a coupon is that panel behaviour can be adequately represented using a thin strip (beam behaviour). This approximation is valid for panels with large aspect ratios where edge effects in the long direction of the panel can be ignored. Furthermore, in order to subject a test coupon to a pseudo-acoustic load it is assumed that panel/stringer constructions exposed to an acoustic field have predominant modes of vibration associated with symmetrical bending of the panel skin around the substructure. The above assumptions will become invalid if panel aspect ratios are small and different structural modes are excited, owing to spatial effects of the acoustic field. Consequently, the coupon's $S/N$ data obtained in *Task 2* must be validated by tests which are more representative of large structural parts of aircraft.

Within subtask 3.2, a total of six advanced composite and metallic panels were investigated in PWTs with respect to their dynamic properties such as natural modes, natural frequencies, damping factors, linearity as well as their acoustic fatigue behaviour. The initial objectives of this were, first, part of the

**Figure 2.37**   Test setup of the progressive wave tube (PWT)

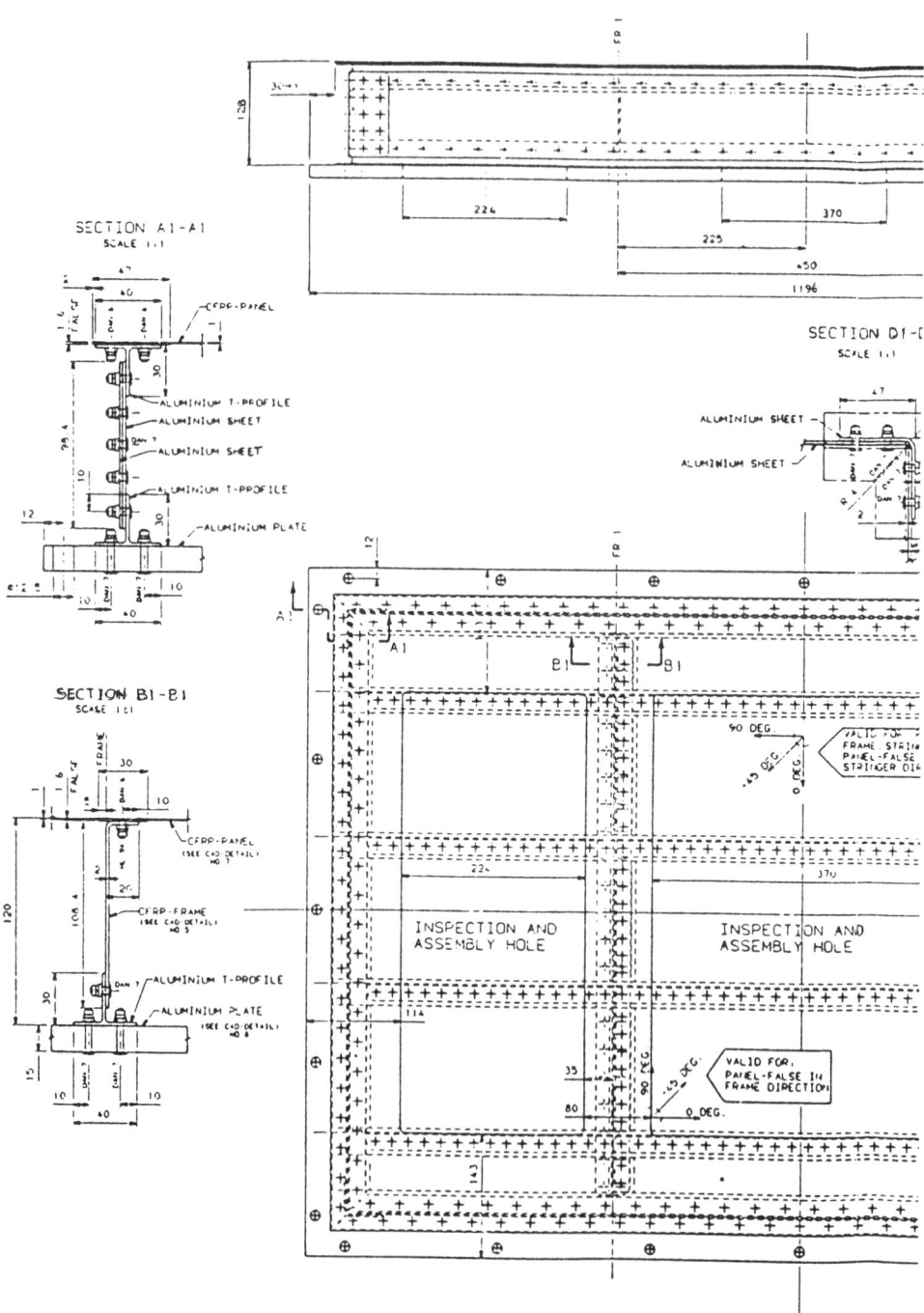

**Figure 2.38**    Dimensions of the HTA/6376 panel

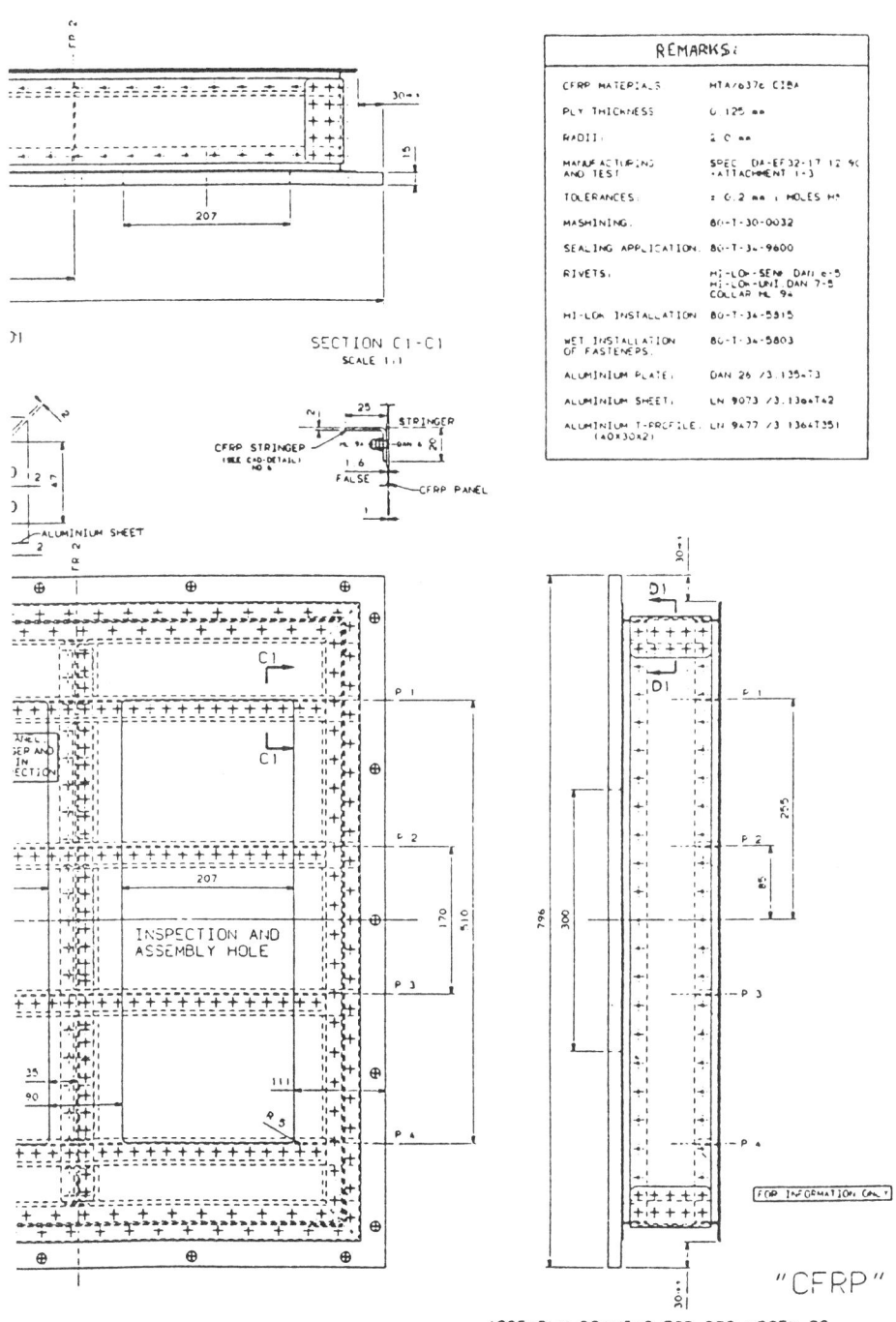

SECTION C1-C1
SCALE 1:1

| REMARKS: | |
|---|---|
| CFRP MATERIALS | HTA/6376 CIBA |
| PLY THICKNESS | 0.125 mm |
| RADII | 2.0 mm |
| MANUFACTURING AND TEST | SPEC DA-EF32-17 12.90 +ATTACHMENT 1-3 |
| TOLERANCES | ± 0.2 mm ; HOLES H9 |
| MASHINING | 80-T-30-0032 |
| SEALING APPLICATION | 80-T-34-9600 |
| RIVETS | HI-LOK-SEN DAN e-5 HI-LOK-UNI DAN 7-5 COLLAR ML 94 |
| HI-LOK INSTALLATION | 80-T-34-5915 |
| WET INSTALLATION OF FASTENERS | 80-T-34-5803 |
| ALUMINIUM PLATE | DAN 26 /3.135-T3 |
| ALUMINIUM SHEET | LN 9073 /3.136+T42 |
| ALUMINIUM T-PROFILE | LN 9477 /3.1364T351 (40x30x2) |

INSPECTION AND ASSEMBLY HOLE

"CFRP"

ASSEMBLY DRAWING FOR PER UDSEN CO.
BRITE EURAM "ACOUSTIC FATIGUE TESTS"

experimental activity to confirm the failures modes and the $S/N$ data gathered during the tests of coupons by shaker excitation (Section 2.3.1), and secondly, to assess the contributory effects of modes other than the fundamental.

As far as the structural response is concerned, this testing activity was in relation to the analytical activity of subtask 4.2, presented in Section 2.5.

*Definition of the six test panels*

The structural concept ('panel' or 'box') of the endurance PWT specimens was defined in *Task 1*, in accordance with the definition proposed for the aluminium specimen dedicated to the WT and PWT tests of subtask 3.1 (see Figure 2.55).

The box concept was selected for these specimens:

- A thin upper panel with or without stringers, on which the acoustic loads of the PWT were applied (see Figure 2.36), representative of aircraft structural parts.

- Substructure components representative of webs or spars or other aircraft structural components; the substructure components divided the upper panel into $3 \times 5$ bays.

- A thick lower plate which gave a high rigidity to the box.

The lower plate had cutouts in each bay to avoid cavity effects and low-frequency modes of the total box. Twenty-four attachment points allowed integration of the test box to the side of the PWT.

Test specimen assemblies are presented in Figure 2.38 (HTA and aluminium–lithium panels), Figure 2.47 (GLARE panels) and Figure 2.50 (T800 panel). For all these specimens, the outer dimensions of the upper plate were 1136 mm $\times$ 736 mm.

Five specimens were defined according to this box concept. The sixth specimen was representative of a real civil aircraft spoiler (see Figure 2.53). In

**Table 2.1** The division of responsibilities among the ACOUFAT partners

| Material | Design | Manufacture | Testing | Calculations |
|---|---|---|---|---|
| Aluminium –lithium | MBB | PUC | IABG | MBB |
| T800/BSL924 | BAe | BAe | BAe | KUL |
| GLARE (2 panels) | Fokker | Fokker | IABG | Dassault |
| SPFDB Ti | BAe | BAe | BAe | Saab |
| HTA/6376 | MBB | PUC | MBB | MBB, KUL Dassault Saab |

the following discussion, all these specimens are referred to as 'panels' (see Table 2.1).

The five materials selected for the second part of the experimental study (panels in PWT) are the same as those investigated in the first part of this study (coupons by shaker excitation). All the panels were associated with one of the coupon types tested in *Task 2*, to provide comparative endurance data between PWT/panel tests and shaker/coupon tests.

All test panels were subjected to similar test programs, which included modal analysis, linearity checks and endurance testing in PWTs.

### Acoustic tests on the HTA/6376 panel

The test article (see Figure 2.38) was instrumented with 12 strain gauges, the location of which can be seen in Figure 40. For modal analysis, 11 additional accelerometers were used. The following investigations were performed.

**Modal analysis**  By means of sinusoidal sweeps produced by a loudspeaker, the test article was excited in its natural modes. In order to avoid interference of the PWT environment, these tests were performed in a separate test setup. Frequency response curves, as shown in Figure 2.39, were taken for each

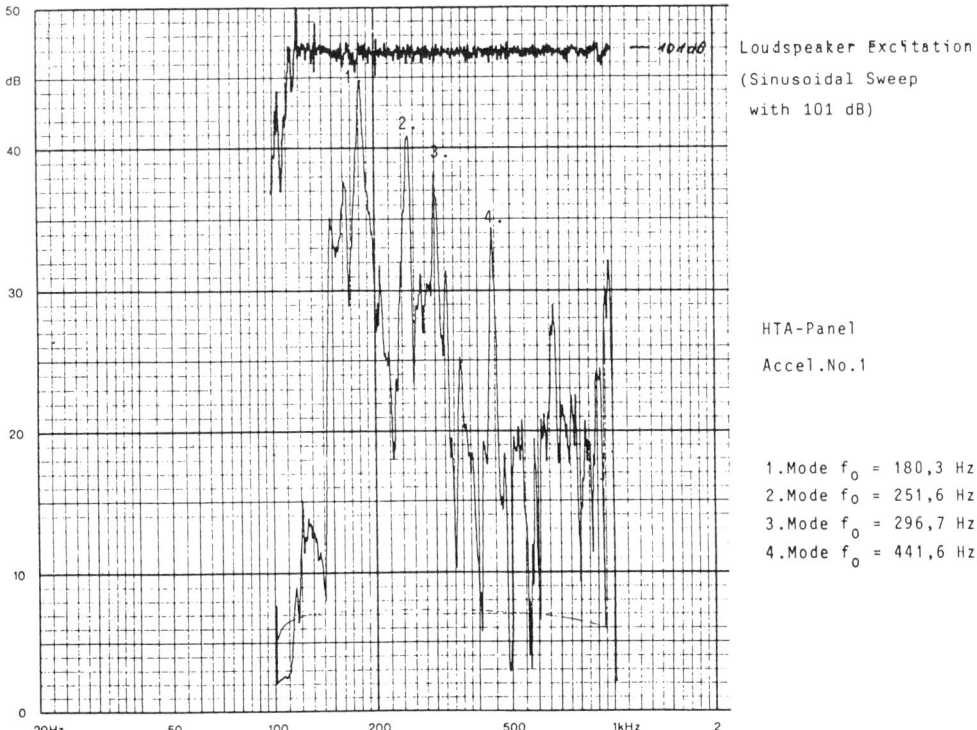

Figure 2.39  HTA/6376 panel: frequency response curve under loudspeaker excitation

individual accelerometer and the dominant natural modes were derived from these curves. In a second step, the natural modes were excited to obtain their mode shapes and the associated loss factors (damping). Further modal analysis was performed by KUL.

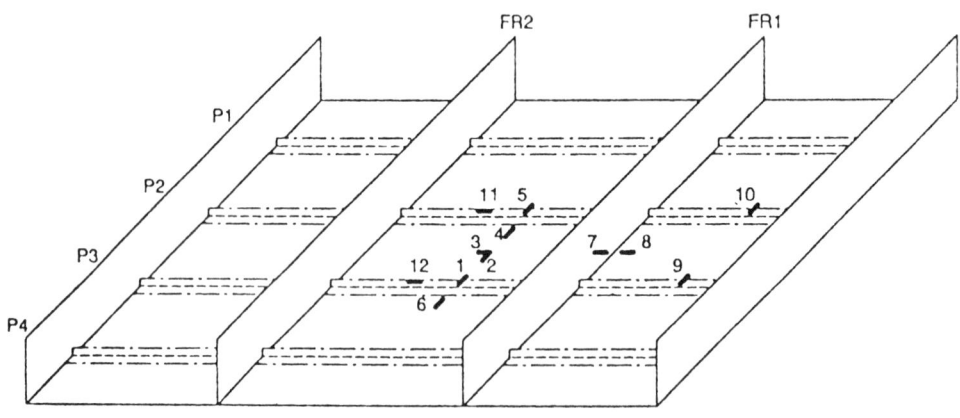

LOCATION OF STRAIN GAUGES

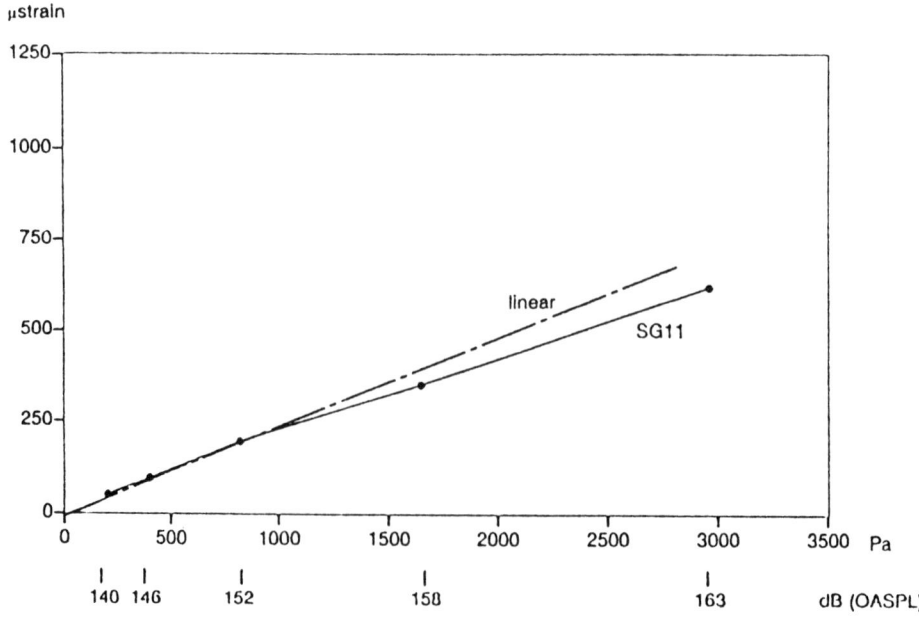

**Figure 2.40**    HTA/6376 panel: plots of strain versus excitation level

**Strain response and linearity**   The test article was then mounted into the PWT and excited by acoustic spectra with increasing sound pressure levels (SPLs). The linearity of structural response versus exciting sound pressures was checked for selected strain gauges (Figure 2.40). Strain-gauge and accelero-

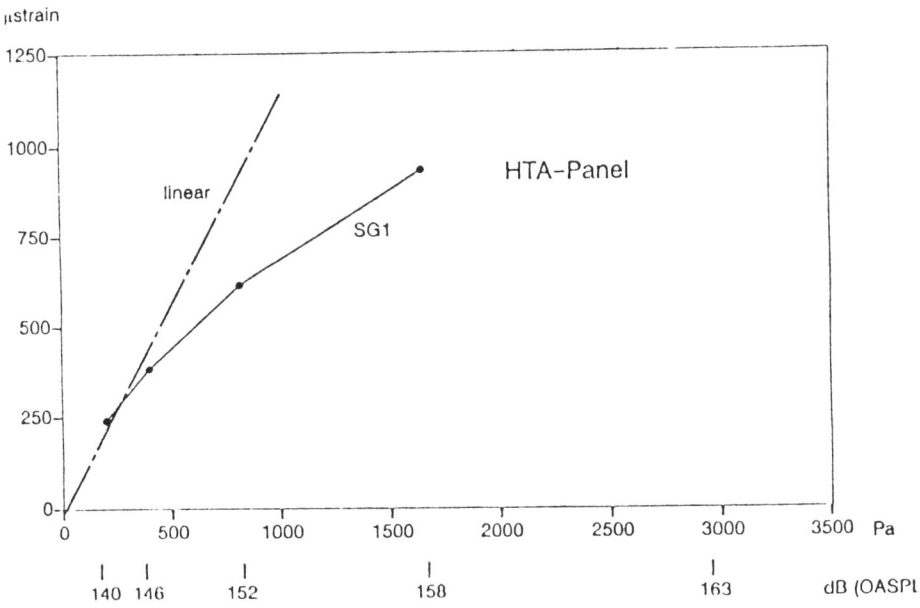

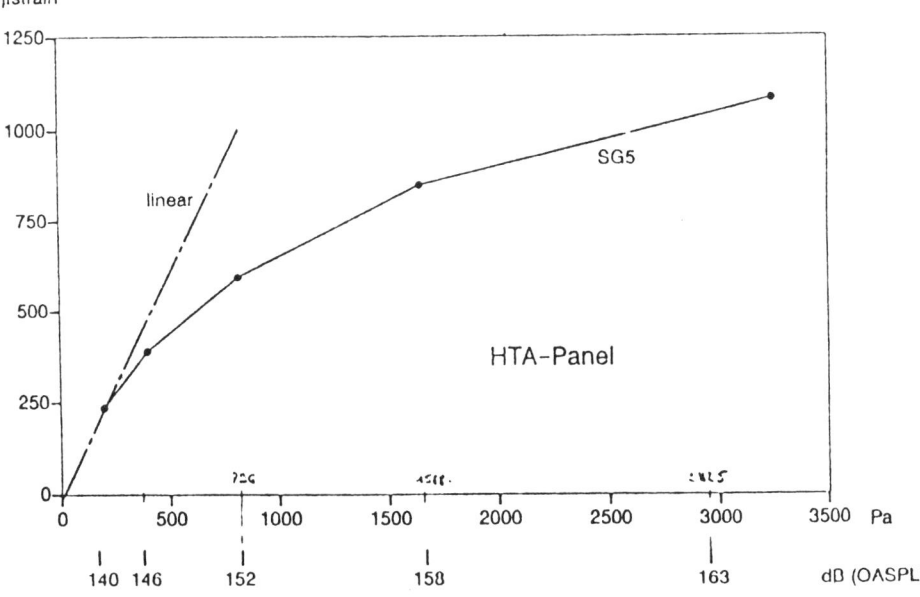

**Figure 2.40**   (*continued*)

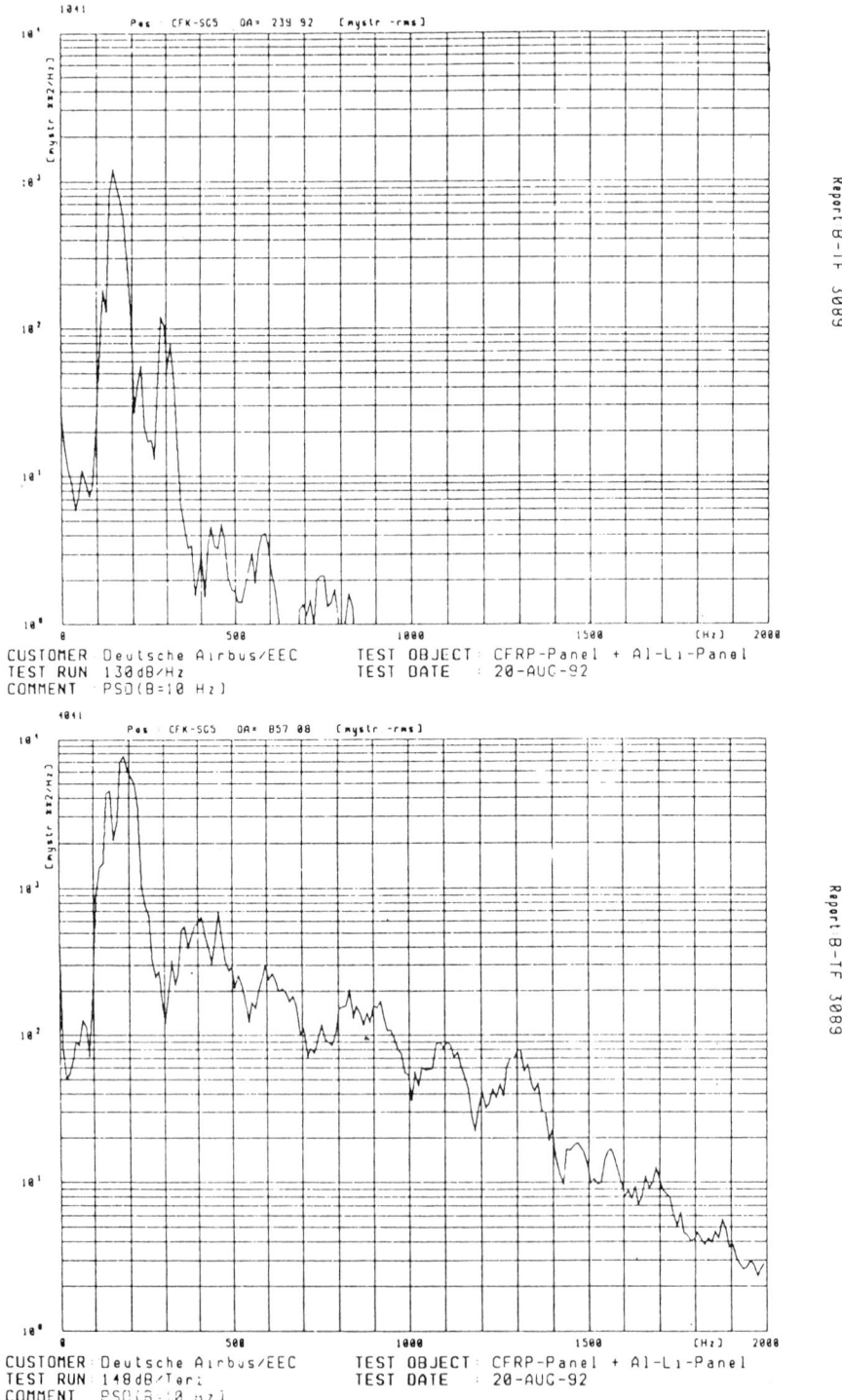

**Figure 2.41**   HTA/6376 panel: PSD plots of strain response

meter signals were also subjected to power spectral density (PSD) analysis, and some typical PSD plots are given in Figure 2.41.

**Acoustic endurance**  The test article was finally exposed to an endurance test which was performed in the sequence shown in Table 2.2.

Up to run 5, no visible damage could be observed. Runs 6, 7 and 8 were performed with intermediate inspections until the damage had developed to the status shown in Figure 2.43. Major damage occurred to the landing along

**Table 2.2**  Acoustic endurance tests on the HTA panel

| Run | OASPL (dB) | Duration (min) | Observation |
|-----|-----------|----------------|-------------|
| 1 | 140 | 13.25 | No visible damage |
| 2 | 146 | 7.45 | No visible damage |
| 3 | 152 | 6.90 | No visible damage |
| 4 | 158 | 4.22 | No visible damage |
| 5 | 163 | 3.15 | No visible damage |
| 6 | 162 | 15.00 | Damage |
| 7 | 162 | 15.00 | Damage |
| 8 | 162 | 15.00 | Damage as in Figure 2.43 |

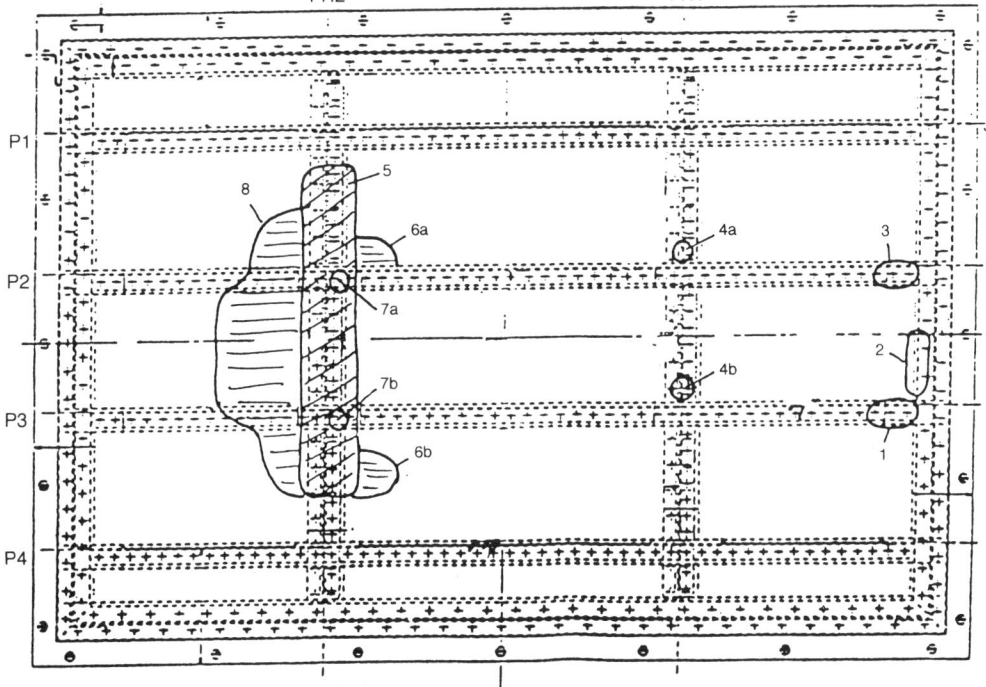

**Figure 2.42**  Areas of damage on HTA panel

Figure 2.43  Damage on HTA panel

the frame rivet line which had separated from the skin. Other damage included pulled rivets.

*Acoustic tests on the aluminium–lithium panel*

The design of the aluminium–lithium panel was similar to that of the HTA panel. The same investigations with the same test procedures were performed for the two panels. For the modal analysis, see Figure 2.44; for the strain response and linearity, see Figure 2.45.

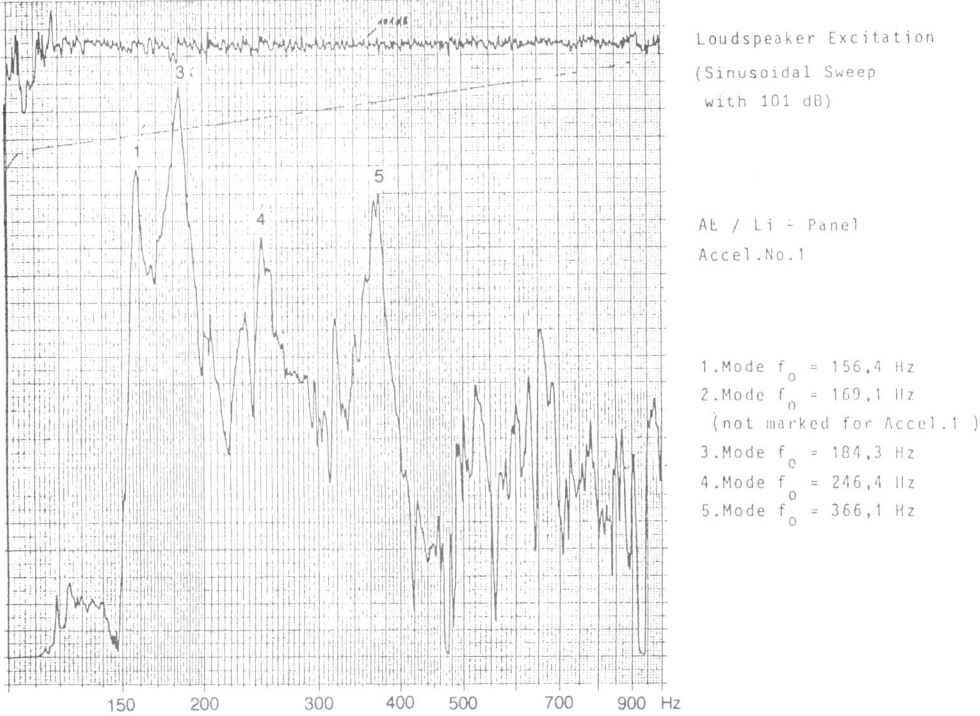

**Figure 2.44** Aluminium–lithium panel: frequency response curve under loudspeaker excitation

Table 2.3   Acoustic endurance tests on the Al–Li panel

| Run | OASPL (dB) | Duration (min) | Observation |
| --- | --- | --- | --- |
| 1 | 140 | 13.25 | No visible damage |
| 2 | 146 | 7.45 | No visible damage |
| 3 | 152 | 6.90 | No visible damage |
| 4 | 158 | 4.22 | No visible damage |
| 5 | 163 | 3.15 | Damage as in Figure 2.46 |

**Acoustic endurance**  The endurance test was performed in the sequence shown in Table 2.3.

Up to run 4, no visible damage could be observed. However, after run 5 the panel was severely damaged as shown in Figure 2.46.

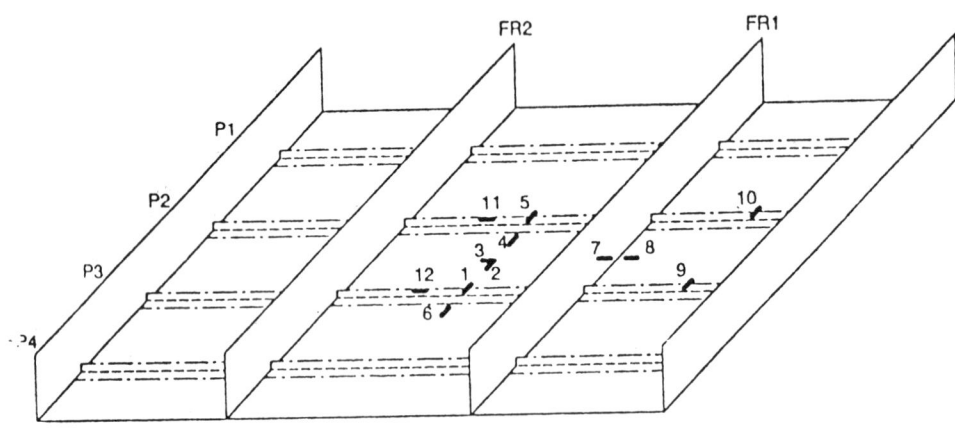

LOCATION OF STRAIN GAUGES

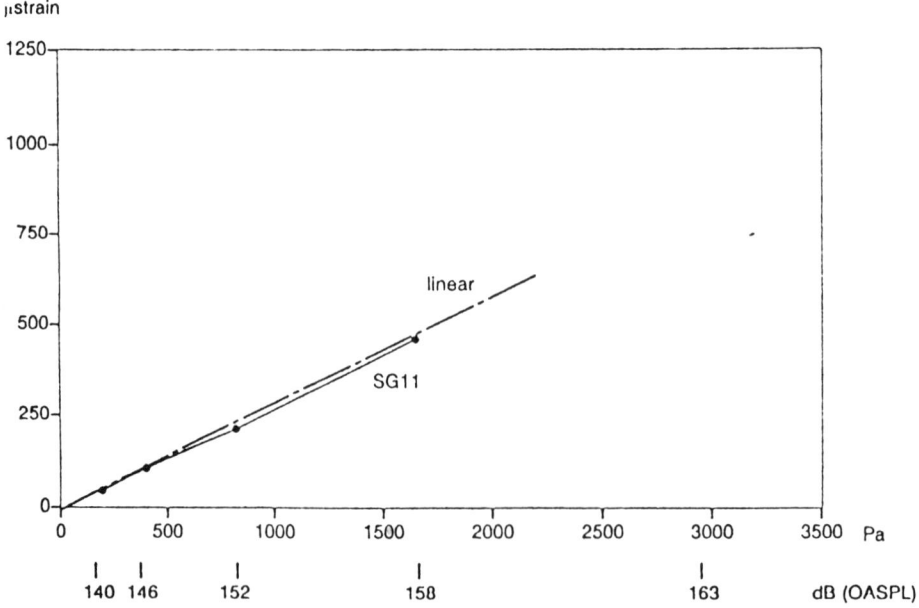

**Figure 2.45**   Aluminium–lithium panel: plots of strain versus excitation level

*Acoustic tests on two GLARE panels*

The two test articles (see Figure 2.47) had different dimensioning of the upper GLARE panels). Each of the test articles was instrumented with 12 strain gauges, the location of which can be seen from Figure 2.48. For modal analysis,

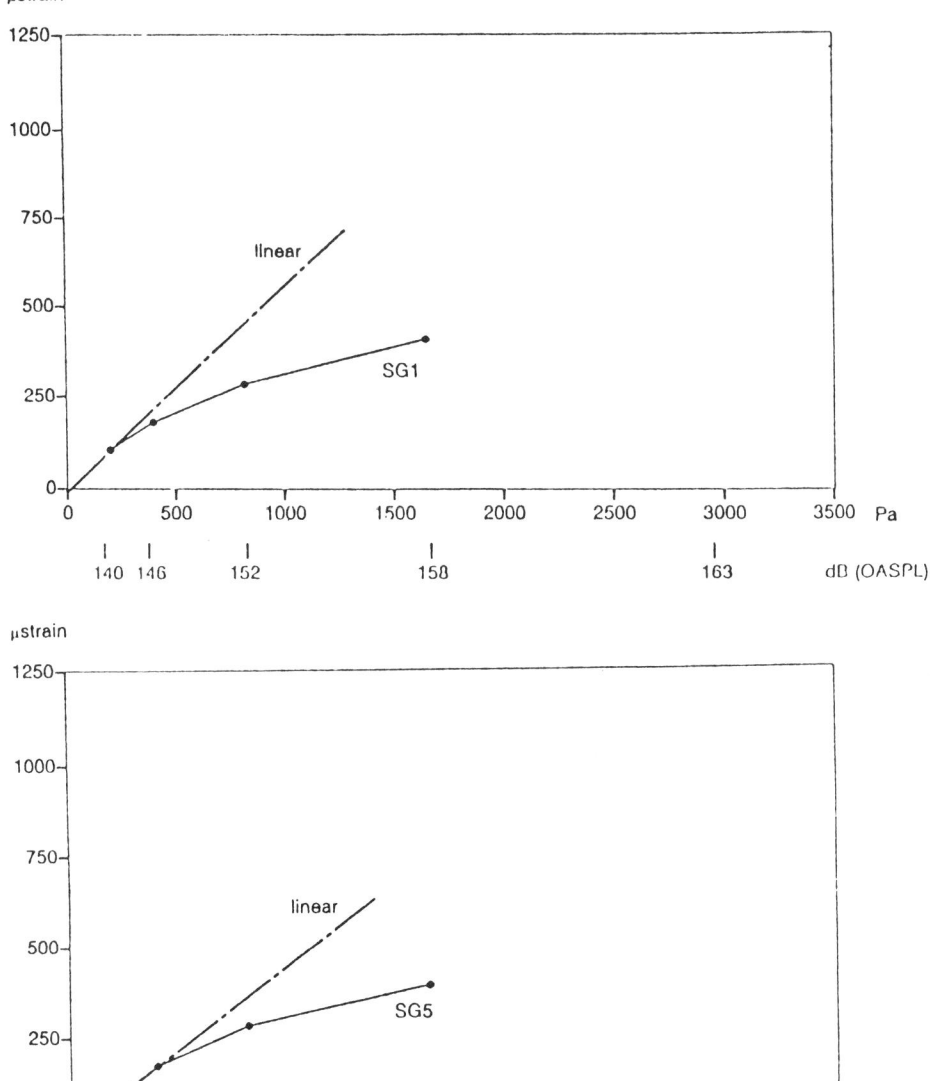

**Figure 2.45** (*continued*)

five additional accelerometers were used. The investigations and the test procedure were similar to that adopted for the HTA and the aluminium–lithium panels. For strain response and linearity, see Figure 2.48.

**Acoustic endurance**   The test article was finally exposed to an endurance test which was performed in the sequence shown in Table 2.4.

Test runs 1–7 were performed to study the structural response of the two articles under increasing acoustic excitation. After each run, the measured strains were evaluated before it was decided to proceed to the next higher loading. Visual inspections were performed after each run to detect the first occurrence of failures.

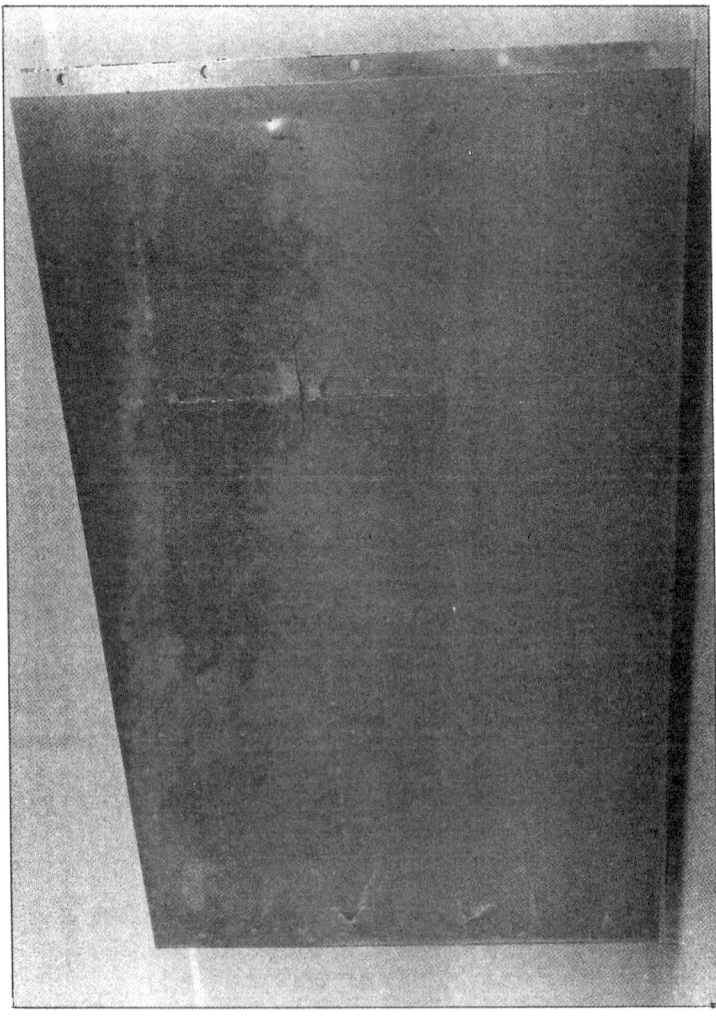

**Figure 2.46**   Aluminium–lithium panel: areas of damage after endurance tests

After run 8, the skin/rib joints failed as shown in Figure 2.49. Both panels were repaired and strengthened by adding a counter flange.

Endurance testing was then continued at an OASPL of 159 dB for 4 hours. Microcracks developed relatively early and increased in size and number, so that after run 15 all flanges had cracks.

### Acoustic tests on the T800/BSL924 panel

The test article (Figure 2.50) was instrumented with 18 unidirectional strain gauges and four rosettes, the location of which can be seen from Figure 2.52.

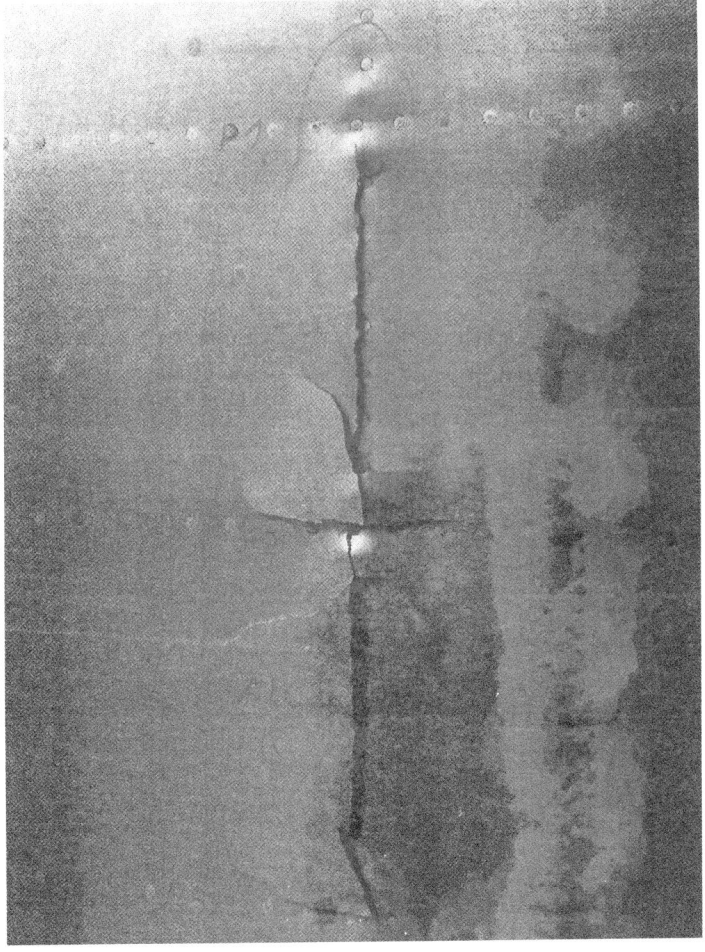

**Figure 2.46** *(continued)*

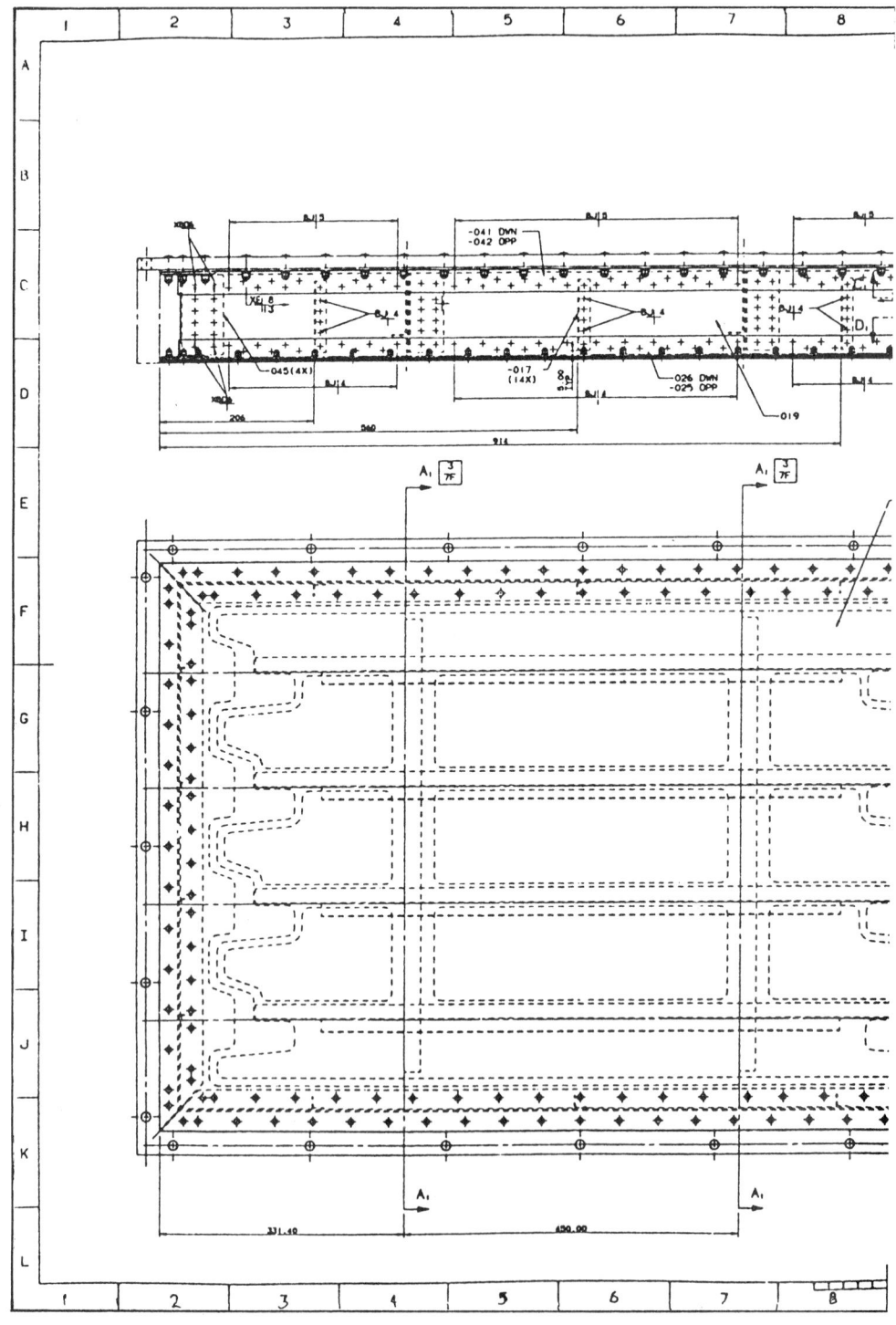

**Figure 2.47** Dimensions of the GLARE panels

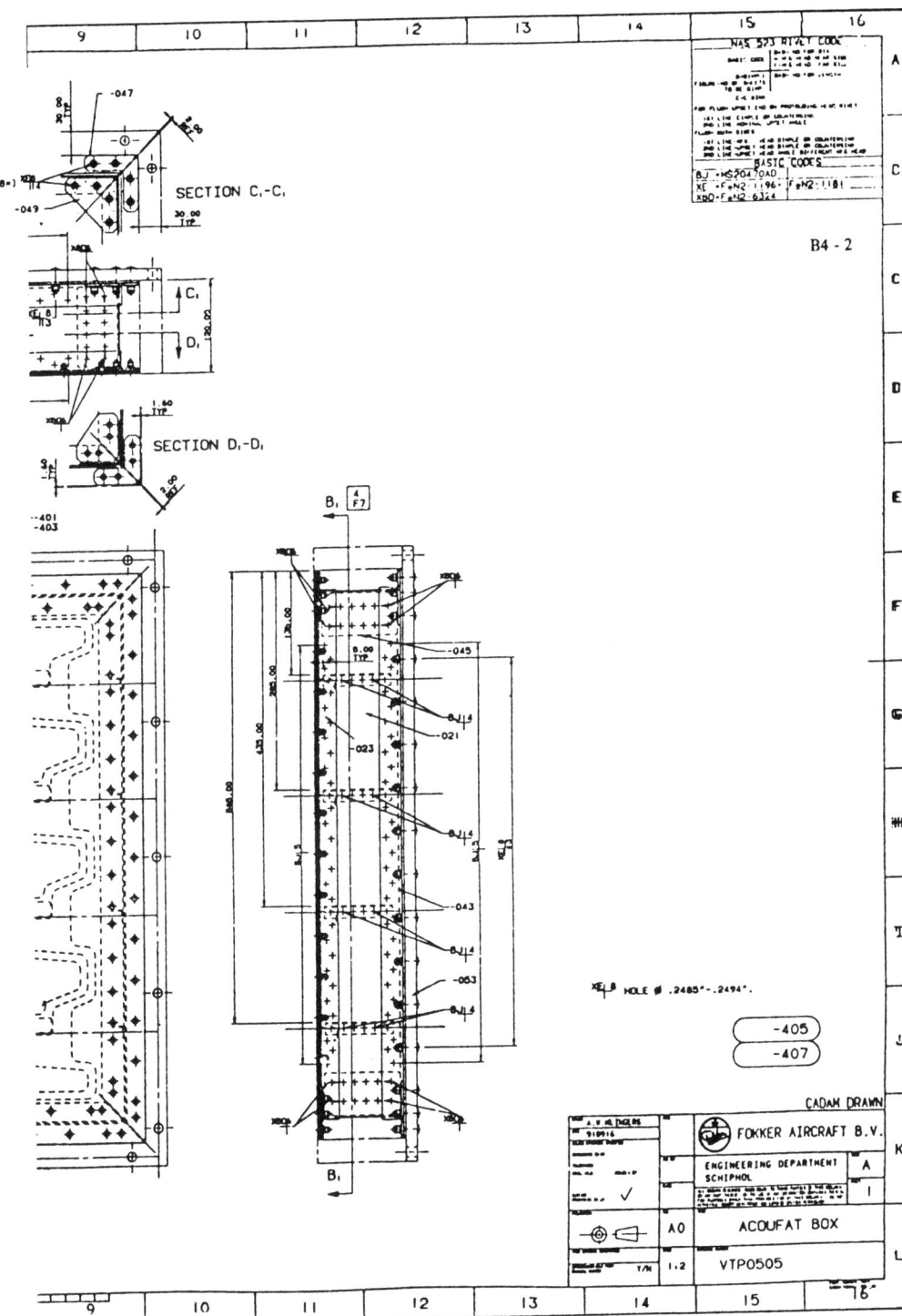

SECTION C₁-C₁

SECTION D₁-D₁

B₄ - 2

XE HOLE ⌀ .2485″-.2494″.

-405
-407

CADAM DRAWN

FOKKER AIRCRAFT B.V.

ENGINEERING DEPARTMENT
SCHIPHOL

ACOUFAT BOX

VTP0505

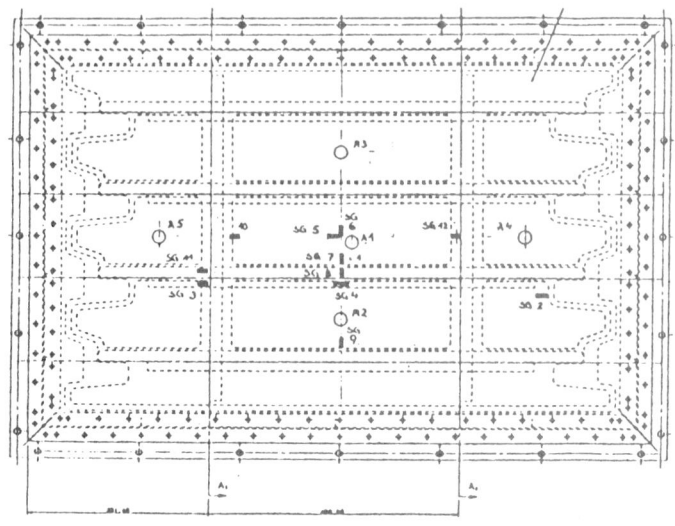

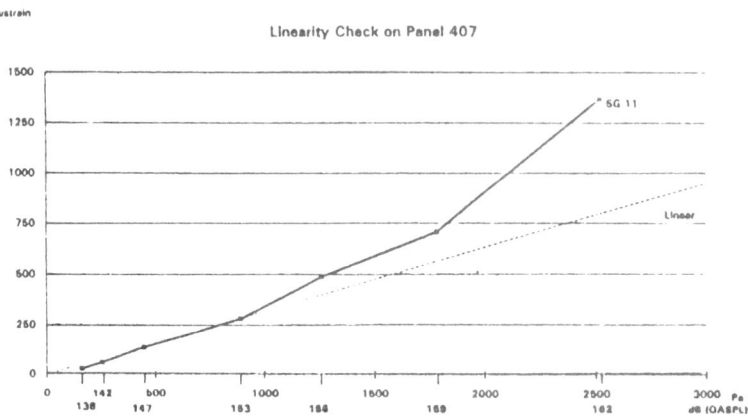

**Figure 2.48**  GLARE panels: plots of strain versus excitation level

For modal analysis, additional accelerometers were used. The following investigations were performed.

**Static load test**  In order to obtain a cross-reading between a static load acting on the panel surface and the resulting strains in the panel structure, hydrostatic load boxes were clamped to the test panel. The data obtained from these loading tests (see Figure 2.51(a)) were needed for the analytical predictions performed in subtask 4.2.

**Strain response and linearity**  The test article was then mounted into the PWT and excited by acoustic spectra with increasing SPLs. The linearity of structural response versus exciting sound pressures was checked. Strain-gauge signals

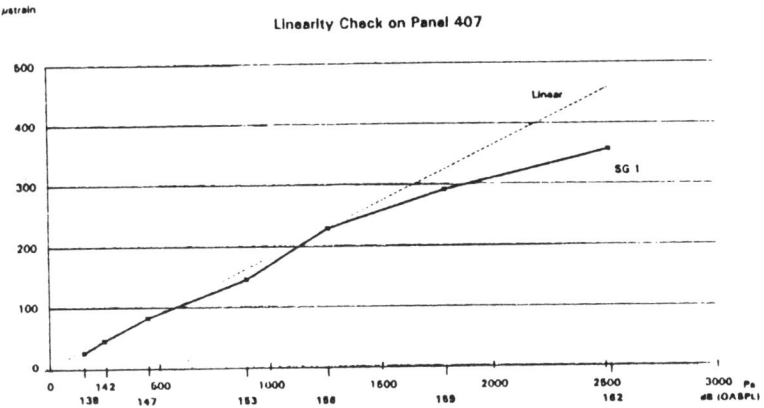

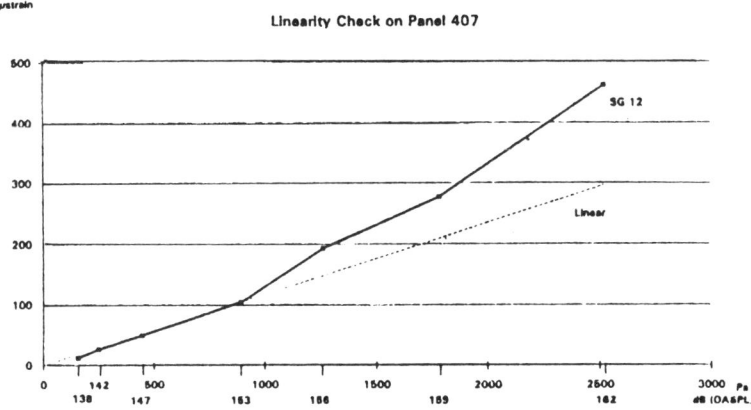

**Figure 2.48**   (*continued*)

were also evaluated in terms of power spectral density (PSD) analysis and some typical plots are given in Figure 2.51 (same strain gauge for increasing excitation levels).

**Modal analysis**   The test article was excited by broadband noise with an OASPL of 145 dB in the PWT. A total of 241 accelerometers, applied in small groups of 20 at a time, were used to measure the structural response and to derive mode shapes, resonant frequencies, damping, etc.

**Acoustic endurance**   The test article was finally exposed to a 20-hour endurance test at 160 dB OASPL broadband (152 dB SPL/160–250 Hz maximum) (see Figure 2.52(b)). Although cracking of the reinforcing cleats occurred, damage

Table 2.4  Acoustic endurance tests on the GLARE panels

| Run | OASPL (dB) | Duration (min) | Observation |
|-----|-----------|----------------|-------------|
| 1 | 138 | 6.50 | No visible damage |
| 2 | 142 | 3.93 | No visible damage |
| 3 | 147 | 2.70 | No visible damage |
| 4 | 153 | 4.17 | No visible damage |
| 5 | 156 | 2.87 | No visible damage |
| 6 | 159 | 3.20 | No visible damage |
| 7 | 162 | 2.00 | No visible damage |
| 8 | 159 | 10.00 | Cracks in flange (Figure 2.49) |
| Repair of frame attachment | | | |
| 9 | 159 | 15.00 | No visible damage |
| 10 | 159 | 45.00 | First microcracks in flange |
| 11 | 159 | 60.00 | First microcracks in flange |
| 12 | 159 | 60.00 | Five microcracks in flange |
| 13 | 159 | 60.00 | Eight cracks in flange |
| 14 | 162 | 28.00 | Eight cracks in flange |
| 15 | 165 | 30.00 | Nearly all flanges with cracks |

propagation was slow, and after 20 hours of testing the connections between the ends of the CFRP stringers and the I-section boundary members were still effective. The failure of the CFRP stringer between bays A and C is probably a consequence of the ineffectiveness of the stringers to stiffen the panel. The disbonding of the CFRP skin was progressive through the skin thickness and not confined to the surface ply.

*Acoustic tests on the SPF/DB titanium panel*

The test article (Figure 2.53) was very different from the five other panels: it was strictly representative of a real civil aircraft SPF/DB spoiler. It was instrumented with 20 strain gauges. The following investigations were performed.

**Strain response and linearity**  The test article was mounted into the PWT according to Figure 2.53 and the lower skin structure was excited by acoustic spectra with increasing SPLs. The linearity of structural response versus exciting sound pressures was checked. Strain-gauge signals were also subjected to PSD analysis.

**Modal analysis**  The test article was excited by broadband noise with an OASPL of 150 dB in the PWT. The survey was performed in three parts: first an

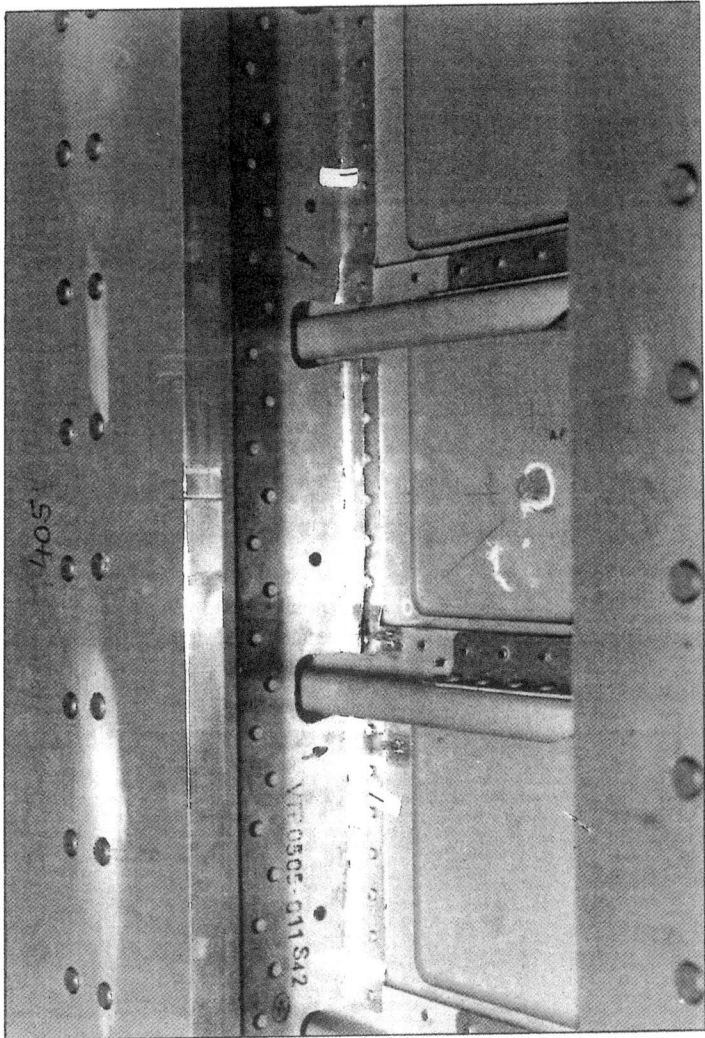

Figure 2.49    GLARE panels: damage after endurance tests

investigation of the modes of the complete panel, and then modal tests of two areas (quarters of the panel area). Mode shapes, modal frequencies and corresponding damping for the three different sections were investigated.

**Acoustic endurance**    The test article was finally exposed to an endurance test at 160 dB OASPL broadband (149 dB SPL/160–250 Hz maximum). The locations of actual fatigue damage after a 30-hour endurance test are given in Figure 2.54. Extensive damage was sustained by the panel skin during the tests and some cracking of the internal structure was detected. Although the upper skin was not directly exposed to the noise environment, extensive damage occurred

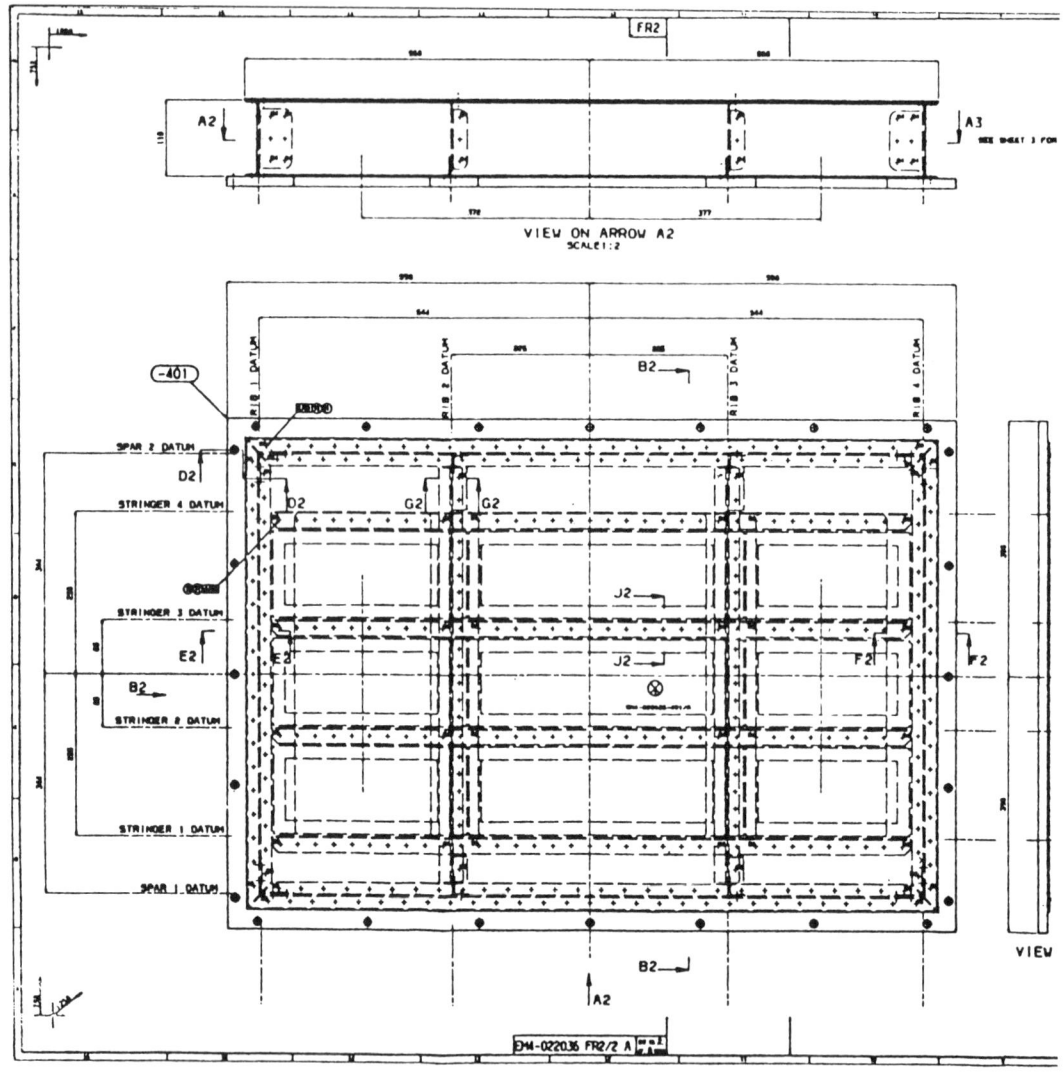

**Figure 2.50**  Dimensions of the T800/BSL924 panel

on this upper surface. This skin is, however, considerably less stiff than the lower skin structure, particularly adjacent to stiffener terminations.

*Conclusions from the acoustic tests on panels*

The main objectives of subtask 3.2 were to investigate built-up panels of advanced composite and metallic materials with respect to their structural response characteristics and to their acoustic fatigue properties. These data

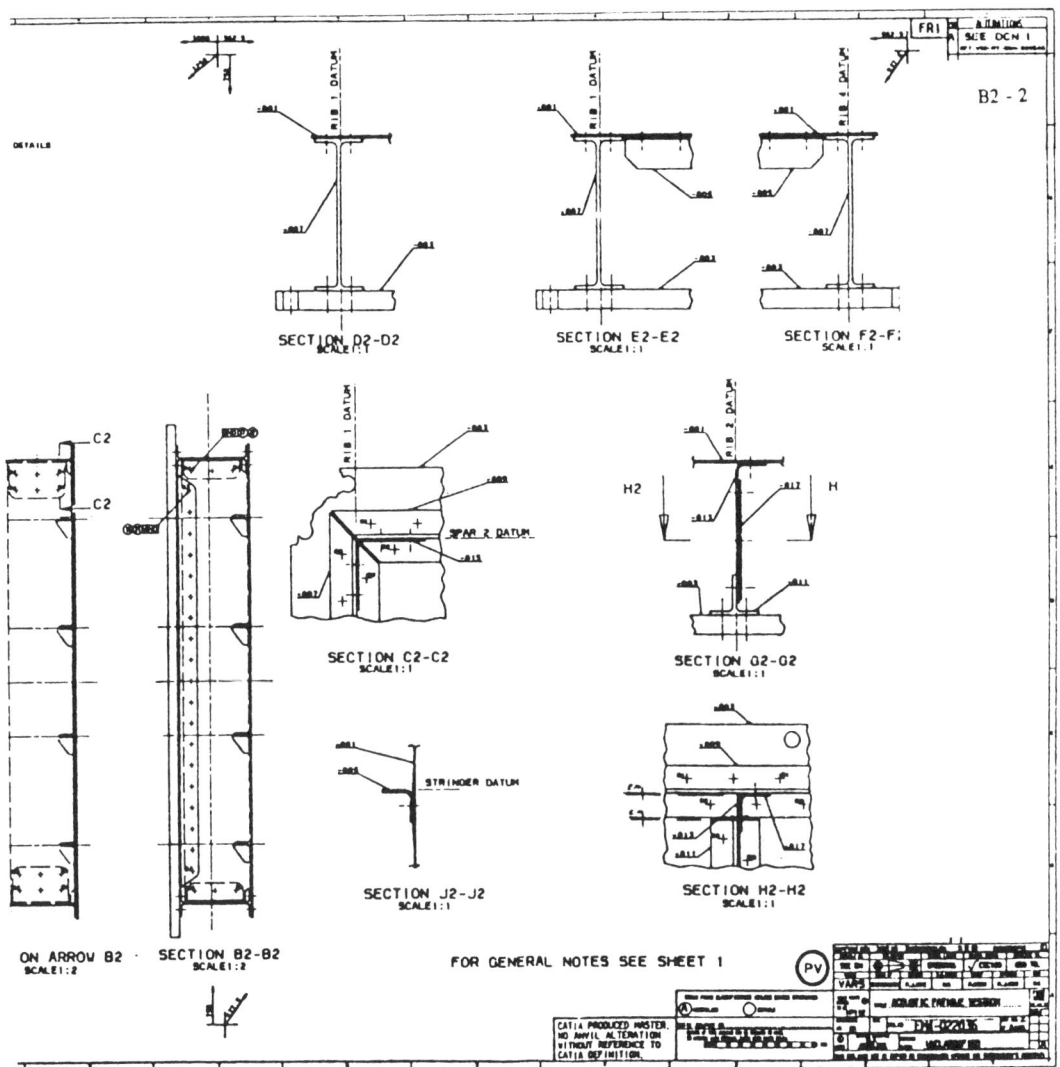

were needed for comparison with the fatigue data obtained with the shaker tests of *Task 2*. As far as the structural response is concerned, this testing activity was also in relation with the analytical activity of subtask 4.2, presented in Section 2.5. From the tests performed, the following points were established:

1. The type and number of structural modes excited in a PWT are governed by the characteristics of the PWT acoustic pressure fields. This becomes clear when the panel response under loudspeaker excitation outside the PWT is compared with the PWT measurements. The natural frequencies

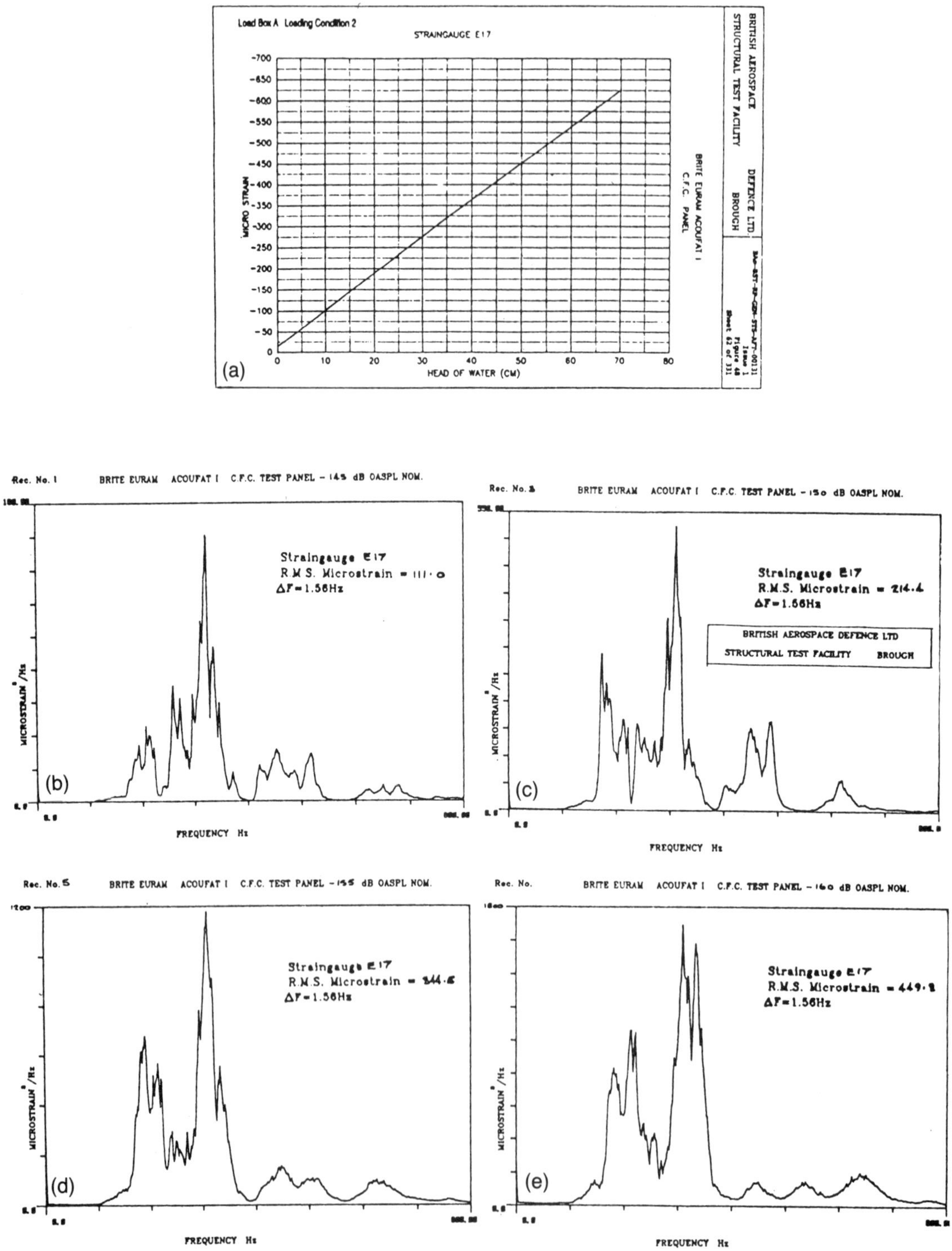

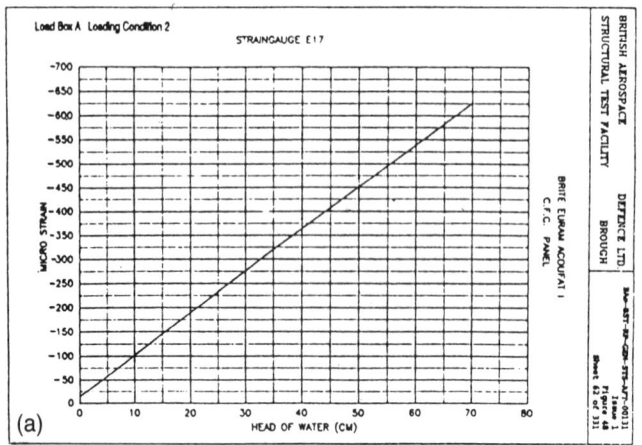

Figure 2.51 Strain response of the T800/BSL924 panel: (a) linearity check under static loading; (b)–(e) PSD plots with increasing excitation levels

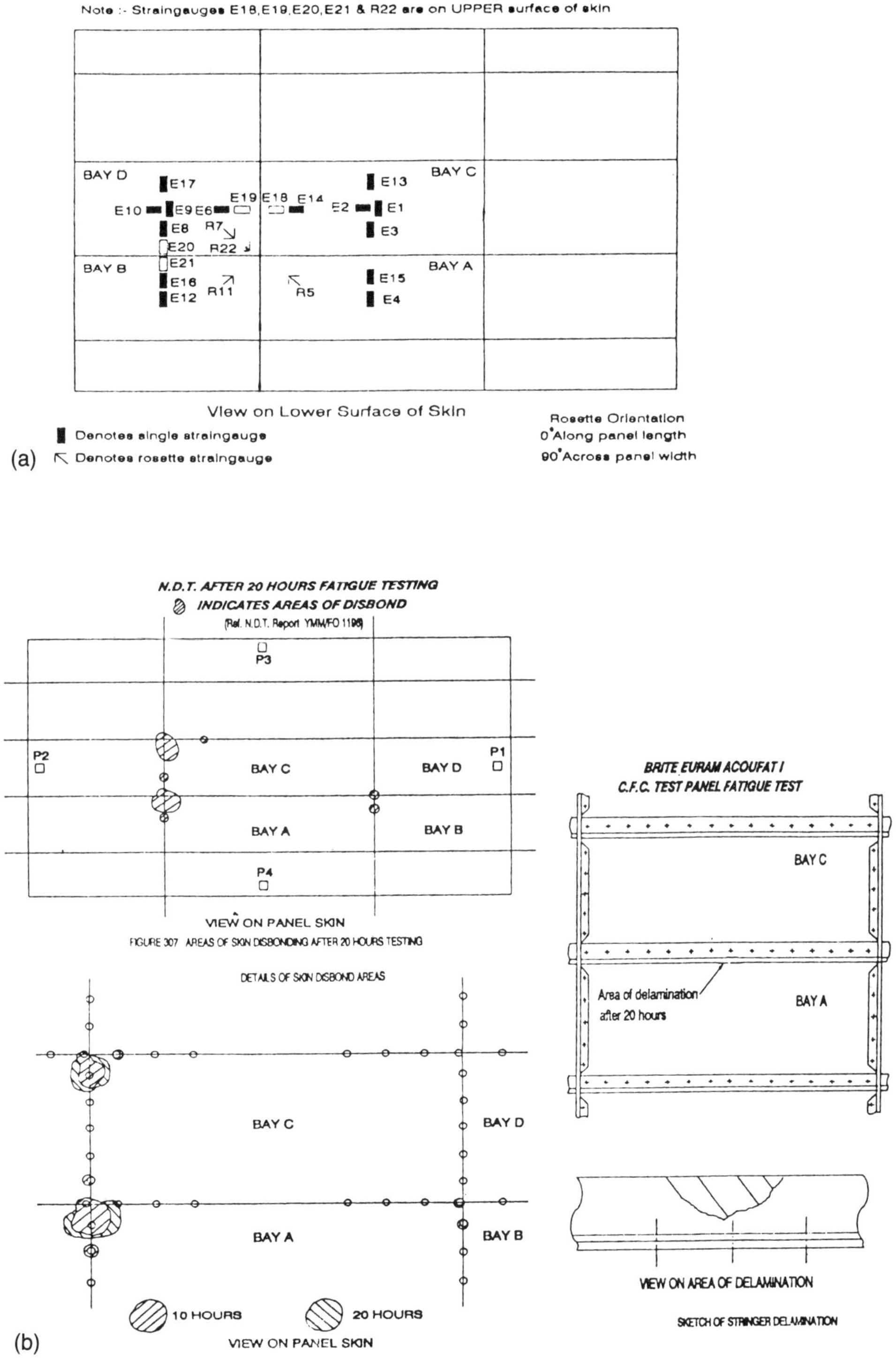

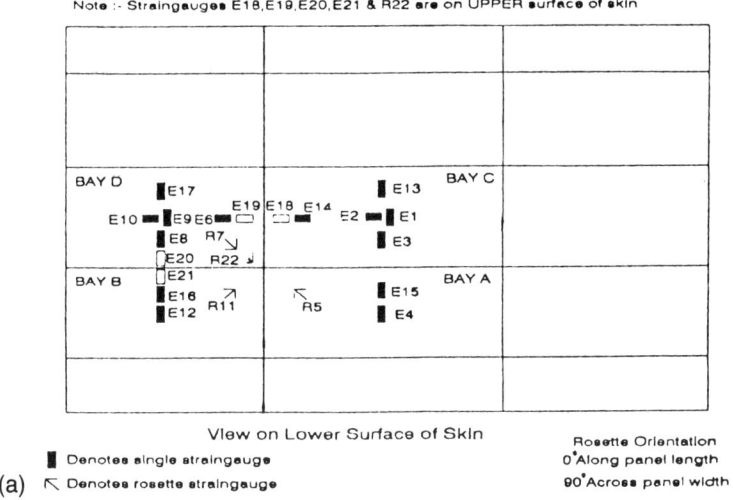

(a)

(b)

**Figure 2.52** T800/BSL924 panel: (a) strain-gauge locations; (b) damage after endurance tests

BRITE EURAM  ACOUFAT I  SPFDB TITANIUM SPOILER

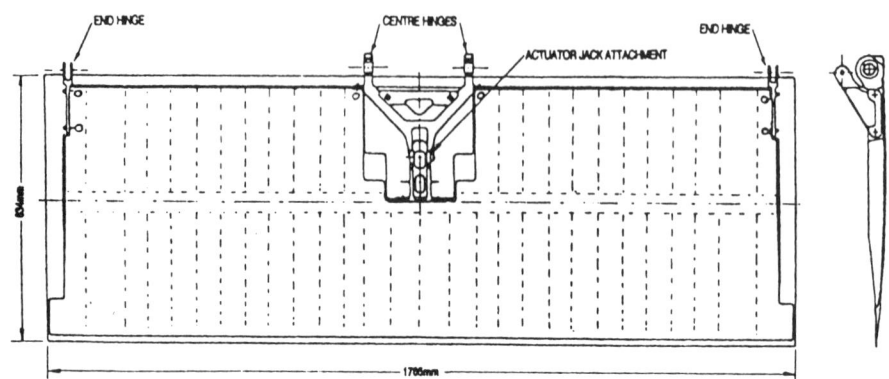

FIGURE 1  TEST PANEL STRUCTURE

MOUNTING CONFIGURATION

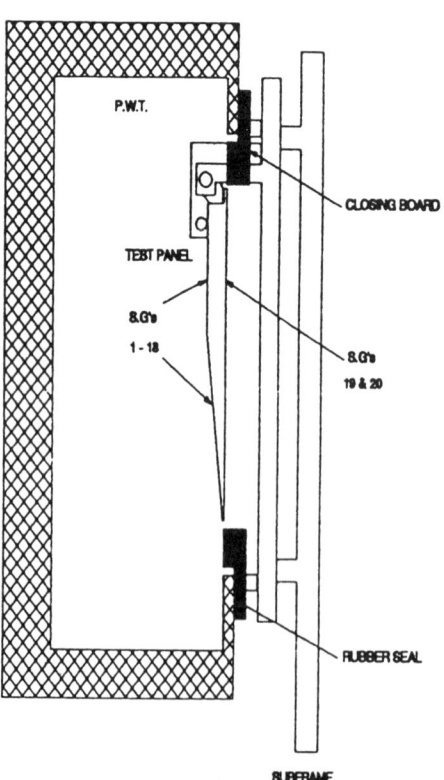

**Figure 2.53**   Details of the SPF/DB titanium spoiler panel

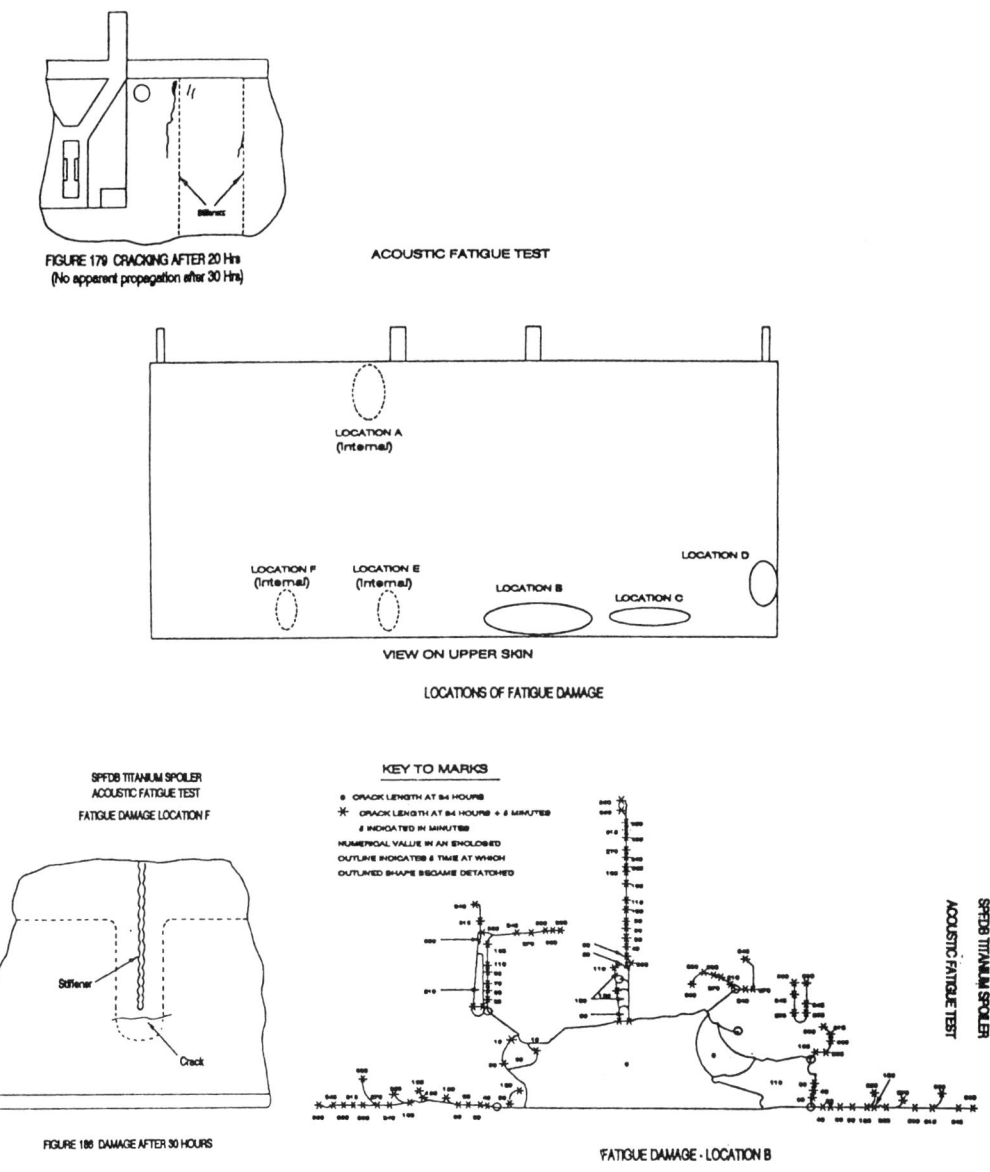

ACOUSTIC FATIGUE TEST

**Figure 2.54**   SPF/DB titanium spoiler panel: damage after endurance tests

under loudspeaker excitation and the resonant frequencies measured in the PWT are different.

2.  Exciting the panels in the PWT with increasing SPLs leads to nonlinearities. This can be seen from the spectral response where the type and number of the dominant modes changes with the level of excitation. Normally the number of dominant modes is reduced when the level of

excitation is increased. For very high excitation levels the response is greatly influenced by forced vibration.

3.  The overall response in terms of r.m.s.-values of microstrains also becomes nonlinear with increasing SPLs. For values greater than about 300–400 microstrain r.m.s., the induced strains in most cases are lower than the linear prediction. The nonlinearity can be attributed to membrane stresses which become effective with higher vibration amplitudes.

4.  From the strain surveys, it was found that the highest strains normally occur along the stringer/skin attachments. This is especially true for all panels of the stiffened skin design (HTA, Al–Li, two GLARE, T800). However, for the two GLARE panels extremely high strains were found at the flanges of the frames besides the stringer cutouts.

The SPF/DB panel which was actually a spoiler structure had the highest strain concentrations at the rib terminations and somewhat lower strains along the continuous skin/rib attachment.

5.  The failures observed in the endurance tests fall into two categories: those which can be attributed to inadequate design of the boundary conditions and those which are typical for the type of structure under investigation.

Failures of the first type are shown in Figure 2.43 for the HTA panel, where the rivets failed because of high inertia loadings of the nonattached stringer ends. Similar failures were encountered on the Al–Li panel at the same locations.

Owing to the panel design (no interconnection between stiffener and rib), high skin shear stresses are suspected to be generated in the corner of each bay, caused by vibration of the stiffener.

This type of failure was unexpected, and consequently, the coupons tested in *Task 2* were not representative of this type of damage. No *S/N* curve is available for this type of damage because the coupons were designed to study the in-plane bending effect, not the out-of-plane strains effects. Consequently, no endurance comparison could be done between the PWT/panel tests and the shaker/coupon tests.

The cracking of the reinforcing cleats of the T800 panel also falls into that category. Additionally, it is remarkable that the GLARE panels were free from failures at the boundaries.

The remaining failure locations are considered to be typical for the corresponding type of construction. This is true for the Al–Li panel, were the most severe damage was caused along the rivet lines of the individual skin panels, and for the GLARE panels with cracking in the frame flanges. According to Fokker, the same failures have been observed on actual airframes. As these flanges were made of conventional aluminium, no statement can be made about the failure mechanisms of GLARE itself.

The fact that very similar failures occurred on the two composite panels

(HTA, T800) is also very remarkable. The areas of skin delamination were nearly identical on the two panels.

### 2.3.3 Damage initiation and crack propagation analysis

This section summarises all the analytical activity performed within sub-task 4.3.

*Introduction: subtask 4.3 presentation*

The design and dimensioning process of aircraft structures involves three main tasks:

- Verification of structural strength and stability.

- The assurance of an adequate fatigue life.

- For fail-safe structures, the proof of sufficient damage tolerance behaviour.

Based on fundamental research and an enormous amount of empirical data, approved and established procedures have been developed to solve all these tasks for conventional, metallic structures. With the development of fibre-reinforced composite materials and their application in primary aircraft components, how could the air ministry requirements be satisfied quickly? Treatment of the strength and stability problem is nowadays well-known and generally accepted. However, fatigue life prediction and proof of damage tolerance for composite aircraft structures still remains a major problem during the certification procedure. The current philosophy is based on 'no damage growth', and this is achieved by conservative strain limitations and verification testing on components and built-up structures. Any extension of that philosophy must take into account current static methodology and current NDT techniques.

The application of classic empirical fatigue approaches to composite materials had only very limited success. Compared with metals, composites show complex damage mechanisms, with different damage modes occurring at the same time and interacting with each other. Another problem encountered with composite materials is that every new layup constitutes a new material which has to be tested in its own right. To reduce the quantity of testing, a theory based on the fatigue characteristics of the individual laminae would be useful.

*Analytical work performed*

Two principal tasks were performed in the course of the investigations concerning fatigue life prediction.

First, existing calculation techniques for the problem of fatigue life prediction, most of them based on 'traditional approaches' (Wöhler curves, Haigh–

Soderberg diagram, Palmgren–Miner rule, fracture mechanics in conjunction with the Paris equation for crack growth) had to be compiled and checked for their applicability. During this scanning of the open literature, it became obvious that there is no 'ready to use' model for the life prediction of composites, which would cover an engineer's needs concerning simplicity and test expenditure for the establishment of the required database.

As a first step, an accumulation model, disregarding any initial flaw or damage propagation, was considered. This model would be able to predict fatigue of a plain specimen on a 'first lamina failure' basis. Possible failure modes are then investigated separately and, with respect to the most critical failure type, the fatigue life is predicted.

The second step was the development of a general, easily used, iterative prediction technique, capable of simulating fatigue damage growth and predicting life or residual strength of dynamically loaded composites. It is based on classical laminate plate theory and ply-by-ply failure principles. Priority was given to a 'fully worked' methodology, simple to be used in practice, rather than to the general validity of all assumptions made. A basic frame was constructed which would allow more advanced fatigue damage models to be integrated.

The fatigue-life prediction technique must be connected to the analytical and/or numerical determination of the residual strength of a damaged composite structure. Numerous investigations have been published during the last 20 years dealing with the applicability of 'linear-elastic fracture mechanics' (LEFM) methods to layered composite materials.

The energy release rate, in particular, as the critical parameter for the onset of delamination growth, has been used with considerable success. Whether this principle would be applicable also to the residual strength prediction for the ACOUFAT specimens was another question to be answered within this subtask. To this end, three-dimensional finite-element calculations were performed in conjunction with a suitable method for determination of the energy release rate $G$ along the tip of the internal delamination within the specimen. These values, together with an appropriate failure criterion, would allow prediction of the damage growth onset load, which is currently considered as the failure defining event.

Owing to the fundamental character of these different investigations and the obvious lack of basic material parameters, which govern the fatigue behaviour of composites, comparisons with shaker tests from *Task 2* are nearly impossible. However, the experimental results can be used to check some of the predictions from the analytical/numerical calculations and to reveal shortfalls of the methodologies used in terms of failure modes and basic dimensioning parameters.

*Conclusions related to the damage initiation and crack propagation analysis*

Damage initiation and propagation/accumulation analysis for composite structures is not a straightforward matter, even for the simple (ordinary) fatigue of isotropic structures. Scanning the literature, it is obvious that there is no complete 'ready to use' model. What exists are different approaches to solving specific problems regarding different $R$-values, layups, notched/unnotched behaviour etc.

Owing to the variety of different failure modes, and their interactions, it is difficult to find a relevant damage parameter and the associated damage growth. Therefore, as a first step, a model for ordinary fatigue must be formulated and then converted for acoustic randomly loaded structures. The semi-analytical models suggested are for ordinary fatigue (i.e. accumulation models) and are based on fatigue failure functions. In one of the approaches, the fatigue failure functions are not determined individually but are transformed from the laminate into lamina constituents and thus constitute the macroscopic behaviour for a built-up laminate. Later the models are based on in-plane behaviour (i.e classical laminate theory).

Further investigation of the coupling between deterministic and randomly loaded specimens for the possible use of ordinary endurance data was planned for a follow-on study. It should also be emphasised that there may be problems with test coupons versus real structures regarding actual failure modes.

Based on classical laminate theory and ply-by-ply failure principles, an iterative finite-element procedure was developed for life prediction with a basic frame where the semi-analytical models could be invoked. The model, which includes both damage modelling and structural analysis in one prediction technique, is capable of simulating fatigue damage growth and predicting life or residual strength of dynamically loaded composite components.

The onset of delamination growth in aircraft structures is currently considered as the critical event, since the damage growth onset load is identical with the residual strength of the structure. To determine the delamination growth onset load of damaged structures, numerical fracture methods have been explored and show promise.

Two different numerical methods have been considered for the calculation of energy release rates along the delamination front; i.e. the 'virtual crack extension' (VCE) method and the 'modified crack closure' (MCC) method. Overall the MCC method proved to be numerically more stable and to require less modelling effort than the VCE method. In addition, it offers the advantage of a straightforward separation of the individual contributions from the different fracture modes to the energy release rate.

Calculation of the energy release rate along the boundary of the elliptical delamination in specimen D02 (see Figure 2.30) revealed very low values, even at the estimated failure load level. The energy release rate contributions of fracture modes I and III were negligible. Peak values for GII were reached at the ends of the short axis of the ellipse. However, application of a linear failure criterion led to predicted delamination growth onset loads which were approximately 50 times higher than the compressive strength of the laminate, predicted by conventional laminate theory. Therefore, the onset of delamination growth during the residual strength tests could be excluded and the failure load had to be established by a convenient static failure criterion. This prediction was confirmed by the tests performed within subtask 2.2.

Although the numerical predictions have been substantiated by tests, a complete verification of the numerical fracture mechanics approach could not be achieved owing to the specimens' strength limit. Further investigations with components which show stable delamination growth before final fracture are necessary for validation of the onset load predictions.

Also reported is a semi-analytical procedure to calculate interlaminar shear stresses at plate bending. The procedure is very much the same as in ESDU documents but with the extension of interlaminar shear stress calculation. The equations for interlaminar stress calculation could also be invoked in a post-processing program after an ordinary FE analysis. To calculate the threshold

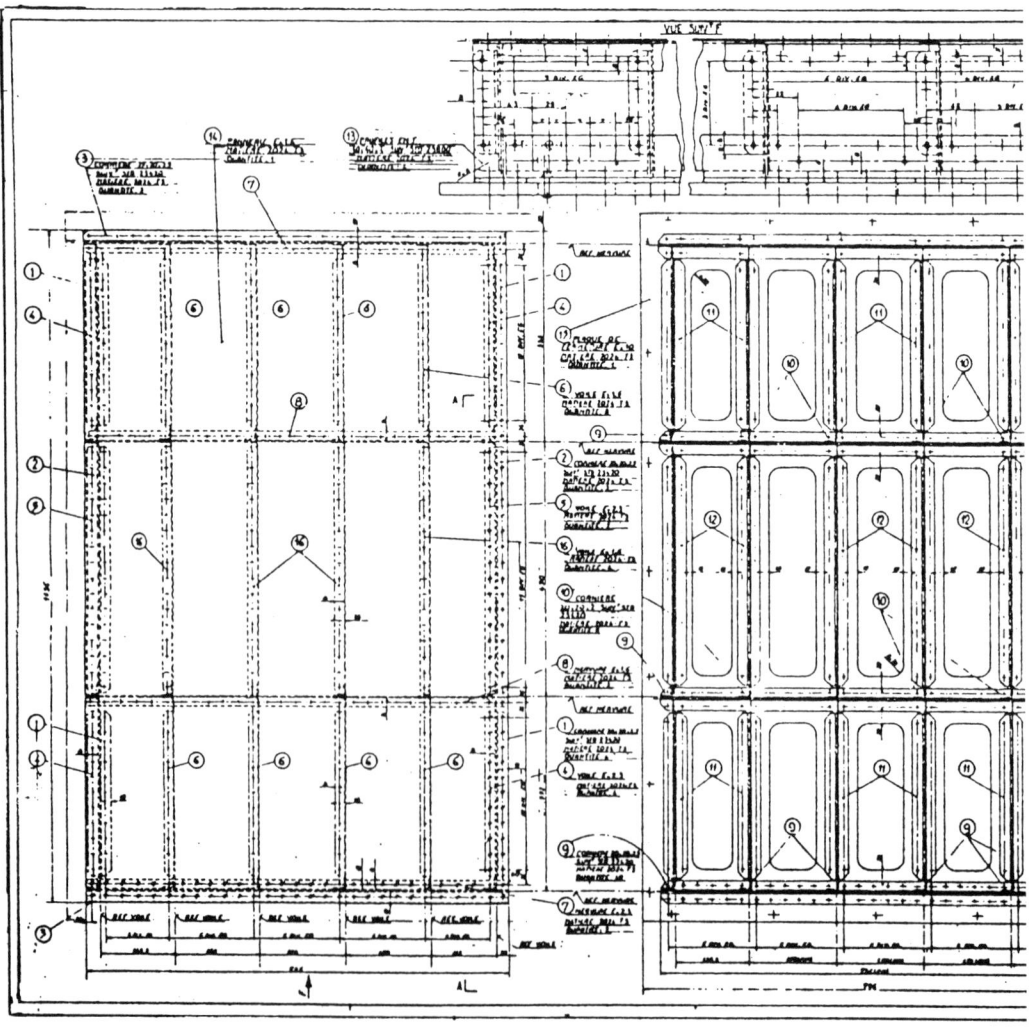

**Figure 2.55** Dimensions of stiffened aluminium panel for WT and PWT tests

for damage initiation, these stresses must be combined into a failure criterion and this will be heavily dependent on test results. Preliminary calculations reveal small interlaminar stresses for actual laminate thicknesses.

## 2.4   Research results: aero/acoustic loads

This section summarises all the results obtained within the aero/acoustic loads study performed as part of the ACOUFAT program. The results of the

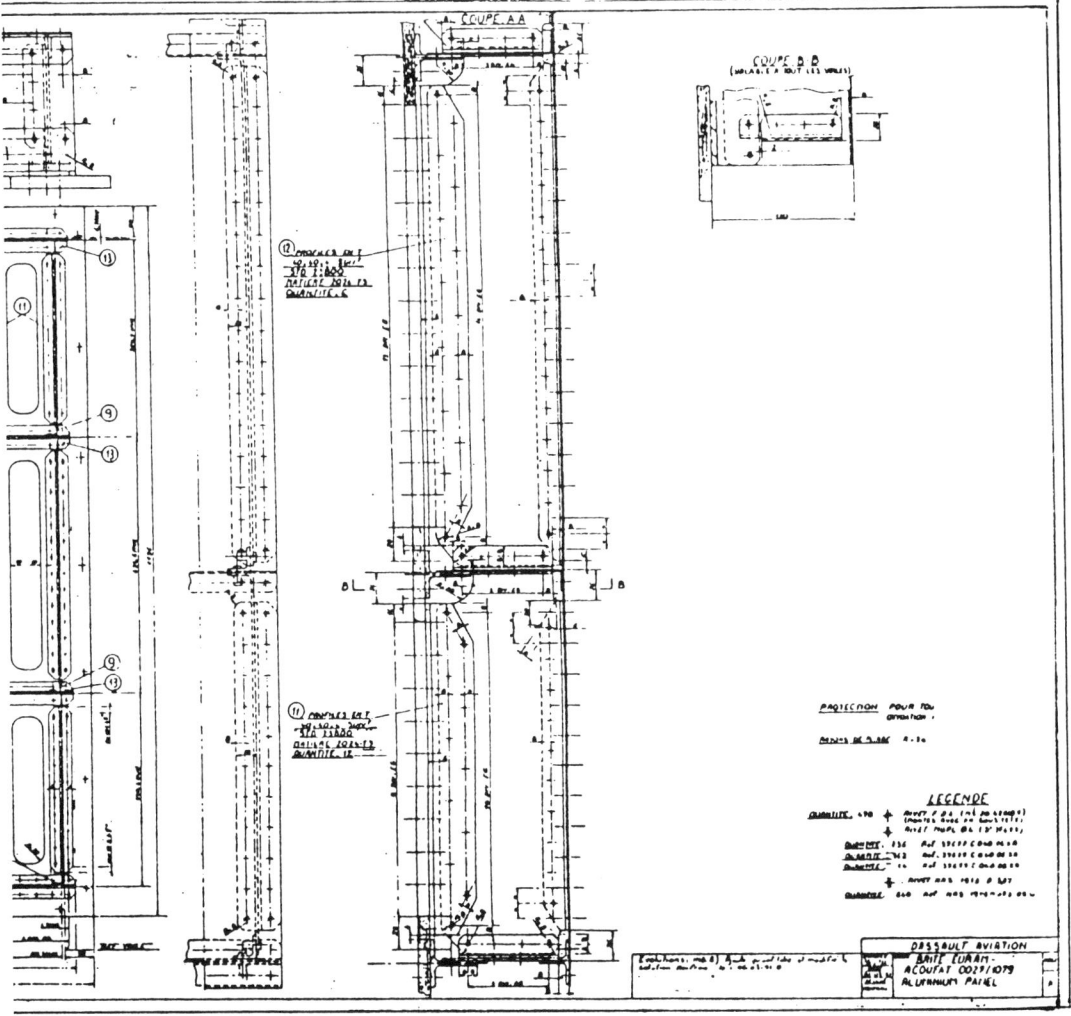

experimental loads study (related to subtask 3.1) are presented in Section 2.4.1 and the results of the analytical study (related to subtask 4.1) in Section 2.4.2.

The initial state-of-the-art, the objectives and the work related to this load study are presented in Section 2.2.2.

### 2.4.1  Experimental results

*Introduction: subtask 3.1 presentation*

The objectives were as follows:

- To determine, in a wind tunnel, the loading actions of separated flows representative of real flight conditions and to identify the structural response of a test panel under these aero/acoustic excitations.

- To calibrate the tests in PWTs by identification of the structural response of the test panel in different PWTs.

- To define a testing strategy in the PWTs, to get the same structural response of the test panel, with the aero/acoustic loads in the WT and with acoustic loads in the PWTs.

For this purpose, a stiffened aluminium panel was designed by Dassault Aviation and manufactured by Per Udsen Co. For this test panel, Dassault Aviation proposed the box concept and a design representative of aeronautical structures (see Figure 2.55). The outer dimensions of the upper panel were 1136 mm × 736 mm. The same structure was adopted for the acoustic fatigue panels (endurance tests in PWT—see Section 2.3.2). To calibrate the structural response of the test panel, it was equipped with a total of 16 active strain gauges, as shown in Figure 2.58.

This test article was first built into the S1-Modane ONERA wind tunnel (diameter 8 m) and tested under separated flow conditions (Figures 2.56 and 2.57). The fluctuating pressure field in the area of the recirculation bubble behind the flap was evaluated for different flap sizes and Mach numbers. The conditions of flap number 3 and Mach 0.8 were used as reference levels for the PWT tests.

In a second test phase the test article was mounted into the PWT at IABG (Figure 2.37). Investigations were conducted to establish the response of the panel at the reference pressure spectrum and, if necessary, to modify the applied acoustic sound pressures in the PWT to match the strain spectral densities measured on the panel in the WT.

Finally, in a third test phase, the PWT tests were repeated in the acoustic facility of BAe, Brough.

In the following, the results of the three test campaigns are evaluated in terms of loading actions and structural response characteristics.

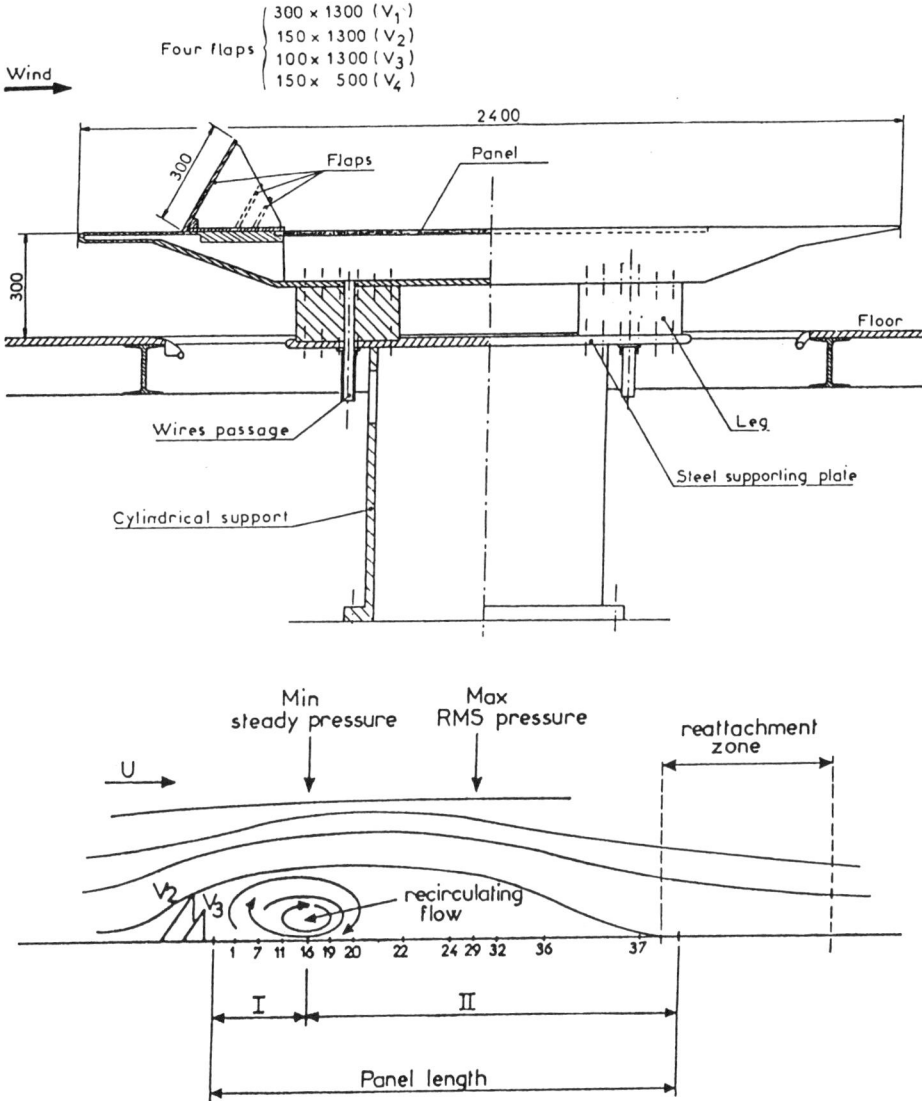

**Figure 2.56**  Wind-tunnel tests: setup and mean flow pattern

*Loading actions*

One of the main objectives was to study the loading actions prevailing in areas of separated flows. These data are needed as inputs for computational methods, as well as for experimental testing.

Whereas the computational aspects are covered in Section 2.4.2 (subtask 4.1), this section compares the results with other methods of acoustic testing.

Separated flows were generated by placing a test setup as shown in Figure 2.56 into the S1–Modane ONERA wind tunnel. At trans-sonic speeds, a

(a)

(b)

**Figure 2.57** Wind-tunnel tests: (a) setup of flap V2 (150 × 1300 mm); (b) setup of flap V3 (100 × 1300 mm)

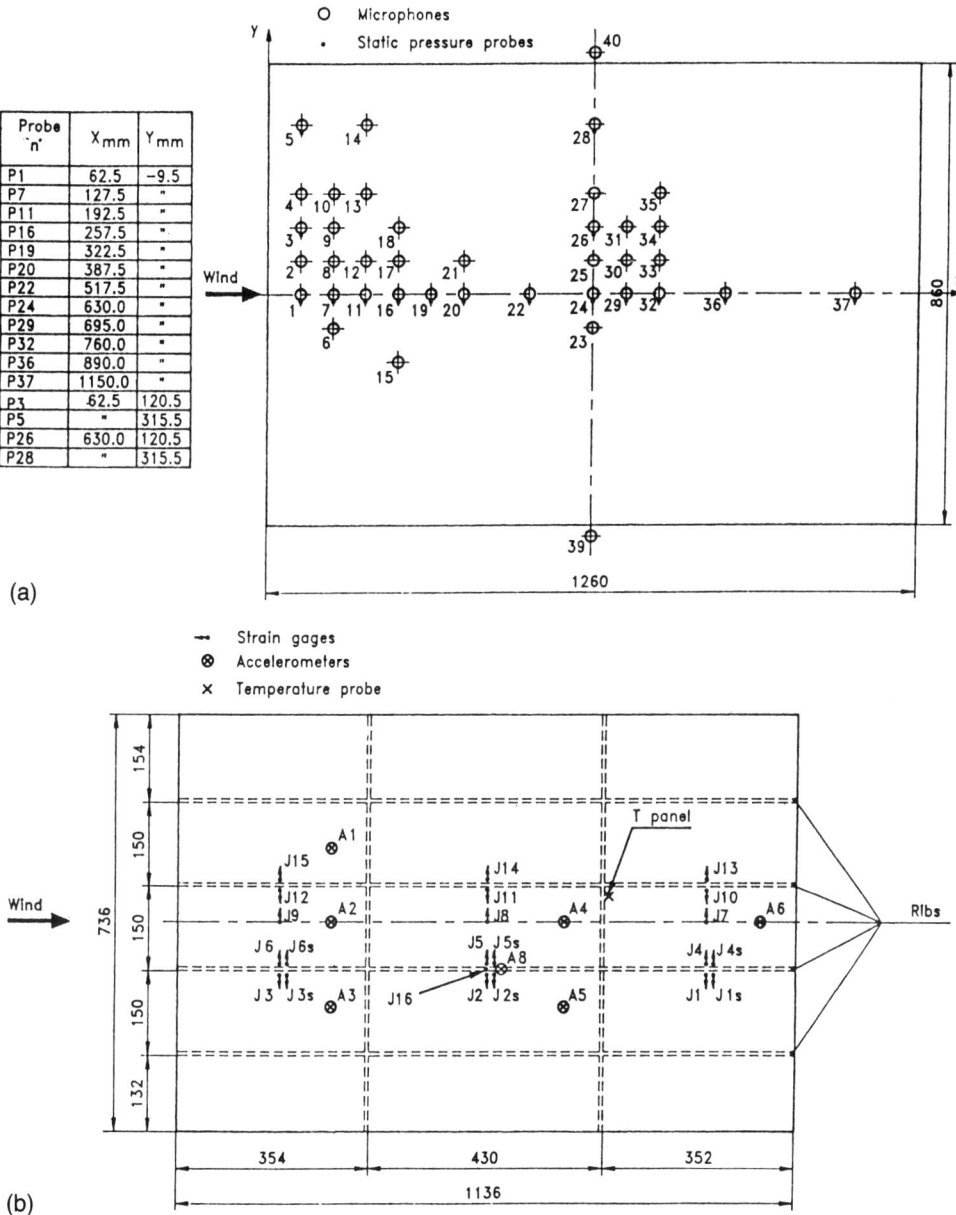

**Figure 2.58** Wind-tunnel tests: (a) location of pressure measurements; (b) location of strain gauges and accelerometers on the aluminium panel

turbulent pressure field is developed behind an elevated spoiler. In the first step, a measuring plate containing 40 pressure sensors (Figure 2.58) was installed behind the spoiler. Each individual sensor signal was evaluated in terms of pressure spectral density and selected pairs of microphones were analysed with respect to their coherence.

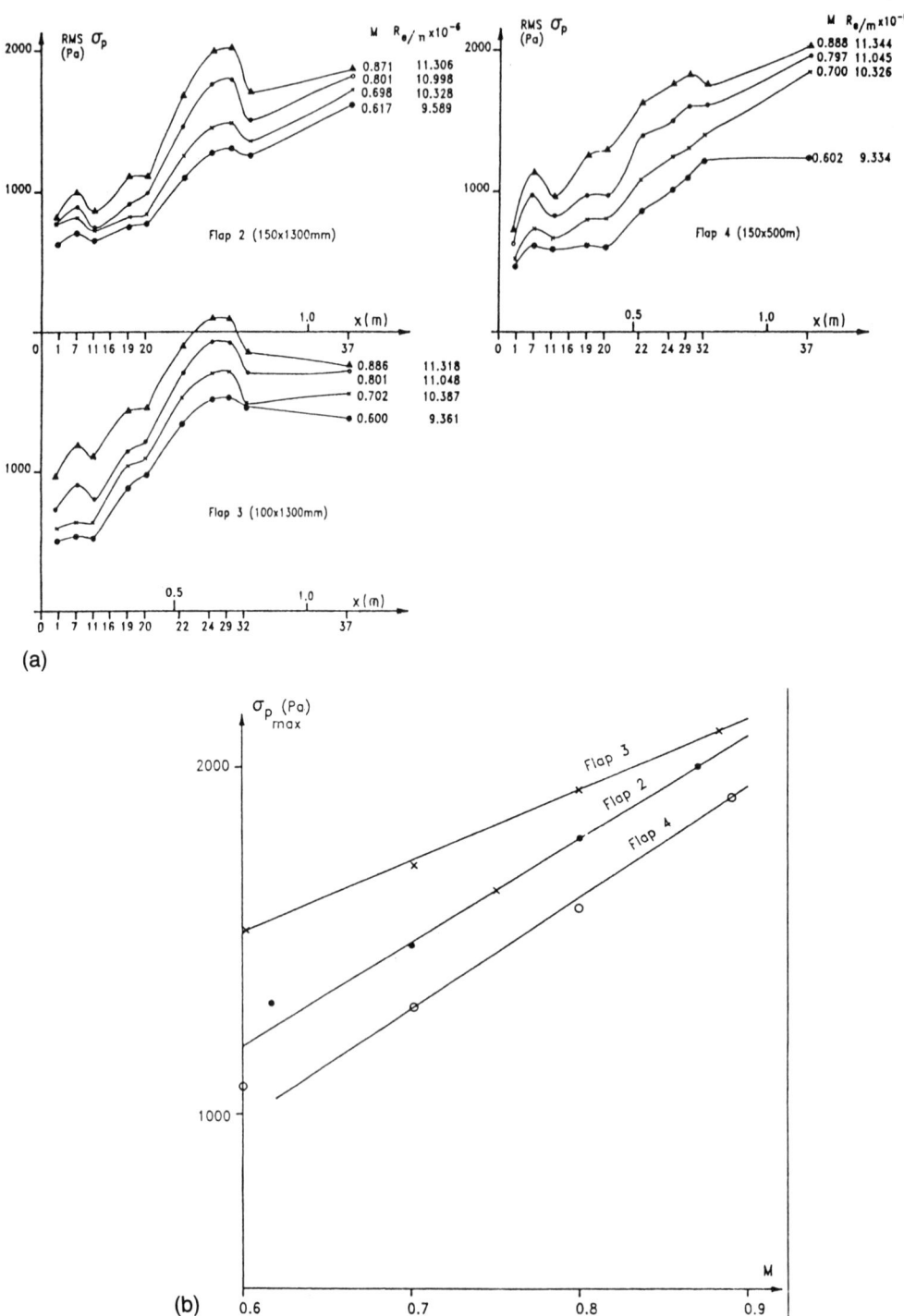

**Figure 2.59** Wind-tunnel tests: (a) overall r.m.s. pressure along the panel length; (b) maximum r.m.s. pressures (point M29) versus Mach number

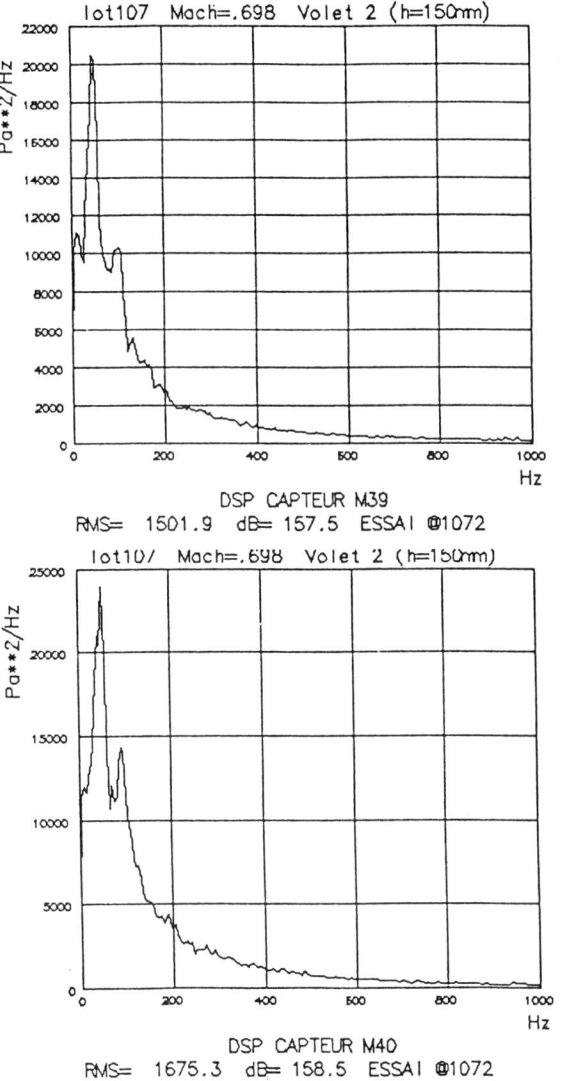

Figure 2.60   ONERA wind-tunnel test results: pressure spectral densities

Typical results are shown in Figure 2.59 for the static pressures and in Figure 2.60 for the pressure spectral density. The plots indicate high intensities in the frequency range around 100 Hz with a rapid decay towards the higher frequencies. The maximum intensity and the corresponding frequencies depend on the test configuration—Mach number, size of spoiler and location of pressure transducers. However, all spectra exhibit a continuous energy distribution due to the randomly fluctuating pressures.

Coherence functions have been produced between selected pairs of microphones and typical plots are shown in Figure 2.61 (real and imaginary parts of the coherence functions). They indicate that the spatial coherence is generally

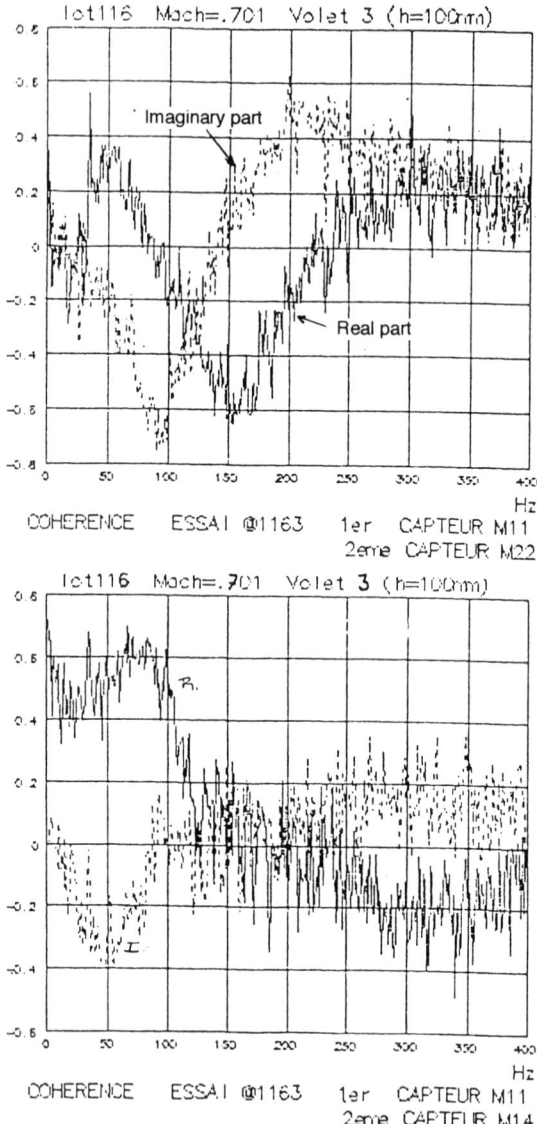

**Figure 2.61** ONERA wind-tunnel test results: coherence functions (selected pairs of microphones)

very low. This implies that the microstructure of the turbulence field contains a high number of uncorrelated eddies.

The comparison between the WT data and the corresponding PWT data is given in Figure 2.65. As can be seen, the spectral densities measured in the PWTs are very peaky compared with the smooth curve of the WT. The peaks in the PWT are due to standing waves associated with the width, height and length of the tube. Differences in the overall sound pressures ($Pa_{rms}$) are not critical as they can be equalised by a more refined control of the excitation.

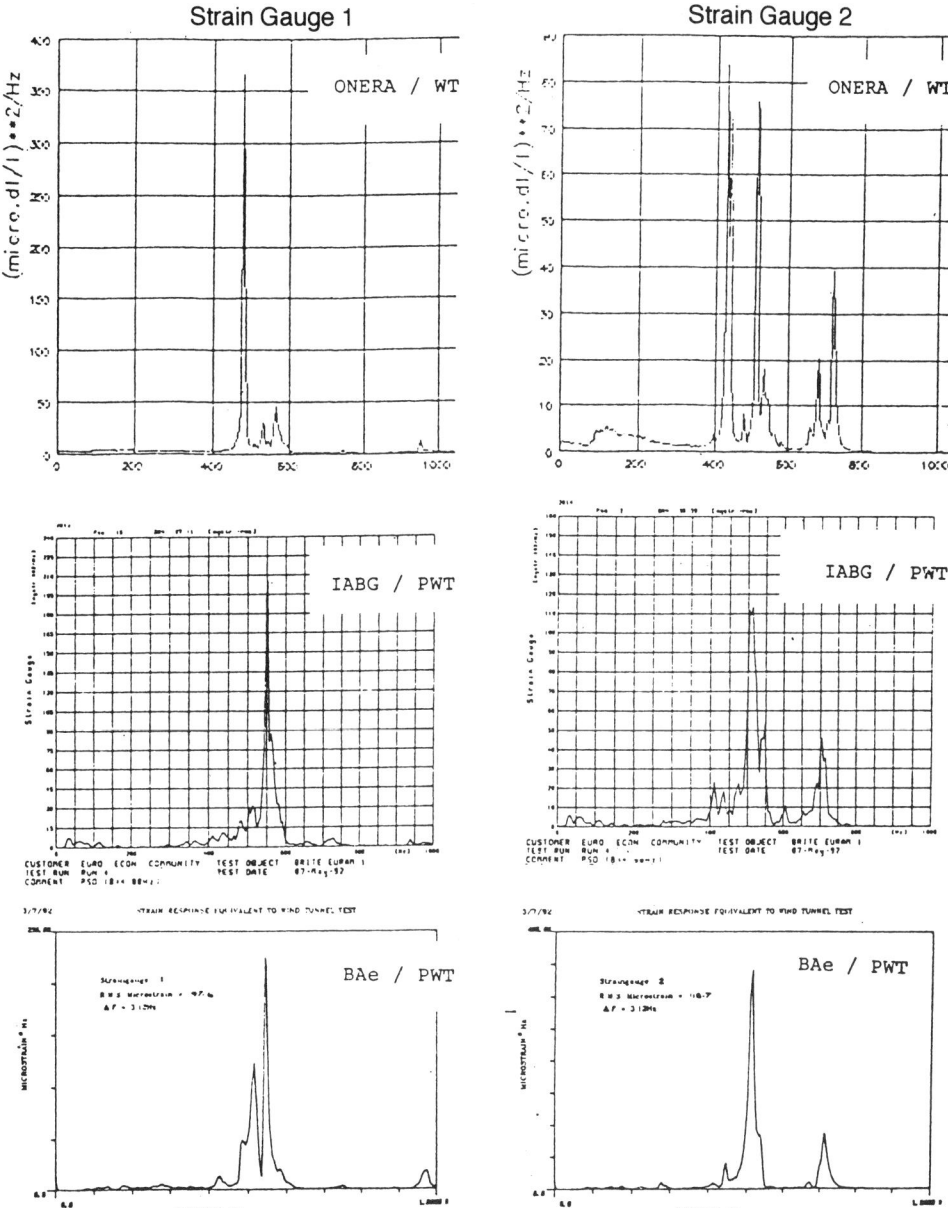

**Figure 2.62**  Comparison of WT and PWT tests of structural response (strain gauges 1 and 2)

The coherence functions obtained within both of the PWTs (see Figure 2.64) indicate that the instantaneous sound pressures are highly coherent, which means that the total pressure field is fully in phase.

The basic experimental data recorded in these WT tests were the reference data for elaborating a mathematical model of the statistical characteristics of the excitations (see Section 2.4.2).

*Structural response characteristics*

Apart from evaluating the loading actions associated with separated flow conditions in the WT, the main objective of subtask 3.1 was to compare the structural response of the test panel under WT and PWT excitations and to

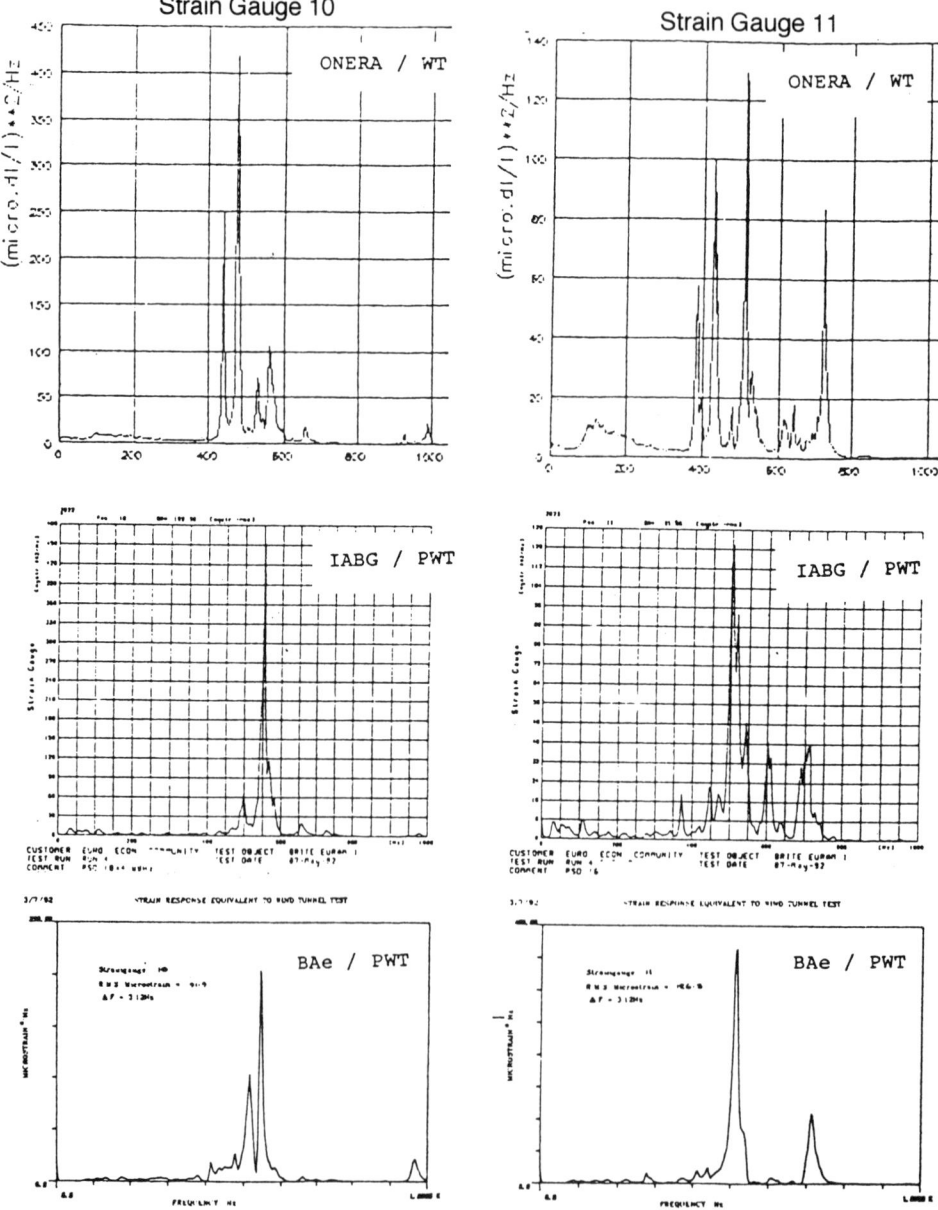

**Figure 2.63**  Comparison of WT and PWT tests of structural response (strain gauges 10 and 11)

define, if necessary, a testing strategy to get the same structural response in both cases.

**General characteristics**   The comparison was performed on the basis of the induced strains measured by a total of 16 strain gauges (see Figure 2.58(b)). The strain signals were analysed to obtain the power spectral densities (PSD), and this was performed for the WT tests as well as for the two PWTs. The PSD diagrams are presented in Figures 2.62 and 2.63 for strain-gauge numbers 1, 2, 10 and 11, and for the three test facilities.

From examination of the strain PSDs it is evident that the calibration test-box has a multimodal vibration response, where the dominant modes occur in the frequency range 400–800 Hz.

When specific comparisons between the responses of reference strain gauges number 10 and 11 (see Figure 2.63) are made, it is apparent for strain gauge 10 that only two of the dominant modes at 515 and 545 Hz can be significantly excited, in comparison with the four being excited in the WT at 440, 475, 515 and 560 Hz. On this gauge the high spectral peak at 475 Hz in the WT could not be excited in the PWT.

For strain gauge number 11, it is also apparent that only two of the dominant modes at 520 and 720 Hz can be significantly excited, in comparison with the four being excited in the WT at 385, 425, 520 and 720 Hz. In this case, however, the highest spectral peak was excited.

**Table 2.5**   Comparison of strain-gauge responses

| Strain gauge | R.M.S. values of microstrains | | |
|---|---|---|---|
| | ONERA/WT (157.8 dB) | IABG/PWT (155.9 dB) | BAe/PWT (156.9 dB) |
| 1 | 83.0 | 87.1 | 97.6 |
| 2 | 68.6 | 90.6 | 118.7 |
| 3 | 59.8 | 172.0 | 101.4 |
| 4 | 50.8 | 39.8 | 51.7 |
| 5 | 43.5 | 51.8 | 65.1 |
| 6 | 38.5 | 46.8 | 61.2 |
| 7 | 91.6 | 66.5 | 91.9 |
| 8 | 115.0 | — | 114.8 |
| 9 | 47.8 | 61.2 | 103.4 |
| 10 | 115.7 | 109.9 | 91.9 |
| 11 | 90.8 | 91.0 | 126.3 |
| 12 | 49.6 | 70.1 | 103.4 |
| 13 | — | 46.6 | 55.5 |
| 14 | 60.8 | 66.3 | 80.4 |
| 15 | 35.8 | 44.9 | 61.2 |
| 16 | 10.2 | 8.7 | 11.1 |
| Means | 64.1 | 70.3 | 83.5 |

Other strain gauges show similar characteristics; i.e. major responses in the PWT occurring in the 500–600 Hz range with individual spikes at 415 and 720 Hz, whereas in the WT major responses occur in the 400–600 Hz range with individual spike at 720 Hz.

(*Note*: From a survey in the finite-element analysis work performed by Dassault Aviation—subtask 4.1—it is apparent that the symmetrical modes of the panel are being excited in the PWT, whereas both symmetrical and antisymmetrical modes were excited in the WT.)

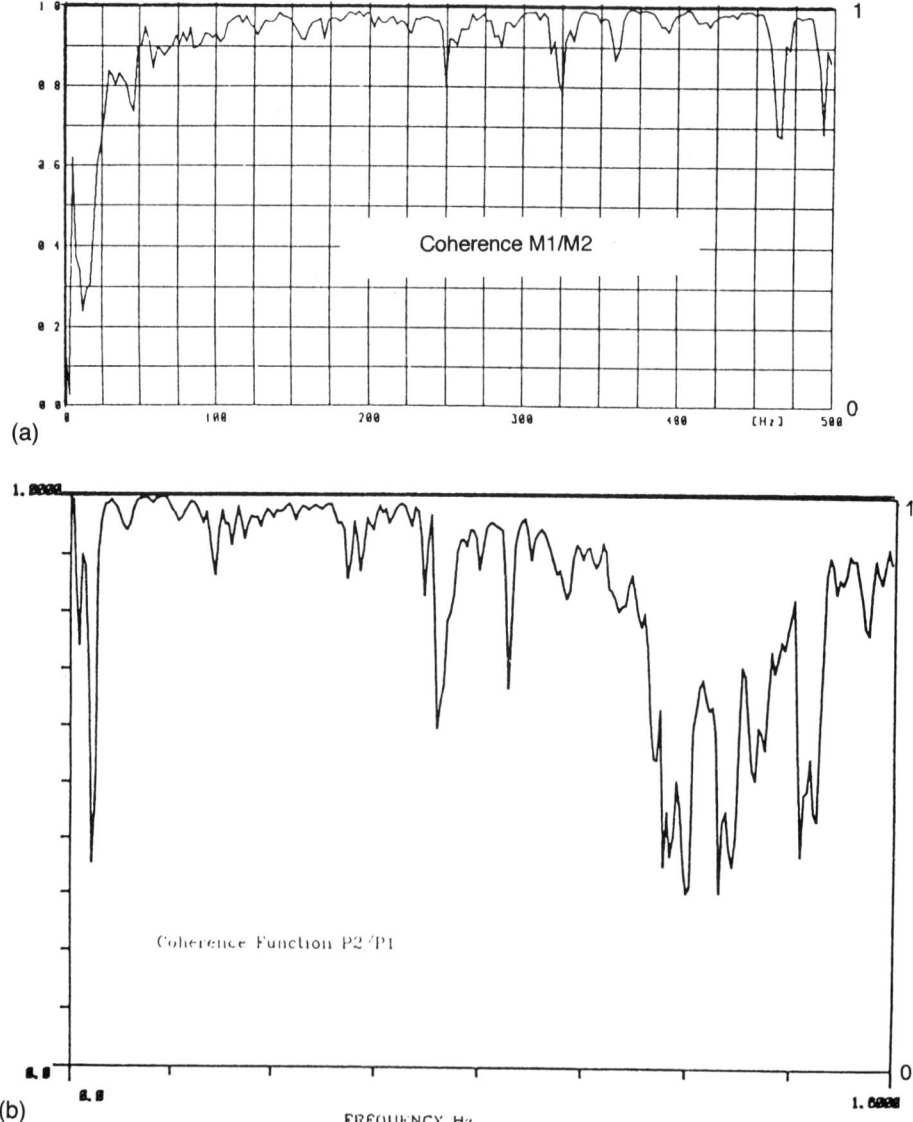

(a)

(b)

FREQUENCY Hz

**Figure 2.64**   Typical coherence functions in the PWTs: (a) IABG; (b) BAe

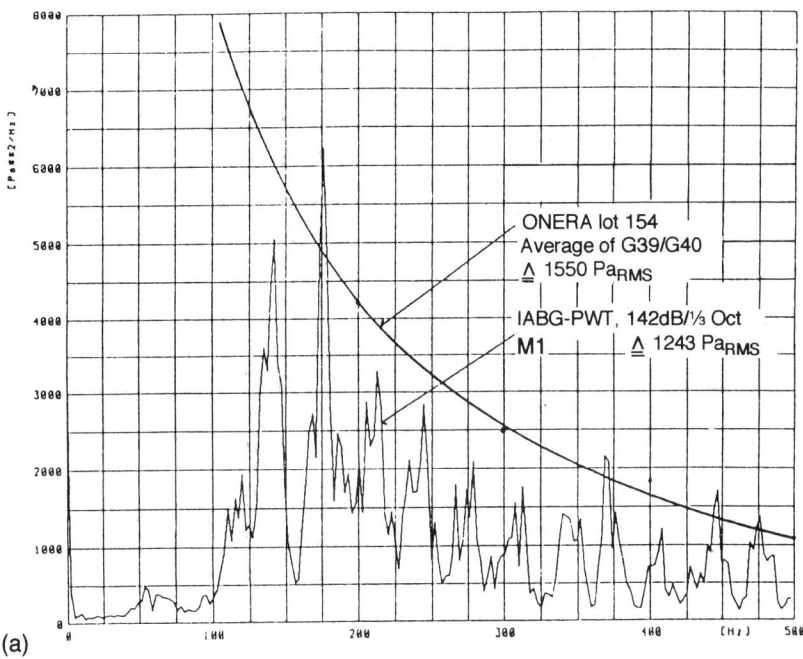

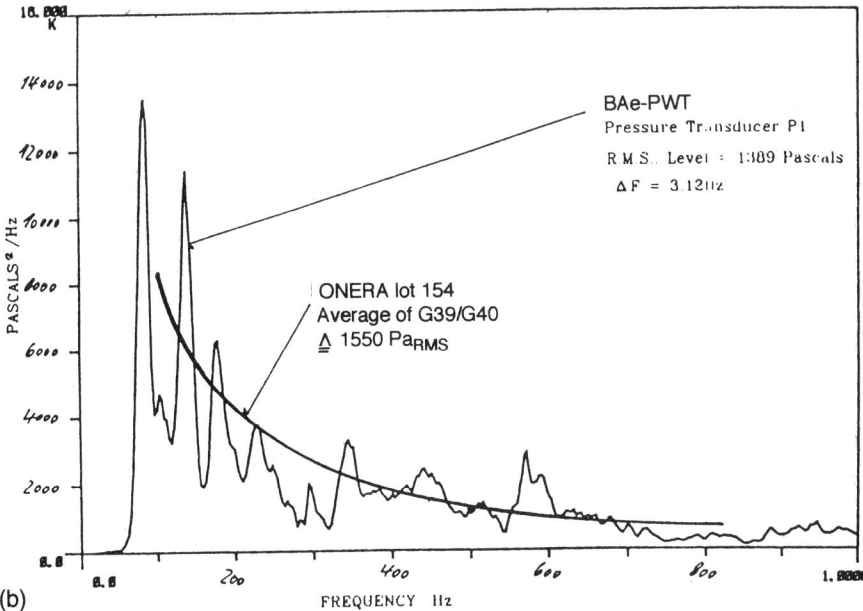

**Figure 2.65** Comparison of WT and PWT tests of pressure spectral densities: (a) ONERA WT and IABG PWT; (b) ONERA WT and BAe PWT

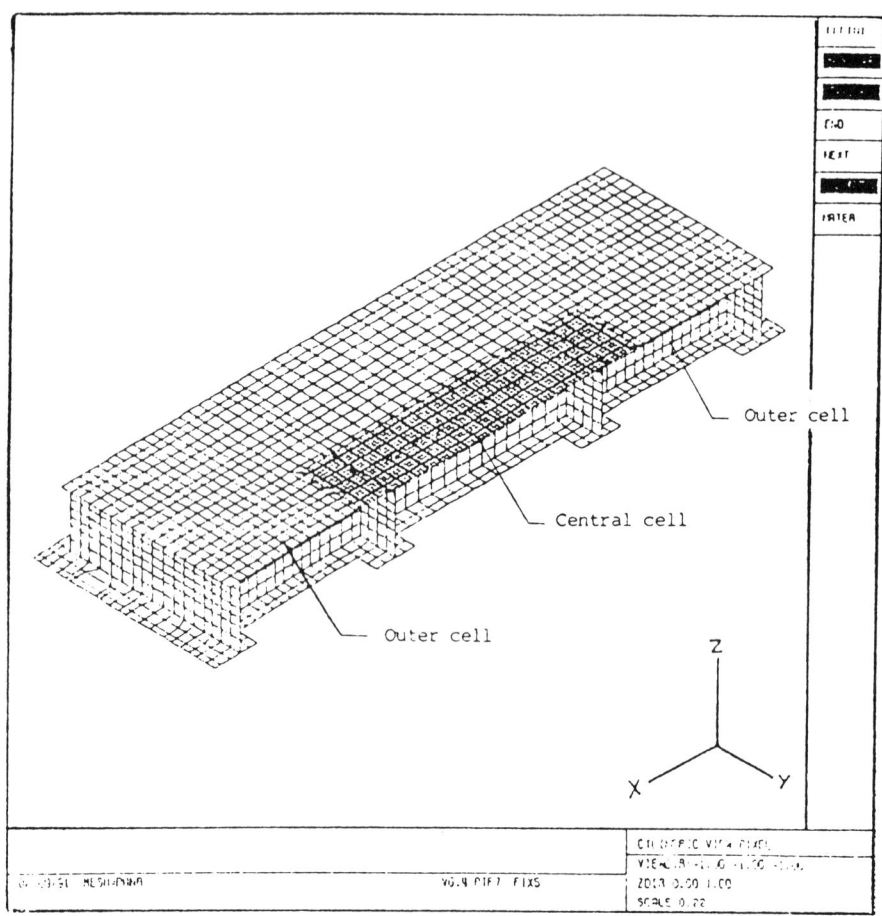

**Figure 2.66** Finite-element calculation of the aluminium panel: mesh and first symmetric mode shapes

**Matched strain power spectrum** The test panel was then subjected to a modified sound pressure spectrum, where the aim was to match the r.m.s. level and strain spectra for strain gauges 10 and 11. The overall responses in terms of r.m.s. values were in reasonably good agreement, as can be seen from the comparison of strain responses in Table 2.5.

The higher values obtained by BAe are most likely due to the higher acoustic excitations. The lowest strains together with the highest excitation level at ONERA are explained by the fact that the WT results are determined mainly by the high intensities at low frequencies where no panel response is present.

The deviations in mean strain should not be considered essential, because they can be adjusted by a more precise tuning of the PWT excitation.

As with the earlier test, at strain gauge 10 two dominant modes, at 515 and 545 Hz, could be significantly excited; but once again, the high peaks at 440 and 475 Hz could not be excited. For strain gauge 11, as with the earlier test, two dominant modes at 520 and 720 Hz could be significantly excited; but once again, the spectral peaks at 385 and 425 Hz could not be excited.

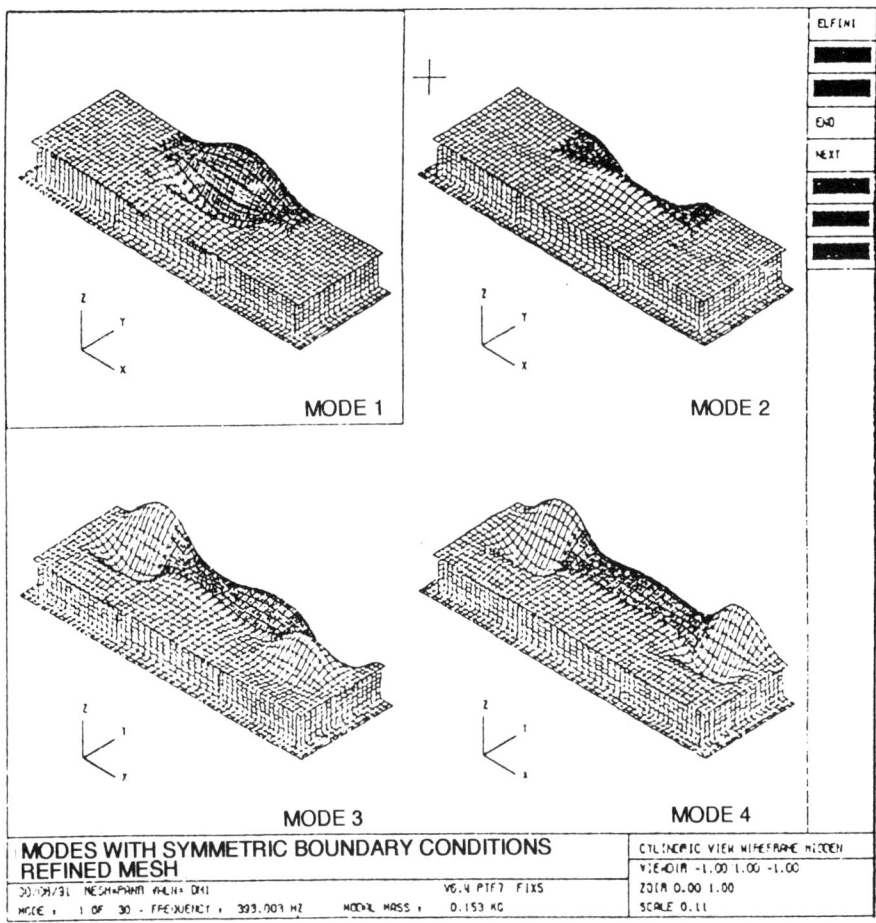

Figure 2.66    (*continued*)

**Narrowband excitation**   To match WT excitations at strain gauges 10 and 11, there is a requirement to excite panel modes at 440 and 475 Hz for gauge 10, and modes at 385 and 425 Hz for gauge 11. For this purpose, narrowband excitation was restricted to between 400 and 592 Hz centre frequencies. However, this resulted in forced vibrations of the panel together with excessive response at resonant conditions.

## Conclusions of the experimental acoustic loads study

The main objectives of subtask 3.1 were to collect basic data on the space–time characteristics of a fluctuating pressure field caused by separated flows and to study the structural response of a test panel subjected to this type of loading in WT and two PWT excitations. Futhermore, if required a testing strategy should be defined to approximate the structural response achieved in the PWT to that obtained in the WT.

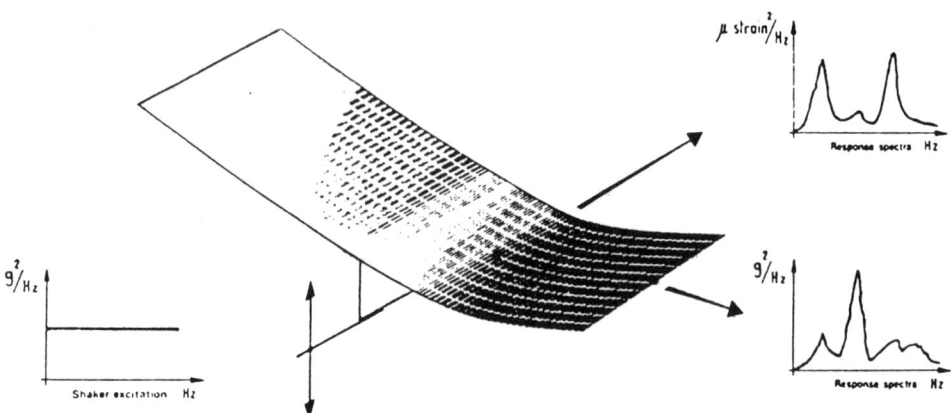

## DYNAMIC TESTS PREDICTION

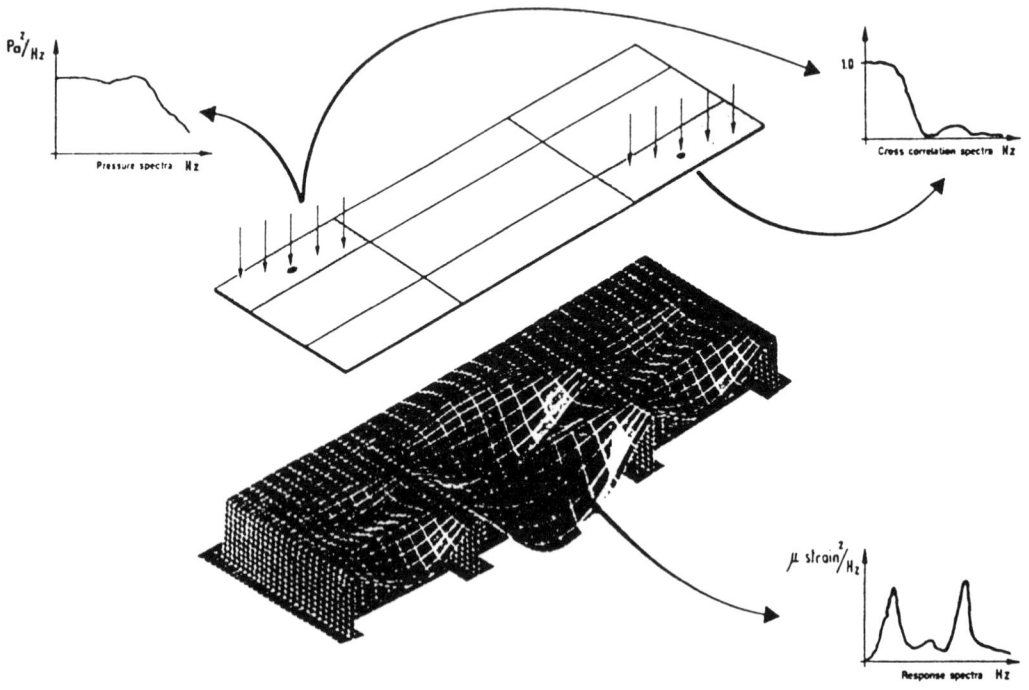

## ACOUSTIC TESTS PREDICTION
## WIND TUNNEL + PWT

**Figure 2.67**  Computational method for simulation of the structural response of the coupons (shaker excitation) and of the panels (in WT or PWT)

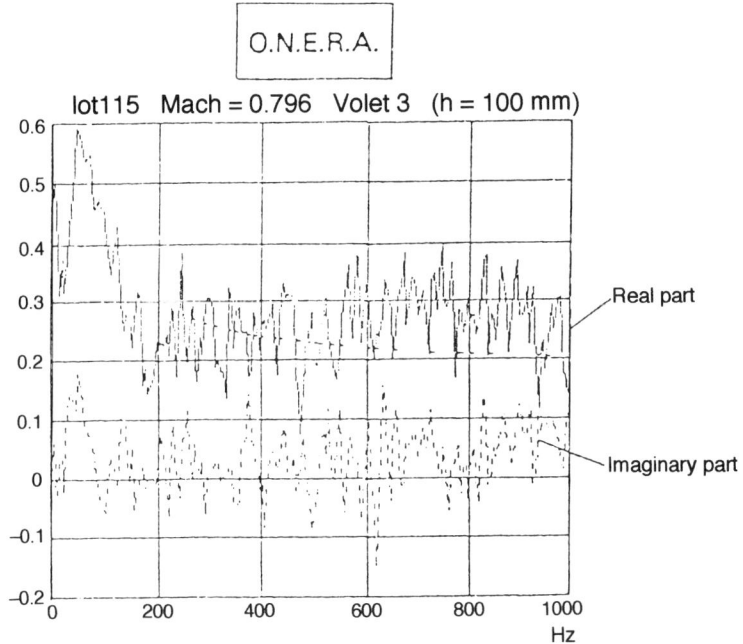

COHERENCE   ESSAI  @ 1152   1er CAPTEUR M1
2erne CAPTEUR M3

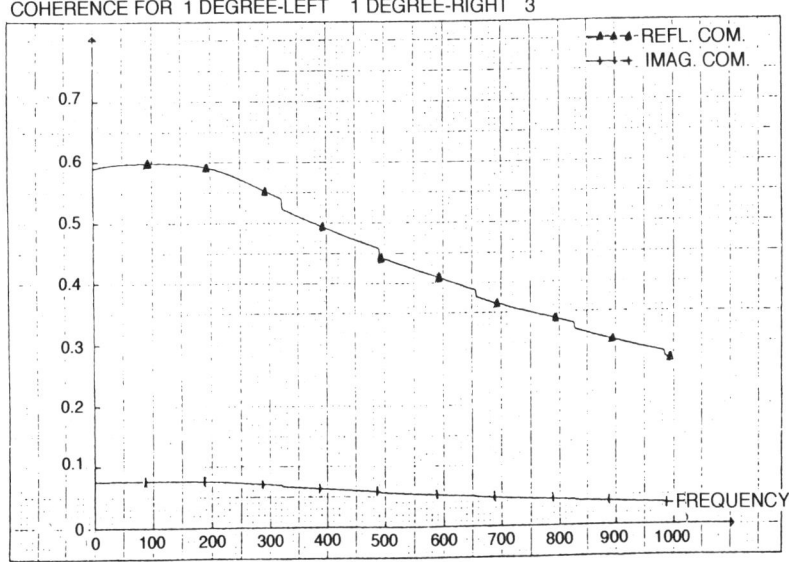

**Figure 2.68**  Typical cross-spectra diagrams in the wind tunnel: comparison between test results in the S1 ONERA WT (top) and from the analytical model of IST

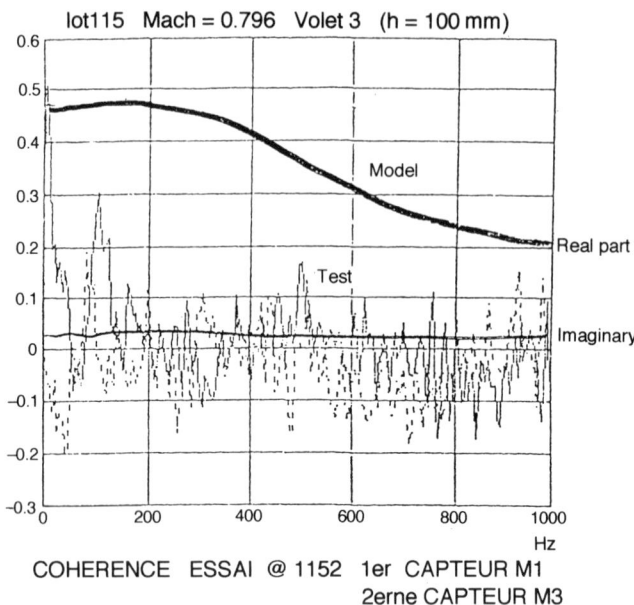

COHERENCE   ESSAI  @ 1152   1er  CAPTEUR M1
2erne CAPTEUR M3

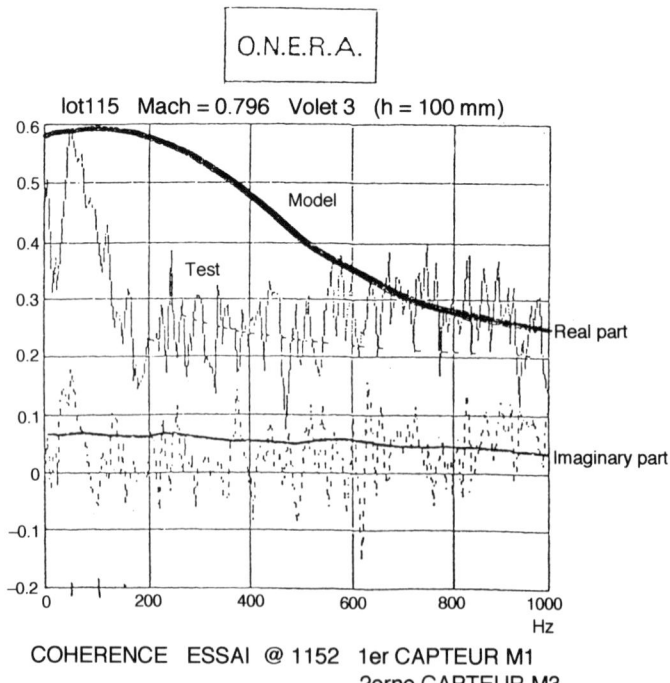

COHERENCE   ESSAI  @ 1152   1er CAPTEUR M1
2erne CAPTEUR M3

**Figure 2.69**   Typical cross-spectra diagrams in the wind tunnel: comparison between test results and the analytical model of IST

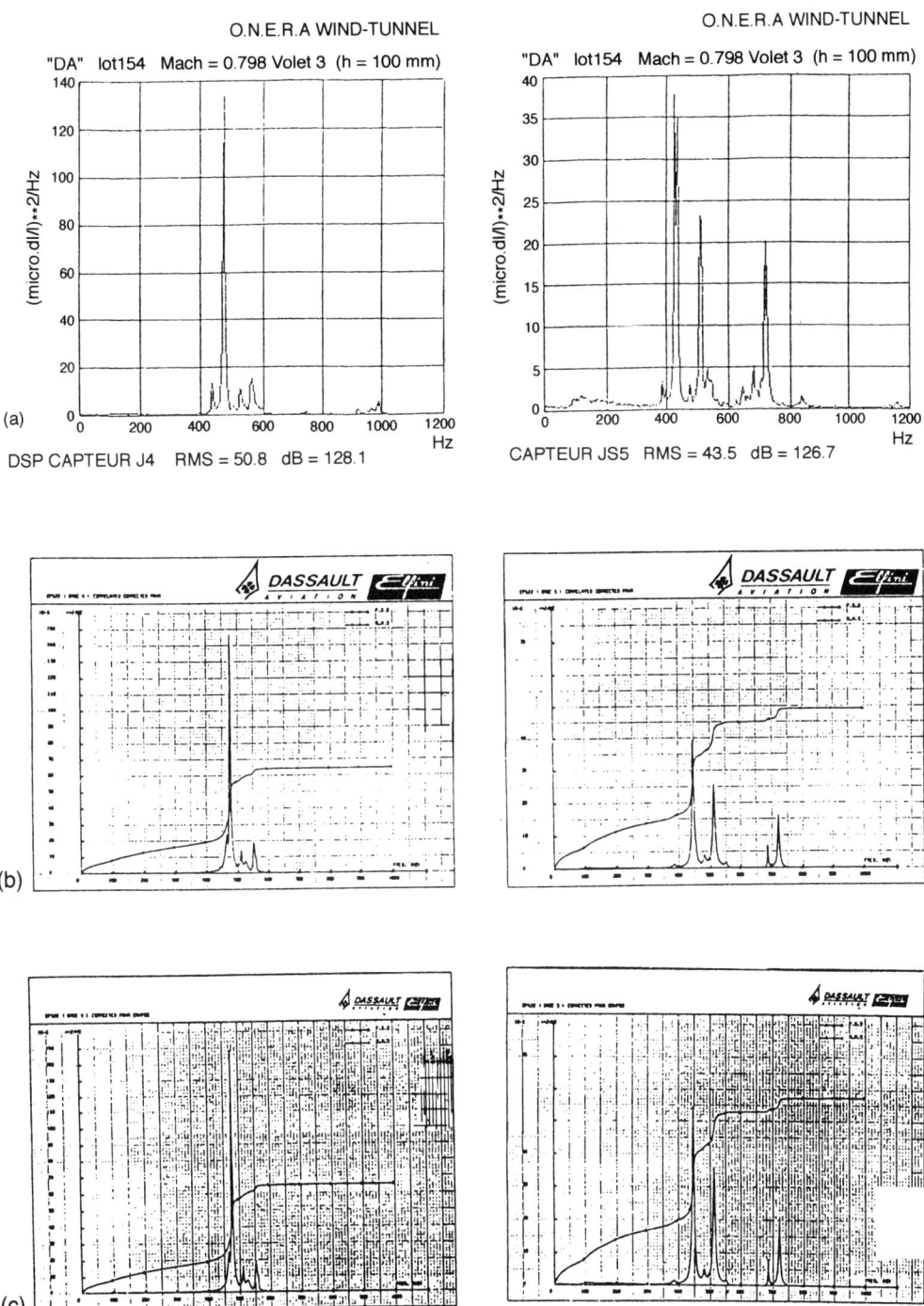

**Figure 2.70**  Wind-tunnel tests of the aluminium panel—PSD of the response of strain gauges J4 and J5: comparison between (a) test results, (b) computations with pressure measurements as load input, and (c) computations with the analytical model as load input

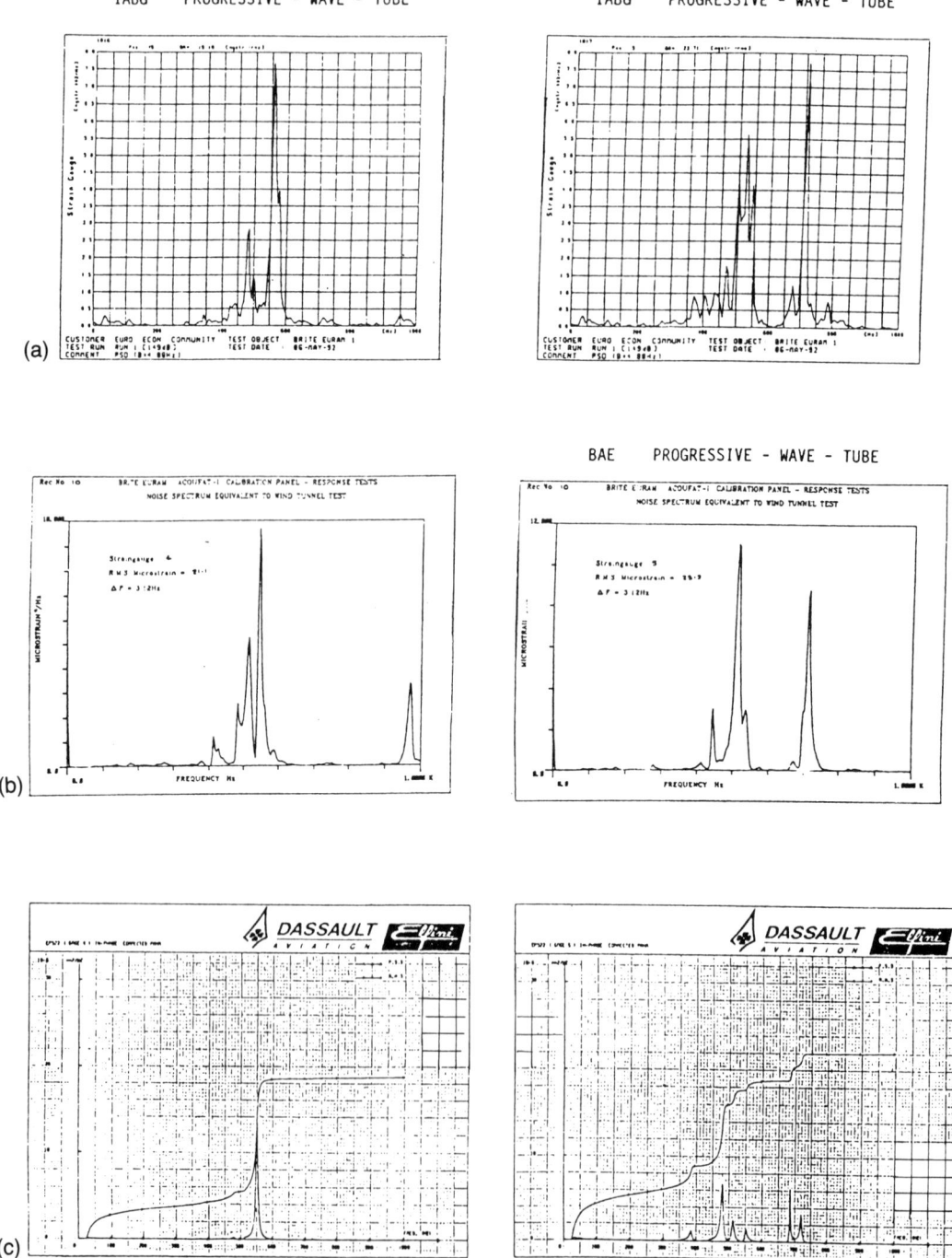

**Figure 2.71** Progressive wave tube tests of the aluminium panel—PSD of the response of strain gauges J4 and J5: comparison between (a) test results in the IABG PWT, (b) test results in the BAe PWT, and (c) computations with an in-phase acoustic pressure

The first two goals have been achieved during the ACOUFAT project, and now more is known about the complex interactions between aero/acoustic excitation and structural behaviour. However, it was not possible to define a testing strategy to obtain the same structural response in the WT and in the PWTs.

The main findings can be summarised as follows:

1.  In the design of a test setup for separated flows in a WT, care has to be taken. Even with a relatively simple structural and aerodynamic configuration the phenomena involved are already rather sophisticated. Several fundamental mechanisms appear to be simultaneously responsible for the fluctuation pattern on the panel, whereby the effect of the internal mean recirculation flow in the separated region cannot be neglected (Figure 2.56).

2.  To represent the statistical characteristics of the pressure fluctuations in mathematical terms it is necessary to extract the most important features from the available experimental data.

3.  On comparing the characteristics of excitation in the PWTs with those in the WT, some significant differences could be observed. Whereas the pressure spectral density in the WT is continuously distributed with high intensities at low frequencies, the PWTs exhibit very peaky spectral densities (Figure 2.65). This is due to internal resonances. Furthermore the coherence of the fluctuating pressures is very much higher in the case of the PWTs (Figure 2.64) compared with the region of separated flows in the WT (Figure 2.61).

4.  Owing to differences in structure acoustic coupling between the progressive wave field of the PWT and separated flow in the WT, different modes are excited and hence spectral levels are different (Figure 2.63). Narrowband excitation has been used to try to excite modes which would otherwise be excited in the WT, but this has resulted in forced vibration of the panel and excessive forcing at resonance.

5.  Matching the one-third octave levels was generally possible within an accuracy of $\pm 1$ dB. However, owing to the characteristics of noise generation, the frequency range in the PWTs is somewhat limited at the low frequency side ($f < 100$ Hz).

It became clear that, even for a simple aerodynamic configuration, the excitation of structures by aero/acoustic loads cannot be simulated by a PWT if only the spectral content is correctly shaped or modified. The effect of the spatial distribution of the loading is clearly different in both cases, which suggests that a theoretical approach, based on correctly predicting responses to both types of environment, is required. If this could be achieved, then it might be possible to design acoustic tests in the PWT (perhaps by applying

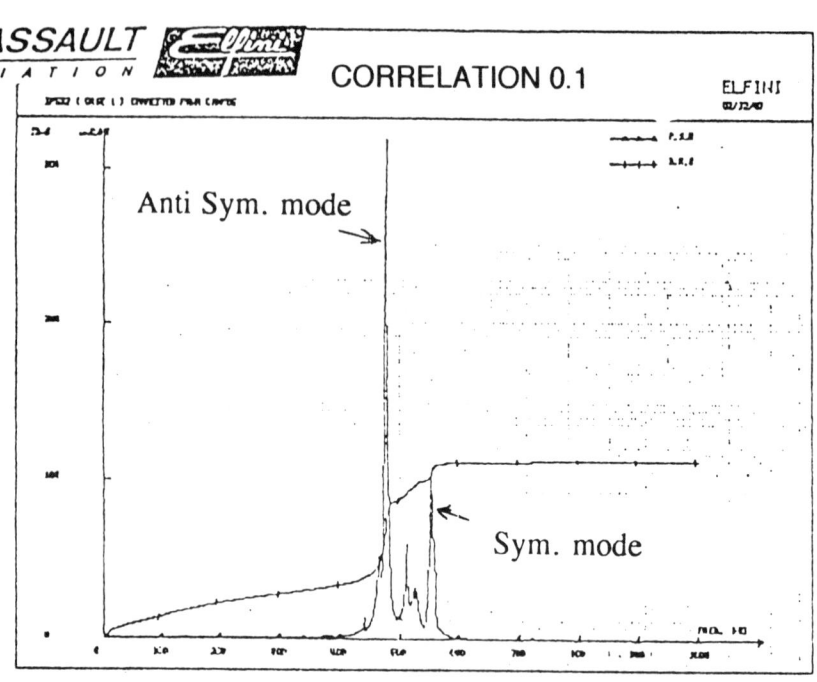

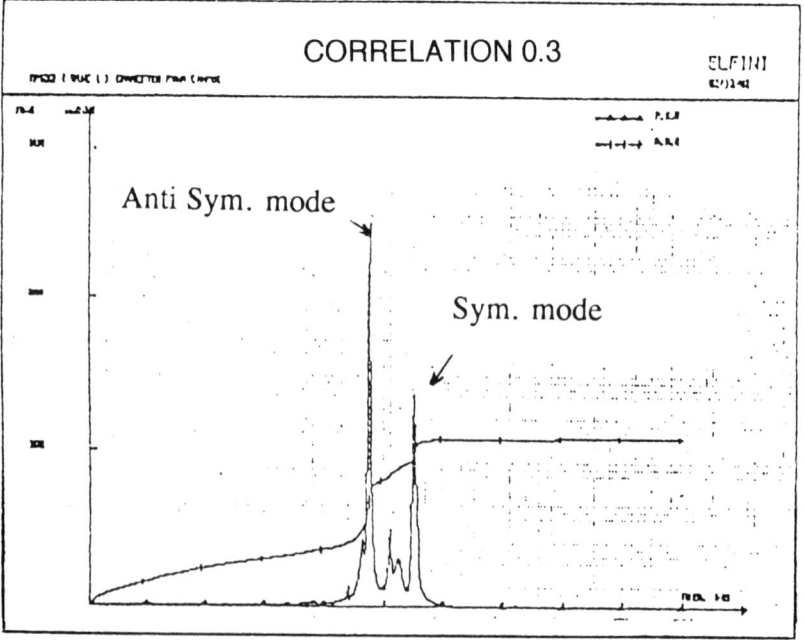

**Figure 2.72** Influence of the correlation factor on the aluminium panel dynamic behaviour (computational study)

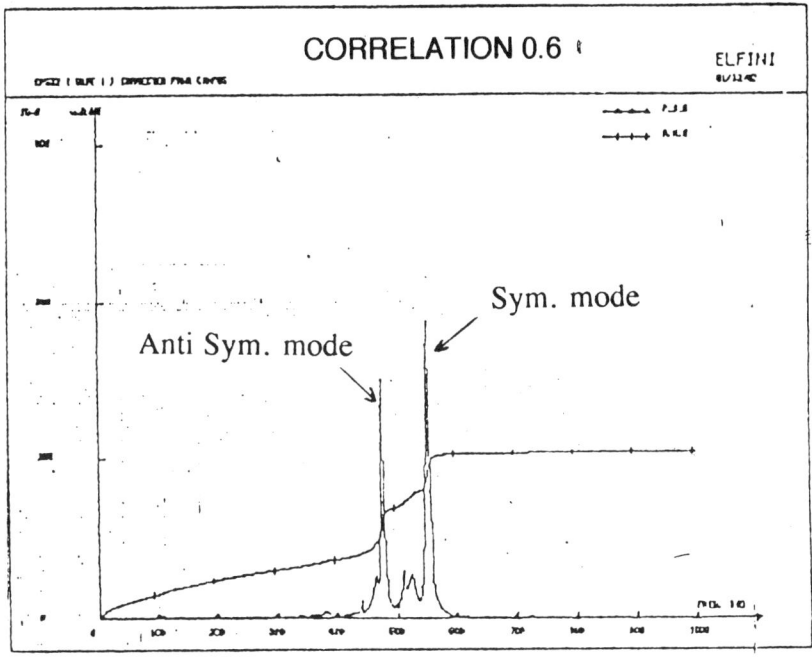

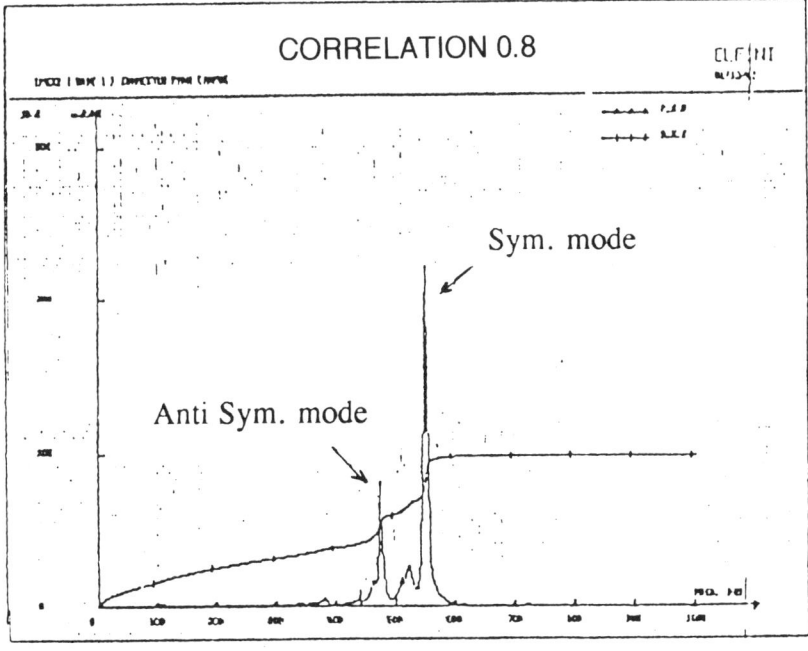

**Figure 2.72**   *(continued)*

narrowband acoustic excitation with additional mechanical excitation) that would excite structures to the same level and spatial stress–strain distribution even in individual modes of the structure.

### 2.4.2  Analytical results

*Introduction: subtask 4.1 presentation*

The objectives were to develop and validate a semiempirical model of acoustic loads, with space distribution and time fluctuation, on the basis of information collected from the open literature and on the pressure measurements obtained by the above experimental study of subtask 3.1. The work related to this analytical study was performed in three steps:

- Development of an analytical model of aero/acoustic loads and comparison with the pressure measurements obtained in the first phase of the WT tests.

- Modelling of the test panel by finite elements, and simulation of the dynamic structural response of the panel: (a) with the WT pressure measurements as load inputs, and (b) with the analytical model as load inputs, followed by comparison with the WT tests results.

- Simulation of the dynamic structural response of the panel in the PWTs with different assumptions related to the PWT load characteristics, and comparison with the PWT tests results.

The analytical model was developed by IST-Lisboa and the FE calculations were performed by Dassault Aviation.

*Analytical model of aero/acoustic loads*

**Development of the analytical model**  The aim was to develop a semi-empirical model to calculate the spectra of the spatial and temporal correlations of the acoustic pressure associated with fatigue loads, for comparison with experiments of panel tests in WT and additionally in PWT. The modelled cross-spectra law can be then used as load input for FE calculations.

For both WT and PWT, a statistical independence is invoked to split the space–time spectra into factors corresponding to temporal spectra and spatial spectra in each dimension (longitudinal and traverse). The simplest case is the test in a PWT which, under the restriction of linear sound, leads to similar spectra in all space–time dimensions. The case testing in a WT leads to progressive complications: a modified Gaussian correlation of phase shifts

**Table 2.6** Equation used to determine the normalised cross-spectra law

$$\gamma(x[=x_1-x_2], y[=y_1-y_2], w) =$$

$$\exp\{j(K_0 x + k_0 y)\} \sum_{n,m=1}^{\infty} D^{n+m-2}\exp(-a)\left\{2\sqrt{\pi}\,\zeta(\Omega) + \sum_{j=1}^{\infty}\left[\frac{bE(x)E(y)}{\sqrt{j}}\right]^j \exp(-\Omega^2/4j)\sum_{p=0}^{j}\frac{(8j)^{-p}}{p!(j-p)!}H_{2p}(\Omega/2\sqrt{j})\right\}$$

$$\sum_{n,m=1}^{\infty} D^{n+m-2}\exp(-a)\left\{2\sqrt{\pi}\,\zeta(\Omega) + \sum_{j=1}^{\infty}(bi/\sqrt{j})\exp(-\Omega^2/4j)\sum_{p=0}^{j}\frac{(8j)^{-p}}{p!(j-p)!}H_{2p}(\Omega/2\sqrt{j})\right\}$$

$$E(x) = (1 - 2x^2/L^2)\exp(-x^2/L^2) \qquad a = \tfrac{1}{2}\sigma^2(n+m)$$

$$E(y) = (1 - 2y^2/l^2)\exp(-y^2/l^2) \qquad b = \tfrac{1}{2}\sigma^2[n + m - (n - m)^2]$$

$$\Omega = (w - w_0)T$$

$$H_{2p}(\Omega/2\sqrt{j}) = 2\{(\Omega/2\sqrt{j})H_{2p-1} - (2p - 1)H_{2p-2}\}, \qquad H_0 = 1,\ H_1 = 2x \ldots$$

leads to spectra which are still stationary, but involve Hermite polynomials, besides Gaussian functions; the positioning of the observer close to a flap of finite span leads to nonstationary spatial spectra, involving error functions as well. All together there is obtained a hierarchy of spectra, the simplest of which depend on the r.m.s. phase shift and correlation scales (time and lengths) and excitation parameters (frequency and wave numbers). Multiple scattering involves the double reflection coefficient, and nonstationary correlations introduce geometric parameters into the exact and asymptotic evaluation of spectral integrals, which generalise the Gaussian type.

The type of final equation used to determine the normalised cross-spectra law, input as load data in FE calculations, is shown in Table 2.6. This model involves eight semi-empirical independent parameters ($K_0$, $k_0$, $D$, $\Omega$, $a$, $b$, $L$, $l$) which must be tuned according to the experimental data obtained by the WT tests of subtask 3.1.

**Comparison of the analytical model of cross-spectra and experimental WT measurements**   On the basis of the transducer responses obtained during the WT tests (see Figure 2.58), the values of the parameters of the above equation were defined in such a way that the computed cross-spectra could be thought to be identical to the measured one.

As Figures 2.68 and 2.69 show, the mean trend of the real and imaginary parts is acceptable. The peaks which appear in the experimental data cannot be modelled. However, the imaginary part seems not to stick to reality as well as the real part does. Along the bandwidth (0–2000 Hz), the cross-spectra value tends to decrease, and this was noticed during the experiments.

*Simulation of the structural response of the panel by FE calculations*

**Mesh characteristics**   The aluminium test panel used for the WT tests is shown in Figure 2.55, and the gauge locations and the numbering are shown in Figure 2.58. A half panel was considered for the FE model; errors in this simplification are considered small.

The FE model is shown in Figure 2.66. A fine mesh of the middle bay was performed to obtain the best idealisation of stiffness, boundary conditions and strain distribution. The model displayed consists in 7393 nodes and 7757 elements, resulting in 38 381 degrees of freedom.

The modal resolution was performed with symmetrical boundary conditions and antisymmetrical boundary conditions at the mid-plane of the panel. The first four mode shapes with symmetrical boundary conditions are shown in Figure 2.66 (nevertheless the first 25, or more, modes with both boundary conditions were computed and visualised).

Gauge locations are accurately idealised in the model, and nine of the sixteen gauge results were obtained (owing to the half model).

To improve the FE model, a static calibration of the gauges was performed by uniform static loading on the real panel. Gauge measurements and gauge predictions were compared and adjustments of the mesh performed to improve the gauge predictions. For each gauge, the experimental results were curve-fitted linearly and the slope difference between this straight line and the computed straight line was considered to be the static coefficient to adjust the static behaviour of the FE model.

**Computational strategy (see Figure 2.67)** To predict the dynamic random response of this structure with the Dassault finite-element program ELFINI, a computational strategy was followed in which computations were carried out in two stages:

*First stage: computational grid* The excited surface of the structure was divided into a number of subareas or elements (the computational grid is a coarse mesh in comparison with the structural FE mesh—see Figure 2.67). To each element of the grid is associated a pressure spectra density definition. The PSDs of pressure are assumed to be uniform over each grid element and perfect spatial correlation is assumed between any two points lying within the same element (in-phase condition). The correlations between grid elements are defined by the cross-spectra. The pressure and the cross-spectra density definitions can be defined by measured data or by an analytical model.

The complete power spectral matrix of auto- and cross-spectra for this multiple-pressure input system is then established, on the basis of the measured (or predicted) auto- and cross-spectra.

*Second stage: projection on the FE model* The pressure load (the power spectral density and the cross-spectra) of each grid element is projected into basic loads on the FE model of the structure. The PSDs of stresses and accelerations at given points are then carried out, from selected elements of the FE model, with the above projected pressure and the associated power spectral matrix as load input.

**Influence of the correlation (cross-spectra)** When defining this ACOUFAT study, it was assumed that the cross-spectra would be an important parameter. A parameter study was performed with a uniform value of the cross-spectra on the whole grid, by increasing this value from 0.1 to 1 (the value 1 is the in-phase condition).

The response of a gauge representative of the panel behaviour is presented in Figure 2.72 for different values of the uniform correlation factor.

The modes do not react in the same way to the same cross-spectra value, and as this value grows the ratio between the modes varies. For a growing cross-spectra value, some modes are reacting more and more (increasing gauge response) while some other modes are reacting less and less (decaying gauge response). For a cross-spectra approaching the in-phase condition $(0 \rightarrow 1)$, the antisymmetrical modes decay and the symmetrical modes grow.

*Simulation of the structural response of the panel with the experimental pressure spectra density (ONERA WT tests) and the experimental cross-spectra (ONERA WT tests)*

**Applied cross-spectra**  It was assumed that the two halves of the computed panel had two independent loading cases but that on each half of the panel there exists a known cross-spectra, the one taken from the wind-tunnel experiment, identical on each half of the panel. The two sides were assumed to be not correlated together (see Figure 2.73); this was confirmed by the fact that the transverse length cross-spectra was experimentally very low. So on a finite-element model which represents half the structural panel, was defined an experimental correlation.

**Predicted strains and r.m.s. results**  Dassault applied the computational strategy of Section 2.3.3 with the ONERA pressure field as input on the grid. The damping levels were determined from an analysis of data which respected the proportions of the different modes for the same gauge, and the ratio of a same mode for different gauges.

The results (PSD) of the gauges J4 and J5, representative of the panel behaviour, are shown in Figure 2.70. The calculated PSD diagrams are in good agreement with the measured diagrams.

(*Note*: Aeroelastic coupling influence. Including aeroelasticity into the model did not change the results at all. Dassault tried to do so with the given ONERA data of correlation and the effect of the aerodynamic forces on the gauge responses was small. It was concluded that, for this panel, there was no effect of such coupling. This was also shown on the flutter curves.)

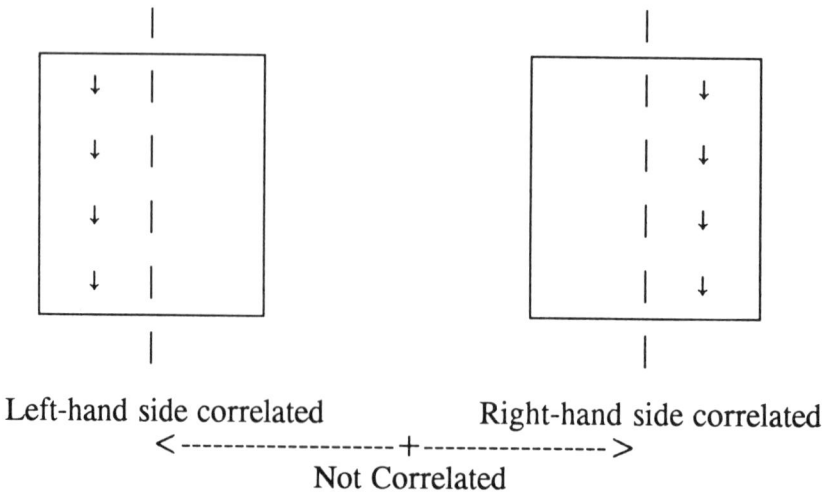

**Figure 2.73**  Independent correlations on the computed panel

*Simulation of the structural response of the panel with the experimental pressure spectra density (ONERA WT tests) and the experimental cross-spectra (IST model)*

**Applied cross-spectra**   The mathematical model from IST was input as a cross-spectra law on the computational grid of the acoustic loads.

**Predicted strains and r.m.s. results**   The results (PSD) are shown in Figure 2.70 for gauges J4 and J5. As one can see, these are very close to the results obtained with the ONERA data. Despite the fact that there is only partial agreement on correlation spectra, one could be surprised at the better agreement on panel response calculations. A possible explanation is that the model and measurements agree in the range of frequencies where the main resonances of the panel are located.

From the panel response, we can infer that the analytical model of pressure field is of good accuracy and allows the behaviour of the panel to be predicted inside the WT.

*Simulation of the structural response of the panel in the PWTs*

**Performed simulation**   To perform the structural response prediction of the panel submitted to aeroacoustics loads in the PWT, Dassault took the same FE model as for the panel response in the WT. The same static, dynamic and damping values were used.

On the other hand it was assumed that the condition of correlation in a PWT was such that the excitation would be in phase. In other words, there would be no phase shift in the excitation between two pressure captors put on the panel.

The computation was essentially the same as before, except for the application of an in-phase correlated pressure field to compute the gauge responses.

**Results from IABG and BAe tests and from Dassault computations**   Figure 2.71 shows a comparison of the PSD responses of gauges 4 and 5. Going from the WT to the PWT computation, the Dassault study on the influence of the correlation predicted that the antisymmetrical modes would tend to disappear (in-phase loading condition) and that, on the contrary, the symmetrical modes would be fully excited. The results show that all the antisymmetrical modes disappeared on the computational solution, while this was not exactly the case on the experimental test although their contribution turns out to be very small.

It was possible that the conditions of cross-spectra inside the PWTs were not exactly in phase. This was experimentally confirmed by the typical coherence functions observed in the two PWTs (see Figure 2.64).

Another comparison is between the two PWT experiments (IABG and BAe). It is observed that the two PWT facilities are different (i.e. size of duct and type of siren (noise generator)) and that the gauge responses are also different. Knowing under which type of cross-spectra the panel is submitted appears to be of primary importance since the gauge response depends on this assumption.

**Conclusion of the computational simulation (PWT)**   As outlined, knowing the conditions under which the panel is submitted is of major importance and this could determine the gauge responses entirely.

From now on, we know that the in-phase condition is probably not exactly true inside the PWT, and that the cross-spectra condition differs from one PWT to another. However, we note that due to an 'almost' in-phase condition, the PWT will not be able to excite the antisymmetrical modes as the WT does. That was confirmed by computation.

*Conclusions of the analytical acoustic loads study*

The main findings can be summarised as follows.

1.   The analytical aero/acoustic field model can be thought to be good enough if we look at it through the panel response for the WT.

2.   It was demonstrated by computations, and confirmed by tests, that different modes are excited due to differences in structure acoustic coupling between the PWT field and the separated flow in the WT. Hence, spectral levels and peaks are different in the structural response of the panel in the two facilities (PWT and WT). As has been seen by computation, the contribution of the antisymmetrical modes in the panel response is very low in the PWT. On the contrary, these seem to be predominant in the WT. That can lead to different failure modes and endurances for the panel in the WT and in the PWT.

3.   As far as the PWT is concerned, we can note that the condition of in-phase correlation, taken as an assumption for the computations, gave predictions that allow a better understanding of differences between WT and PWT. But it appears through the measurements that, for the same panel, two different PWTs do not give the same response. In fact, what comes to light by computation results is that, probably, each PWT has its own spatial distribution of loads. In the PWT tests, it is likely that the cross-spectra condition differs slightly from an in-phase condition (assumed initially). These results were also confirmed by measurements and analysis of the pressure field in PWT.

4.   Simulations were found to be useful to understand the problem and to identify which parameters are influencial. They could also help out in the adjustment of testing facilities to better represent the flight conditions, and so be useful for endurance tests.

5.   An understanding of the pressure field characteristics and influence on panel response has been of benefit in this study.

*Further work to be performed*

The cross-spectra definition is of the utmost importance, and so enhancement of the analytical model is necessary. Having a more accurate modelling for such an influential factor is essential. That could be achieved by additional WT tests with visualisations and measurements in the flow behind the spoiler.

An analytical model of the acoustic field for the PWTs should be produced and then coupled to the finite-element model to improve the predictions inside the PWT (similar to that for the WT).

A generalisation of this type of work to other sources of aero/acoustic excitations (step, jet noise, shock wave etc.) should be performed.


## 2.5  Research results: analytical support

### 2.5.1  *Structural dynamic response evaluation*

*Introduction: subtask 4.2 presentation*

Four computer codes were available for the study of the dynamic structural behaviour of specimens. These codes have existed for years but they still need to be validated by comparison with structural test results.

All the simple specimens for the dynamic tests by shaker excitation, and the seven panels for the acoustic tests in WT and in PWT, have been calculated by at least one partner. Furthermore, to evaluate the ability of the computer codes to deal with the problem, two of these panels have been calculated by the four 'calculating' partners (Dassault, KUL, MBB and Saab) to compare predictions between them and with test results. Thus, the ability of the computer codes to deal with the problem have been evaluated by comparison with each other and with test results.

This section provides an overview of the main conclusions and the major results obtained when comparing the results of the structural response analysis of the ACOUFAT endurance panels with the results of the tests performed in the PWTs.

Because space is limited, the main results presented here relate to the HTA/6376 panel which was the one endurance panel tested in PWT and calculated by the four partners concerned.


*Comparison between computed and measured panel mode shapes and natural frequencies*

A typical FE mesh of a (half) panel is presented in Figure 2.66(a). Typical outputs to visualise the mode shapes are shown in Figure 2.66(b).

The predicted values of the first mode shapes and natural frequencies were compared (when possible) with the different types of modal tests which were available:

- Loudspeaker tests with sweep excitation of the panel.

- Hammer test on the panel.

- PWT test of the panel.

In general, the mode shapes of the different calculations look rather similar to the measured one. Only a few modes appear in a different sequence.

Table 2.7 gives the corresponding natural frequencies computed by the four partners for the HTA/6376 panel, which are to be compared with the measured values of Table 2.8 (test results for the same panel). Differences up to 30% exist between the predicted and the measured natural frequencies. This magnitude of error was confirmed when comparing all the panel computations with the measurements. The main sources of errors of the FE predictions were identified to be:

- In the modelling of the connection between the skin/ribs and between the

Table 2.7   Computed natural frequencies (Hz) for the HTA/6376 panel

|   | Deutsche Airbus | Dassault Aviation | KU Leuven | Saab-Scania |
|---|---|---|---|---|
| 1 | 142 | 137 | 180 | 119 |
| 2 | 166 | 161 | 196 | 145 |
| 3 | 176 | 178 | 234 | 152 |
| 4 | 182 | 186 | 238 | 156 |
| 5 | 188 | 185 | 208 | 171 |
| 6 | 196 | 197 | 245 | 193 |
| 7 | 202 | 203 | 250 | 179 |
| 8 | 216 | 218 | 257 | 200 |
| 9 | 220 | 223 | 261 | 204 |

Table 2.8   Measured natural frequencies (Hz) for the HTA/6376 panel

|   | Sinus test | PWT test | Modal test |
|---|---|---|---|
| 1 | 147 | 142 | 166 |
| 2 | 167 | 166 | 184 |
| 3 | 176 | 174 | 193 |
| 4 | 192 | – | 225 |
| 5 | 180 | 180 | 208 |
| 6 | – | – | – |
| 7 | – | – | 235 |
| 8 | – | – | – |
| 9 | – | – | – |

skin/stringers. As pointed out by many partners, these connections have a nonlinear behaviour (due to contact effects) which make them difficult to model.

- In the thicknesses and the theoretical values of the material characteristics, which need to be adjusted in the FE model by identification and by the use of a modal test.

*Comparison of computed and measured r.m.s. strain levels*

As far as the in-plane r.m.s. strain distribution on the panel skin is concerned, measurements and computations show very little differences in general (see an example in Figures 2.70(a) and (b)). In particular, the maximum strain area (in the middle of the long side bay) is well predicted by the FE models. However, if we compare each gauge separately, the magnitudes of the predicted and the measured r.m.s. strain levels can be quite different (i.e. up to 100% difference).

The main sources of errors identified by the partners in the r.m.s. strain level predictions are listed below in descending order of influence:

- Nonlinearity effect. All the calculations are based on the linear elastic theory and do not take into account the geometrical nonlinear behaviour of the panel skin (membrane effect) when submitted to the high acoustic loading in the PWT. Static nonlinear computations and linearity checks inside the PWT (see Figure 2.40 for the HTA panel or Figures 2.45 and 2.48 for other panels) showed that this effect can more than double the error between a linear and a nonlinear behaviour. Nevertheless, the nonlinear dynamic behaviour, which is apparent in structures under high levels of acoustic loads, can also be simulated. Nonlinear dynamic FE calculations can be performed to evaluate this effect but these calculations are CPU consuming and costly.

- Characteristics of the PWT acoustic pressure field (auto- and cross-spectra). The computations were performed with a perfectly in-phase pressure field which was assumed to be homogeneous and constant on the panel in terms of auto-spectra ('white noise'). Computations with different hypotheses of cross-spectra showed that the cross-spectra inside the pressure field can greatly influence the individual contribution of each natural mode in the structural response of the panel. This was confirmed in comparing measurements of the structural response of the panels excited by loudspeaker with the structural response of the panel in the PWT, or in comparing measurements of structural response of the panel inside the WT with structural response of the panel inside the PWTs (see Figures 2.62 and 2.63). Therefore a precise prediction of the structural response is impossible without very detailed knowledge of the pressure field.

- Damping effect. The computations were performed with a theoretical value of the modal damping, which can be very different from the real damping of the panel inside the PWT. Reliable values are necessary as computational input data.

- FE mesh. A more detailed modelling, or an adjustment of the existing modelling, of the skin/stringer and of the skin/rib connections (including rivets) is necessary, especially in the areas of high strain gradients.

*Failure mode analysis of the endurance panels in PWTs*

As mentioned above, the maximum predicted and measured in-plane strain in the panel skin is located at the middle of the long side bay, just in front of the landing when existing, on the stiff side of the stringer. Despite this fact, for most of the ACOUFAT endurance panels tested in the PWTs the initial damage in the skin was not initiated in this critical area, but just in the corner of the bays, where the stringer crosses the rib (see Figures 2.43, 2.46, 2.49 and 2.52).

Owing to the panel design (no interconnection between stiffener and rib), high skin shear stresses are suspected to be generated in the corner of each bay, caused by vibration of the stiffener. This was confirmed by a finite-element structural response analysis in which the outputs show a high out-of-plane strain concentration in the bay corner where the initial damage occurred.

This type of failure was unexpected. Consequently, the FE models were not adequate to predict representative strain levels in these regions with high strain gradients.

*Fatigue life prediction of the endurance panels in the PWTs*

Fatigue life prediction is usually computed on the basis of random structural response analysis and adequate $S/N$ curves. One of the objectives of the ACOUFAT project was to be able to predict the fatigue life of the panels, under acoustic loading, by the use of $S/N$ curves obtained from the coupon tests of *Task 2*. Nevertheless, the unexpected type of initial damage in the skin of the panels had two consequences:

- The FE models were not adequate to compute representative shear strain levels in these regions.

- The coupons tested in *Task 2* were not representative of this type of damage. No $S/N$ curves were available for this type of damage because the coupons were designed to study the in-plane bending effect, not the out-of-plane strains effects.

Therefore, no endurance prediction could be evaluated from computations for the acoustic fatigue of the panels in the PWT.

*Conclusions on the structural dynamic response evaluation*

One of the main objectives of subtask 4.2 was to investigate the real computational capability to predict the structural dynamic response using the FE method. The comparisons between calculated and test results for the ACOUFAT endurance panels lead to the following conclusions:

1.  Classical random vibration calculations, including the modes of an FE model of a simple structure, can give reasonable results.

2.  Two of the most important FE input data are the pressure field characteristics and the modal damping values. Reliable values are necessary to increase the precision of the computations.

3.  Geometrical nonlinear behaviour of the panel skin (membrane effect) can strongly influence the structural endurance of the panel inside the PWT when submitted to high acoustic loading.

4.  By the FE analysis, a first attempt was made to explain the failure mode which occurred during the PWT endurance tests by the effect of the out-of-plane loads generated by the vibrating mass of the stringer suspended to the skin.

This cooperative work provided a basic comparison between the computational methods, to improve and to extend the existing analytical methods.

## 2.6   Overall summary and conclusions

*The acoustic loads study*

On the basis of wind-tunnel calibration tests, a semi-empirical model of the spatiotemporal characteristics of the aero/acoustic loads exerted on a flat panel by the turbulent field created by a flap (simple configuration of a typical turbulence) has been developed and utilised as 'load data input' for finite-element calculations. The WT tests were reasonably well represented: the development of this semi-empirical model of the spatiotemporal characteristics of the aero/acoustic loads is an encouraging initial success. The results from the initial modelling suggest that this can be extended to modelling the PWT.

Furthermore, with the same panel, investigations were conducted in two PWTs to try to match the strain spectral densities obtained in the WT. By this experimental study, it became clear that, even for a simple aerodynamic configuration, the excitation of structures by aero/acoustic loads may not be simulated fully in PWT, by simply modifying and correctly shaping the spectral content. The effect of the spatial distribution of the loading is clearly different in both cases, and the tested specimen endurance may be significantly different. It is clear that a theoretical approach based on correctly predicting responses to both types of environment is required. If this could be achieved, then it may be possible to design acoustic tests in the PWT (perhaps by applying narrowband acoustic excitation with additional mechanical excitation) that would excite structures to the same level and spatial stress–strain distribution even in individual modes of the structure.

### The structural dynamic response evaluation

Several computer codes were available for this study. Their ability to deal with the problem have been tested by comparison with each other and with test results. All the simple specimens for the dynamic tests by shaker excitation, and the seven panels for the acoustic tests in WT and in PWT, have been calculated by at least one partner. These assessments improve predictive capability, define the analysis assumptions and provide rules of use. They are the necessary support to the testing activity.

This computational activity was essential to develop and to validate the abovementioned semi-empirical model of the spatiotemporal characteristics of the aero/acoustic loads in the WT. Our understanding of the different panel responses in WT and in PWT has been essentially based on the parameter study performed by FE calculations.

Since the ultimate aim is to dimension and to qualify structures subjected to acoustic loads, FE calculations coupled to load input models would be the preferred solution: (a) to improve the representativeness of the test in PWT or in other test facilities (adjustment of temporal and spatial characteristics), and (b) to establish a valid comparison of the structural behaviour in real flight conditions and in PWT or other ground-test environments (i.e. different coupling fluid/structure). For this purpose, FE calculations are the only available and valid tool.

### The acoustic fatigue strength data for the selected materials and the associated designs (riveted, bonded, etc.)

Standard $S/N$ endurance curves have been elaborated for five selected advanced materials (two CFRP + GLARE + aluminium–lithium + SPF/DB titanium) with different designs representative of aeronautical structures ($\sim 550$ coupons tested by shaker excitation + six large panels in PWT). A considerable quantity of data has been produced, related to the five selected materials. This is the reference data required to develop aeronautical structures with advanced materials.

The commonly used methodology for this type of test has been critically analysed. The use of the 'frequency degradation' criterion, which is commonly applied for standard endurance data of classical metallic materials, has been evaluated for CFRP materials. This criterion, as the only parameter, was not considered suitable for determination of specimen 'failures'. Further work is required to investigate the reasons for the 'settling phase' observed in CFRP materials. It has been suggested that a suitable criterion should be based upon the degradation of the mechanical properties of the specimens.

Furthermore, an assessment of coupon failures compared with complex structural failures is required in order to validate the use of current coupon designs and endurance data.

On the basis of these tests, analytical work concerning damage initiation and damage propagation/accumulation has been also performed for CFRP materials.

# 3 Helicopter and tilt-rotor aircraft exterior noise research (HELINOISE)

## V. Klöppel*

This report, for the period January 1990 to December 1992, covers the activities carried out under the BRITE/EURAM Area 5 'Aeronautics' Research Contract No. AERO-CT89-0010 (Project AERO-P1108) between the Commission of the European Communities and the following:

Eurocopter Deutschland GmbH* (coordinator)
Agusta S.p.A
Deutsche Forschungsanstalt für Luft- und Raumfahrt, DLR
CIRA, S.p.A.
ALFAPI S.A.
Brüel & Kjaer
Bristol University
Instituto Superior Tecnico, Lisboa

Contact: V. Klöppel, Eurocopter Deutschland GmbH, Dptm. D/ET2 Aeromechanics and Flight Control, D-81663 München, Germany

*Advances in Acoustics Technology* Edited by J.M. Martin Hernandez. © ECSC-EEC-EAEC, Brussels–Luxembourg, 1994. Published in 1995 by John Wiley & Sons Ltd.

## Abstract

Eight partners investigated the noise radiation of helicopter rotors, including the aerodynamic sources. The dual objectives of the task were to improve rotor noise prediction, and to acquire a deeper understanding of rotor noise phenomena.

To validate the prediction codes, wind-tunnel tests with a model rotor equipped with 124 pressure sensors on the upper and lower sides of one blade were performed in the open jet test section of the Dutch/German wind tunnel DNW.

The model rotor system applied was a 1/2.46 hingeless Mach-scaled BO105 rotor of diameter 4 m. The dynamic and steady blade pressure signals were amplified in the rotating system and transmitted through a slip-ring to the data-acquisition system, so the signal-to-noise ratio could be kept low.

The test rig was mounted by a rear cantilever to the DNW suspension sting. This allowed relatively undisturbed flow around the model. A microphone traverse moved below the rotor plane to record the rotor's footprint by 11 microphones. The traverse moved continuously at low speed and recorded the noise signature synchronously with the unsteady blade pressure distribution. This procedure is mandatory for tracing the correlation between the blade pressure as noise source and the noise radiation itself.

The test program covered the flight envelope of the BO105 with special emphasis on descent conditions, with strong interactions between main rotor blade tip vortices and blades—a particularly annoying source called blade/vortex interaction (BVI) noise.

Prediction of BVI noise involves determining the structure and strength of the vortex wake leaving the blades. The blade tip vortices were visualised by smoke injection and a laser light sheet method available at the DNW.

The results obtained validated the first part of the theoretical work (viz. description of the noise sources by determining the blade pressure or load distribution, being strongly influenced by the rotor wake). To simulate the aerodynamic behaviour of blade flow and wake, the partners developed and improved different analytical methods.

In the second part of the theoretical work, aeroacoustic codes for the prediction of rotor noise radiation were addressed. Different sources such as thickness, loading and quadrupole noise were considered. As with the aerodynamic codes, the partners partially addressed different phenomena (complementary work) and partially applied different methods (comparative work). In the latter case, validation by the test results of a follow-on program will identify the superior method and enable the partners either to adopt this approach or to improve their existing codes accordingly.

## 3.1  Introduction

Noise is one of the most prominent constraints on helicopters that are to operate near populated areas. Hence, the noise levels of a new helicopter must be known in the early design stage since improvements to a flying prototype are very expensive if not impossible. Therefore, reliable noise prediction is of great importance for helicopter manufacturers.

The noise radiation of a helicopter is very complex. The rotor noise consists of discrete-frequency and broadband components. Recent experimental programs, undertaken individually by some of the partners, showed that helicopter noise can be reduced for example by means of a lower tip speed, larger rotor blade area and by aerodynamic improvement especially of the blade tips. All these noise-reduction techniques are effective, but might not be acceptable in civil helicopters because of their weight and performance penalties.

The relative intensities of the noise sources depend on rotor configuration and flight conditions. At moderate Mach numbers, adequate predictions can be made while ignoring high-speed terms. Of the remaining sources, thickness noise is by far the more tractable and computer programs accounting for blade section and planform distribution have been written successfully.

The calculation of rotor harmonic noise based on available methods is well in hand for the prediction of noise from propellers and rotors in hover. The limitation of existing methods lies in the calculation of high-resolution airloads required as input to the acoustic theory. Up to now, the fluctuating blade loads have been calculated with sufficient accuracy only for rectangular planform blades.

The current prediction of broadband noise is based on empirical procedures using rotor geometry and flight operational data. The correlation with measurements is acceptable, but the method is not able to model the detailed aeroacoustic phenomena.

Blade/vortex interaction (BVI) impulsive noise, which is one form of so-called 'blade slap', being most annoying in descending flight, in particular requires detailed knowledge of the unsteady blade pressure distribution for its successful prediction. If these unsteady aerodynamic blade pressures cannot be measured in a wind tunnel or big full-scale experiments, they must be calculated by high-resolution airload codes, which however do not yet exist to the desired degree of accuracy and thus have to be improved considerably.

High-speed (HS) impulsive noise, or compressibility noise, is correlated with the unsteady, transonic flow field around the blade tips. Prediction and possible reduction by optimum tip shape design, for example, again requires detailed information on the blade pressure and velocity distribution obtained either by experiment or by advanced three-dimensional, unsteady transonic rotor codes currently under development in some of the EU countries.

In view of our limited knowledge, the USA has over the past few years conducted a substantial research program on helicopter noise, entitled 'National Rotor Noise Reduction Program'. Cooperative European research in this area began with the project HELINOISE, bringing together the following partners:

|  | Role | County |
|---|---|---|
| Eurocoter Deutschland GmbH | CO | G |
| G. Agusta | CT | I |
| DLR-Institut for Aerodynamics | CT | G |

|                                      | Role | County |
|--------------------------------------|------|--------|
| Cira                                 | CT   | I      |
| ALFAPI                               | CT   | GR     |
| Brüel & Kjaer                        | CT   | DK     |
| I.S.T. (Instituto Superior Tecnico)  | CT   | P      |
| University of Bristol                | CT   | UK     |
| Subcontractors were:                 |      |        |
| DNW                                  | SC   | NL     |
| DLR-Institutes for Flight Mechanics  |      |        |
| and Aeroelastic                      | SC   | G      |

Co-ordinator (CO), contractor (CT), subcontractor (SC).

The research was originally scheduled to last for two years. However, owing to severe difficulties with construction of the model rotor, which was supplied by an external company, the contract time had to be extended to three years.

## 3.2  Objectives

Civil helicopters are an important part of the European aircraft industry. To compete against the substantially larger US helicopter industry, it is vital to have a technically and environmentally superior product. Noise, in particular, has raised public concern to an extent that it will be next to impossible to sell a noisy aircraft in the future. Moreover, the current noise limits for helicopters, as internationally agreed within the ICAO member states, are likely to become more stringent, so helicopter development requires substantial research efforts in aeroacoustics and, because of the inherent relationship, in aerodynamics and aeromechanics too.

The long-term objectives of any external noise research program must be to establish a basis for the following:

- The accurate prediction of helicopter and tilt-rotor aircraft noise characteristics during the design stage.

- The definition and optimisation of flight conditions for low noise emission ('fly neighbourly aspects') as guidelines for pilots.

With these long-term objectives in mind, the more immediate objectives to be approached within the planned research effort were:

- Improvement of our understanding of noise generation mechanisms, especially at high speeds and under descending flight conditions, by means of simultaneous measurements of blade pressure distribution and radiated noise, as well as by the identification of blade vortex positions in the vicinity of the rotor blades.

- A substantially improved prediction capability for helicopter external noise and related unsteady aeromechanical phenomena, by providing a reliable database to validate the aerodynamic and aeroacoustic prediction codes applied by the partners.

Research should then continue towards the use of the database for intensive validation of the aerodynamic and aeroacoustic prediction codes, and their further improvement; and application of the theoretical tools to design rotor blades with the specific aim of noise abatement.

## 3.3  Experimental activities

### 3.3.1  Work share

ECD was responsible together with DLR for the procurement of the model rotor (see Figure 3.1 and Table 3.1). The design and manufacture of this rotor for aeroacoustic tests in the DNW was performed by DEI, Virginia, USA. For this task, ECD defined the required dynamic, aerodynamic and geometric properties.

In order to approve the model rotor blades' dynamic quality and compatibility with the wind-tunnel model support, ECD and DLR tested in the DNW two configurations of a model rotor blade set, fabricated by DEI with the same dynamic layout as the one of the HELINOISE model rotor but without pressure sensors. Ground resonance stability and load tests were conducted. During the load tests, the rotor failed twice due to improper design. This led to a one-year delay of the task.

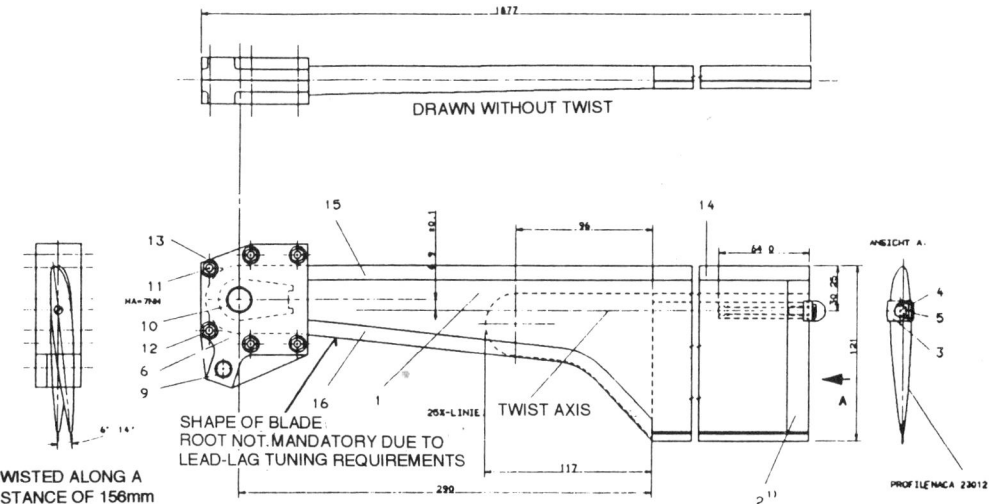

**Figure 3.1**   Main dimensions of the model rotor blade

**Table 3.1**  Model rotor blade data

| | |
|---|---|
| Scaling factor | 2.455 |
| Number of blades | 4 |
| Rotor diameter | 4 m |
| Blade chord (rectangular) | 0.121 m (larger than geometrically scaled due to Reynolds considerations) |
| Twist (outboard of $r = 0.44$ m) | 6° 14' |
| Airfoil | NACA 23012 modified |
| Rotation | Counterclockwise seen from top |
| Nominal rotation speed | 110 rad/s |
| Maximum rotation speed | 130 rad/s |
| Centre of mass | 24.5% |
| Shear centre (elastic centre) | 20% |
| Tension centre | 21% |
| Design thrust | 3900 N |
| Maximum thrust | 11000 N |

DLR defined the sensor characteristics and location (Figure 3.2) and led the experimental group 'XG'. The tasks of XG were basically threefold:

- To generate an optimum test plan that would yield the 'best return' within the financial and time constraints (Figure 3.3).

- To provide the test hardware for the wind-tunnel (WT) experiments— specifically the rotor test stand, the rotor instrumentation system, the

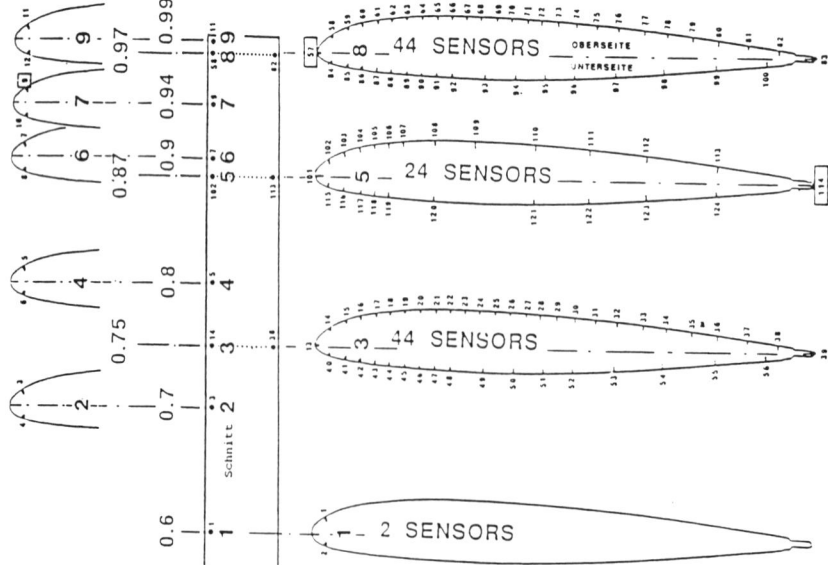

**Figure 3.2**  Location of the pressure sensors

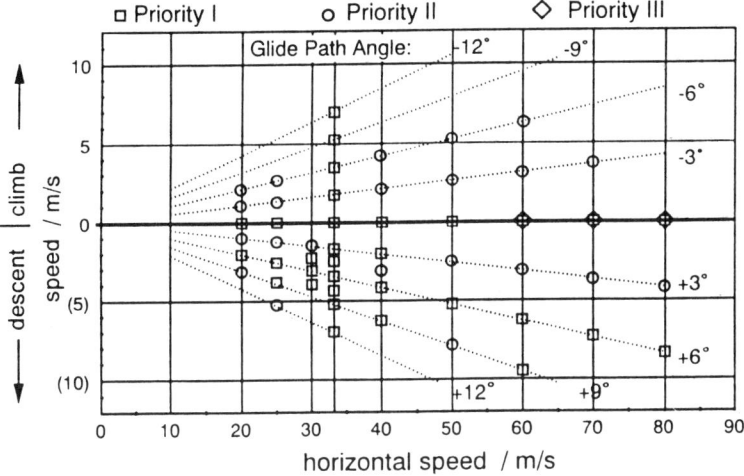

**Figure 3.3** Test plan for the wind tunnel campaign

data-acquisition and reduction systems for the rotor operational data, the aerodynamic blade pressure and acoustic data.

- To provide appropriate software for the tunnel experiments—specifically relating to the data-acquisition and data-reduction systems, with emphasis on an efficient 'quicklook' analysis capability.

- To perform the WT tests and to analyse the results together with ALFAPI.

ALFAPI acquired or developed all the required hardware and software facilities for the data-reduction task and, where appropriate, performed evaluation tests. The main hardware configuration for this task consisted of a Sun SPARCstation 2 together with all the necessary peripherals, whilst the primary software was the PV-WAVE programmable graphics package employed for data analysis, visualisation and final presentation purposes. A communication link was established between ALFAPI and DLR Göttingen for ease of data transfer between the two partners via the BITNET–INTERNET.

After the HELINOISE wind-tunnel tests, ALFAPI analysed all corresponding pressure data. These data were filtered, averaged, controlled with respect to plausibility and reproducibility and then documented in time and frequency domains. A representative part of the analysed results is presented in Section 3.4.

The HELINOISE wind-tunnel tests were performed in the German–Dutch wind tunnel (DNW). The following were the subcontractor's responsibilities [1]:

- Provide 12 consecutive tunnel occupation days of 8 hours each.

- Provide 8 m × 6 m open jet test section and acoustic testing hall (Figure 3.4).

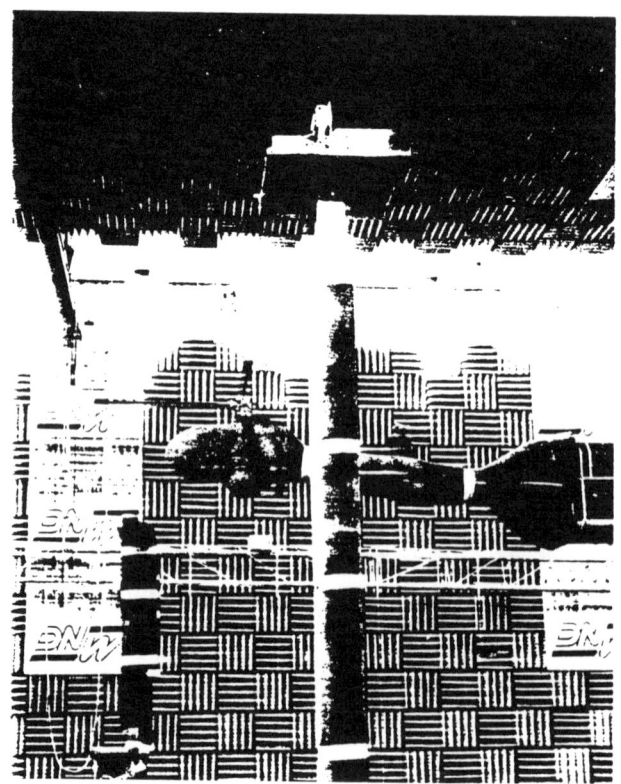

**Figure 3.4** HELINOISE test configuration in DNW

- Provide necessary operational and personnel support for the execution of the test program.

- Provide a microphone wing and associated traversing mechanism.

- Programme and operate the traverse.

- Implement the computer network between DNW computers and those of DLR-FM and DLR-EA.

- Provide timely transfer of tunnel, sting and traverse data via computer during the test.

- Execute the test program for flow visualisation and provide the necessary hardware of laser light sheet, traversing smoke probe, stroboscope flash and high-speed video camera.

- Provide a stiff mast for blade tip video recording.

- Analyse data of flow visualisation and video recording.

- Provide general electrical, mechanical and computing support.

### 3.3.2 Preparation and performance of DNW tests

The experimental setup is shown in Figure 3.5. It comprises the instrumented main rotor above a relatively small 'fuselage-body', containing rotor drive system and balance. This allows the measurement area to be extended below the 4-metre-diameter rotor 6 m upstream and 4 m downstream over a width of at least 4 m. Constant slow traversing of the microphone array saved up to 60% data-acquisition time.

The quality of the results of the HELINOISE project depended largely on the quality of the blade pressure sensors used. Here, it was important that the

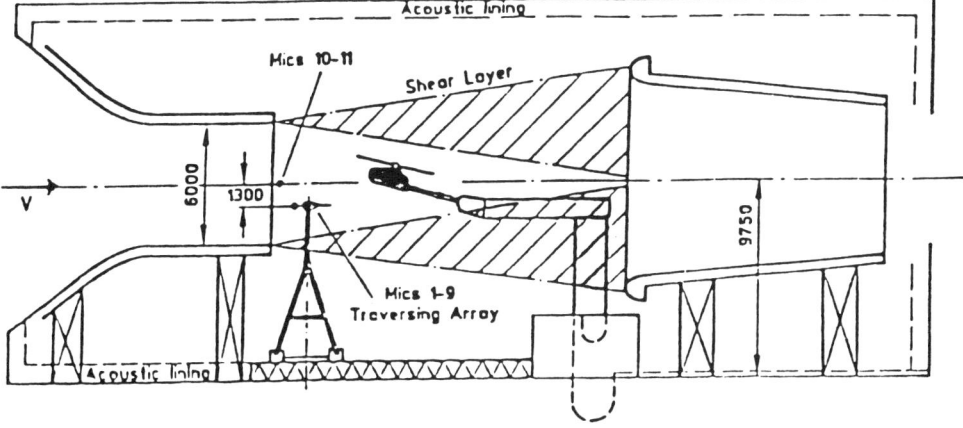

**Figure 3.5**  Test configuration

sensors exhibit sufficiently good pressure sensitivity, acceptably low electronic self noise, smooth frequency response characteristics over a sufficiently large frequency range (0 to 10 kHz) and acceptably low vibration sensitivity. DLR therefore developed special miniature low-noise amplifiers (DLR type 533) to be used in conjunction with KULITE pressure sensors. Preliminary frequency calibration (conducted by DLR Berlin with a specially developed calibration technique) on a number of representative sensors indicated a satisfactory frequency response. Additional calibrations were performed on all sensors in their installed versions.

### 3.3.3 Data-acquisition system

For storage and analysis of the test results, a large data-acquisition system called TEDAS (Transputer-based Expandable Data Acquisition System, Figure 3.6) was developed. Starting with 180 channels, the data-acquisition system had to be expandable for an envisaged continuation of the project with a more sophisticated rotor. The sensors—very small pressure transducers—were integrated flush into the rotor blades. The data were sampled at about 35 kHz with 16-bit resolution for 5 seconds. This gave a sum data rate of 13 Mbyte/s. As this

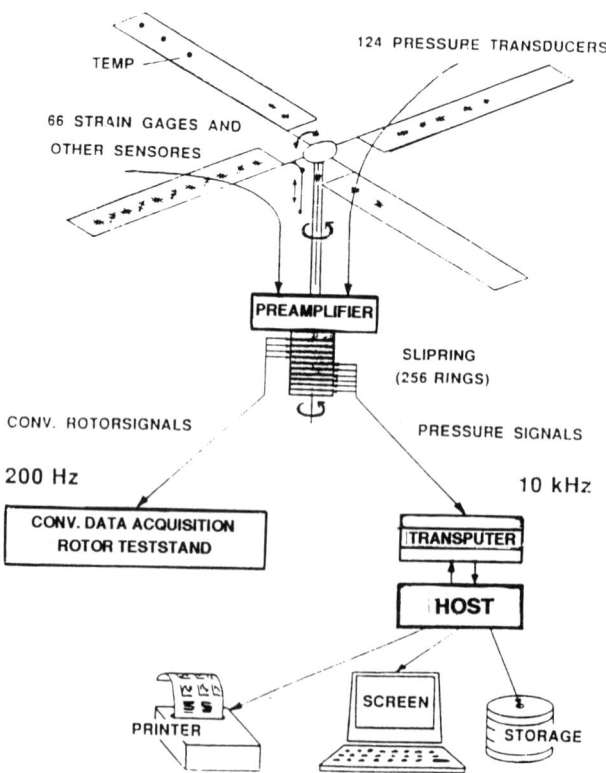

**Figure 3.6**    Rotor data-acquisition system

amount of data could not be stored online, a concept was used which stored the data locally and then transferred it to a mass storage device.

Transputer modules, serving eight dual A/D converters each and equipped with 4 Mbyte of local memory, sampled the data, stored it in local memory and performed signal processing on it. FFT, RMS, DC offset and other algorithms were performed by the transputers in parallel. Then the raw data and the results of the calculations were transmitted to a host computer, to be stored on a disk. Because of the large number of channels, the required anti-aliasing filters for 16-bit resolution would have required a significant part of the cost. Therefore, no traditional successive-approximation A/D converters were used, but rather delta/sigma converters, sampling the data internally at 64 times the desired word rate and producing a 1-bit datastream that was bit-density modulated. A decimation filter then reduced the word rate to 35 kHz at 16-bit resolution. Delta/sigma converters, originally designed for audio purposes, map most of the analogue filtering into the digital domain. The built-in decimation filter had a definitive linear phase, a very flat passband and a stopband rejection of $-90$ dB.

DLR performed the wind-tunnel tests supported by partner experts from ECD, Agusta and Cira. The Institute for Design Aerodynamics, Braunschweig, led the tests and measured the noise signature below the rotor with 11 microphones on the moving traverse (Figure 3.7).

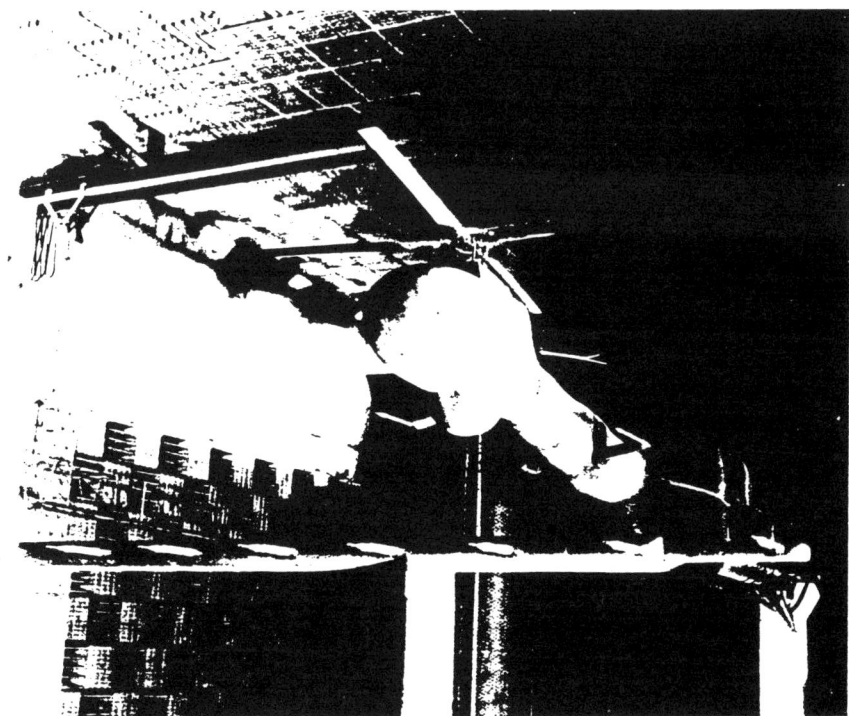

**Figure 3.7** Noise measurement below the rotor

The Institute for Flight Mechanics, Braunschweig, controlled the model rotor and analysed the flight mechanical data. It measured also the blade tip vortex loci at the position of the BO105 tail rotor by two up-and-down moving dynamic pressure probes (Figure 3.8).

The Institute for Aeroelasticity, Göttingen, measured and analysed the blade pressure data. In addition, the absolute blade tip position was recorded for validation of the flight-mechanical codes to be applied for the prediction of the blade motions.

### 3.3.4  Flow visualisation

In order to determine the misdistance between the blade and the blade tip vortices, a special flow visualisation technique was applied. With the aid of the same technique the vortex trajectories could be evaluated. A detailed description of this technique can be found in reference [1]. The most important aspects of this technique can be summarised as follows.

Using a 5 W argon laser and an optical package, a continuous thin light sheet was erected in the visualisation plane perpendicular to the rotor plane. For flow visualisation, oil smoke, generated by a smoke generator, was introduced

**Figure 3.8**   Wake capturing device at the position of the BO105 tail rotor

into the flow at the wind-tunnel nozzle by a smoke probe, which could be traversed in the vertical plane. In order to get a focused picture of a tip vortex, a stroboscopic, triggerable high-speed video camera equipped with a light intensifier was used. The shutter time of the camera was $1/10\,000$ s. The laser sheet could be traversed in such a way that the vortices to be visualised became illuminated by the laser light. Simultaneously, a high-power stroboscopic light source was used to illuminate the rotor blade. The light source and the video camera were triggered by the r.p.m. of the rotor.

Thus, the vortex and the blade could be visualised at the same time. Furthermore, in order to determine the position of the vortex in space, a grid was recorded with the same video camera. However, the rotor and the wind tunnel were then turned off. The grid was placed in the plane of the laser light sheet. Later, the recorded picture was digitised and used as an overlay on the

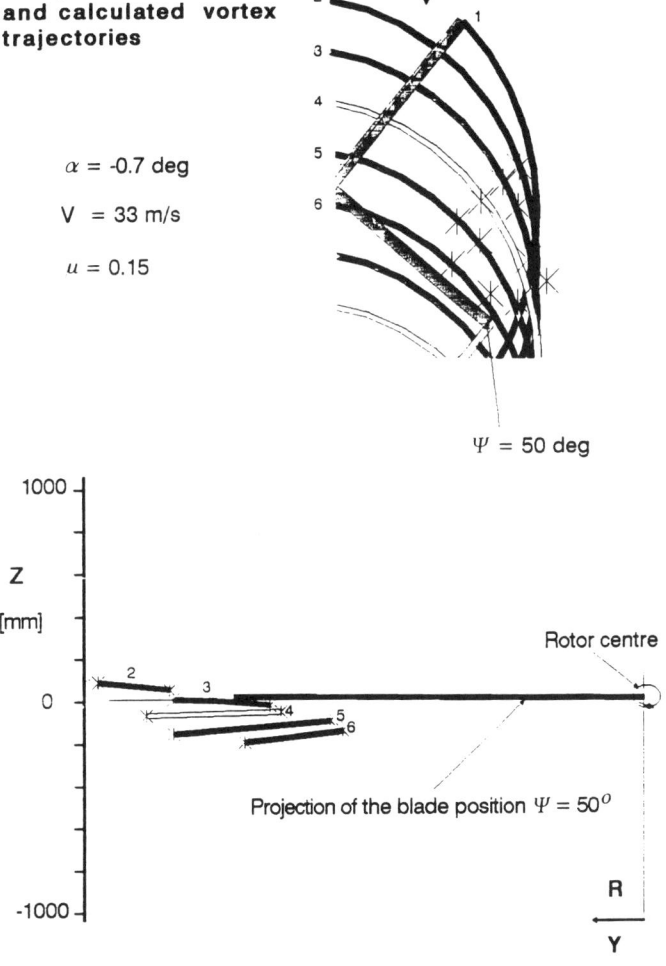

**Figure 3.9**  Measured and calculated vortex trajectories for a moderate descent flight case

actual flow recording, which then provided the relative positions of vortices and blade.

Figure 3.9 displays the top view of the measured vortex trajectories at a tunnel speed of 33 m/s and an angle of attack of −0.7°. The recording took place whenever the azimuthal blade angle was at 50°. Furthermore, the measured results were compared with theoretically calculated trajectories. The calculation is based on a simple flat wake model. Apparently, the agreement between theory and experiment is very good. Figure 3.9 also contains the frontal view of the vortex trajectories. The rotor blade and the measured sections of the trajectories are now projected into a vertical plane erected at an azimuthal angle of 90°. It is interesting to note that although the angle of attack of the rotor plane was slightly negative, trajectory 2 was located above the rotor plane, whereas trajectory 6, which was closest to the blade, was already far below the rotor plane. For this rotor condition the blade crossed trajectory 3 at an azimuthal angle about 90°. It can be expected that at this location the highest level of impulsive noise due to blade vortex interaction was radiated.

Figure 3.10 displays results of the same type of measurements, but with the angle of attack changed to +5.3°. Again, the top view of the vortex trajectories

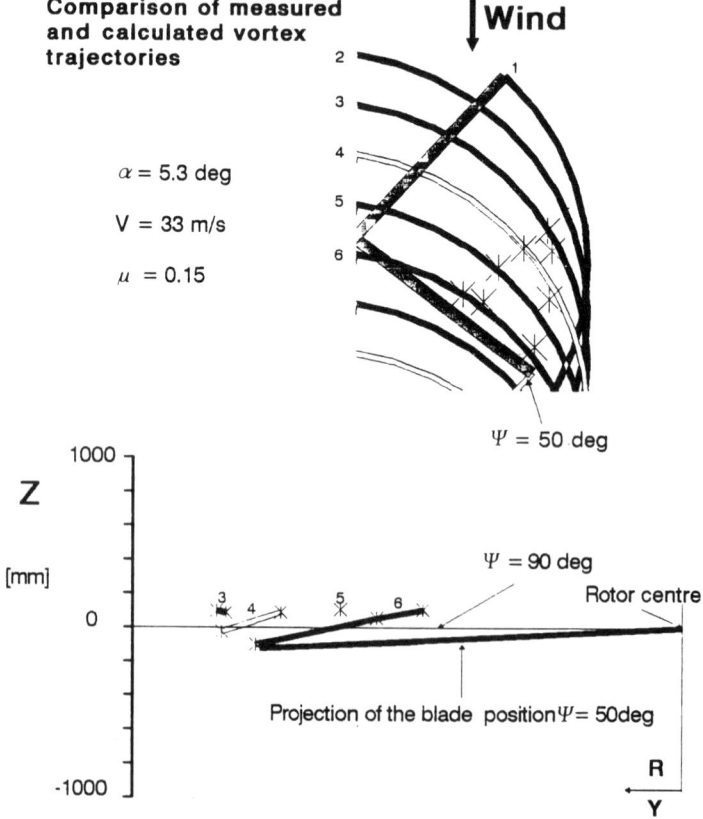

**Figure 3.10**  Measured and calculated vortex trajectories for a steeper descent case

reveal the good agreement between theory and experiment. However, owing to the large angle of attack all trajectories on the advancing side of the blade were now located above the rotor plane. It is most likely that for this rotor condition the blade tip had crossed the vortex trajectory 5 at an azimuthal blade angle of 50°. But also later, when the blade advanced to 90°, for example, trajectory 4 will be crossed by the blade. This can be recognised by the thin line which is the projection of the blade at an azimuthal angle of 90°. Again, it can be expected that impulsive noise will be generated at these locations. For convenience, Table 3.2 shows the measured vortex positions. Coordinates in $x$- and $y$-directions indicate the position with respect to the rotor centre. The $z$-coordinate is the distance of the vortex to the blade passing plane.

### 3.3.5   Blade tip position measurements

In order to improve the quality of aerodynamic prediction codes, knowledge of the blade tip deflections during different flight conditions is of significant importance. Usually, these quantities are measured with strain gauges pasted to the surface of the blade. It is however difficult to analyse such data, because the measured signals of the different bending modes (flapping, lagging, torsion) of a loaded blade are usually not decoupled, and one strain gauge measures several modes simultaneously. In addition, the hingeless BO105 rotor system allows no direct measurement of the first (rigid) blade flapping mode, a measurement which for articulated rotor systems is performed at the flapping hinge. Therefore, it was envisaged during this test campaign to record the tip of the rotor blade with the help of a high-speed video camera and analyse the position of the image of the tips with the grid technique described above. For this purpose, special identifiers were marked at the surface of the

**Table 3.2**   Measured vortex position (misdistance vortex/blade path plane) [1]

| Conditions in Figure 3.9 | | | Conditions in Figure 3.10 | | |
|---|---|---|---|---|---|
| $x$ | $y$ | $\Delta z$ | $x$ | $y$ | $\Delta z$ |
| 120 | −1691 | 111 | −238 | −1770 | 53 |
| 606 | −1110 | 113 | −103 | −1609 | 0 |
| 100 | −1455 | 100 | 74 | −1398 | −46 |
| 281 | −1238 | 133 | 263 | −1172 | −100 |
| 532 | −939 | 158 | 130 | −1760 | 0 |
| −70 | −1656 | 83 | 445 | −1385 | −62 |
| 585 | −1671 | 33 | 663 | −1125 | −152 |
| 1037 | −1547 | 0 | 672 | −1857 | −79 |
| | | | 935 | −1490 | −210 |
| | | | 1016 | −1757 | −171 |
| | | | 823 | −2037 | 75 |

$x$ and $y$ are measured with respect to the rotor centre, and $\Delta z$ is the measured vertical position to the blade path plane (all in millimetres).

blunt blade tip, and recording took place with the rotor and the wind tunnel in operation as well as turned off. However, in the latter case, the blade tip was replaced by a grid board and the recording was used as an overlay on the actual recording of the blades set in motion. The video camera and the stroboscope were installed on a rigid mast in the rotor plane outside the jet (see Figure 3.4). The trigger mechanism of the camera and the stroboscope were linked to the r.p.m.-signal of the rotor (1024 pulses/s). Thus, for each rotor revolution one image of a blade was recorded. Although the shutter time of the camera was set to 100 $\mu$s, the tip was visible, and thus recorded, over a period of only 15 $\mu$s owing to the preselected short duration of the stroboscope flash. (The flash time period is based on the half-life span, which is defined by the time period during which the light intensity of the flash has decreased by 50% of its maximum value.) During this period the blade was moved by approximately 3 mm in the circumferential direction, indicating the lower limit of the resolution capability of this measuring system. With a nominal rotor speed of 1050 r.p.m., 17 images of each blade were recorded, whereby the pressure instrumented blade was used as a reference blade.

Table 3.3 shows the flap deflection ($\Delta z$) at azimuthal angles of 270° and 300°. For this test the rotor thrust was increased from 2600 N to 3600 N at a tunnel speed of 33 m/s. In one case the shaft angle was 0° and for another test case the shaft angle was increased to 5.3°. Additionally, the table contains information about the variation of the blade angle when the rotor thrust was increased by 1000 N. With the knowledge of the adjusted collective pitch angle of the rotor, the torsional angle of the blade can be calculated. The collective pitch angles fed in at the blade root were measured by DLR-FM, Braunschweig, and secured in their data files.

In order to control the new measurement technique described above, DLR Göttingen performed parallel blade tip measurements with a laser Doppler method.

### 3.3.6 Intensity measurement technique

Partner B&K conducted sound intensity measurements on the DLR tail rotor test stand in the Braunschweig acoustic wind tunnel to check the technique. Details of the intensity measurement technique and some specific results, as obtained during the experiment on the DLR tail rotor, are discussed in the following.

**Table 3.3**   Measured blade-tip flap deflection

|  | $\psi = 270°$ | $\psi = 300°$ |
|---|---|---|
| Thrust: 2600–3600 N<br>$V = 33$ m/s, $\alpha$ (shaft) = 0° | $\Delta\alpha$ (tip) = 4.66°<br>$\Delta z$ (tip) = 80.3 mm | $\Delta\alpha$ (tip) = 6.14°<br>$\Delta z$ (tip) = 67.9 mm |
| Thrust: 2600–3600 N<br>$V = 33$ m/s, $\alpha$ (shaft) = 5.3° | $\Delta\alpha$ (tip) = 4.46°<br>$\Delta z$ (tip) = 73.2 mm | $\Delta\alpha$ (tip) = 6.07°<br>$\Delta z$ (tip) = 71.9 mm |

In order to capitalise on the potential of the new technique, it is necessary to perform such intensity measurements on rotors at a distance of 200–300 mm from the rotor blades. Until now, most acoustical measurements on rotors have been made in the acoustic far-field. The purpose of the test on the DLR tail rotor was to identify any problems associated with near-field measurements.

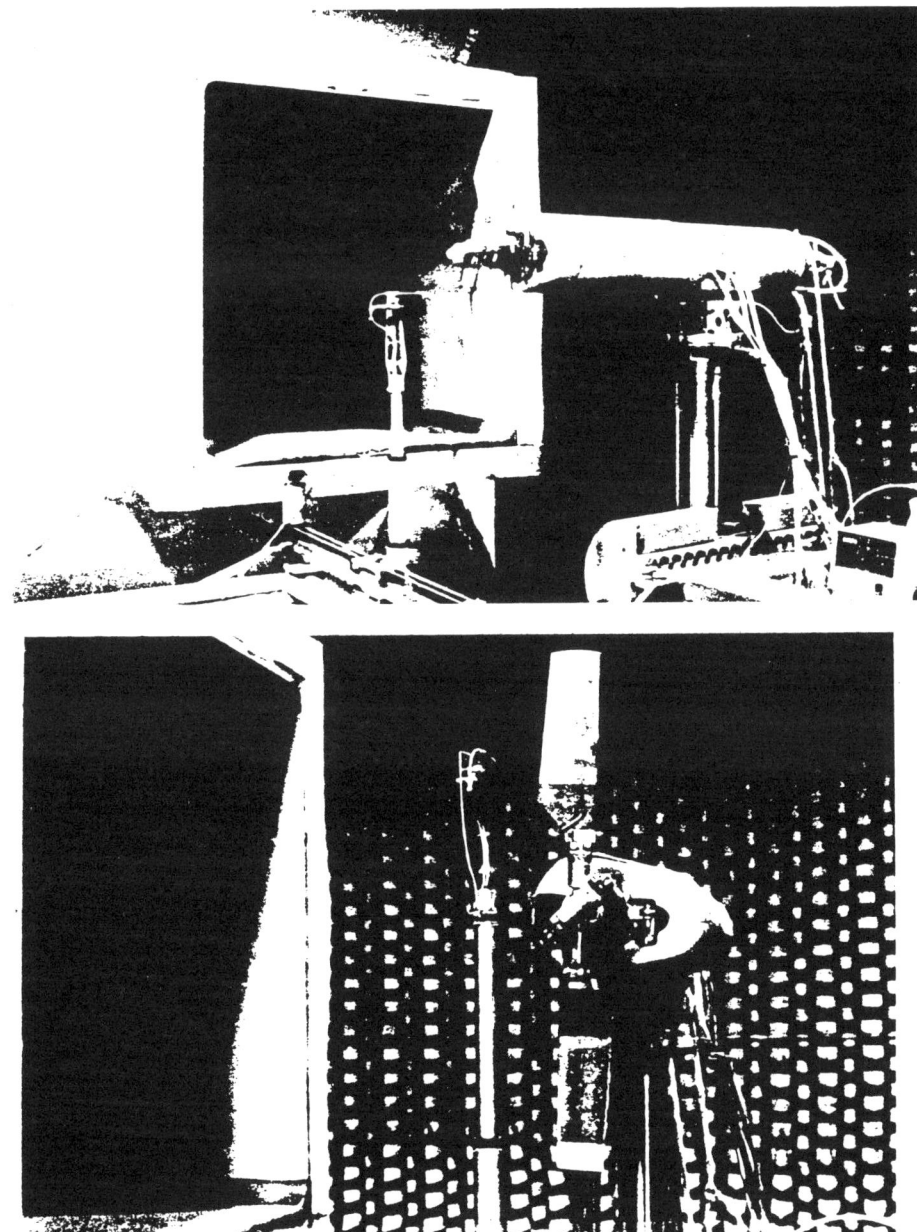

**Figure 3.11**  Test setup to check the acoustic intensity measurement technique in the DLR wind tunnel using a small model rotor

For the pilot tests, the DLR tail rotor was mounted in the anechoic part of a wind tunnel (Figure 3.11). Tests were made at wind speeds up to 40 m/s. When measuring in the far-field, the wind direction at the microphone positions is well-defined, as it is determined by the wind tunnel. The effect of the wind on the microphones can therefore be reduced by using nose cones on the microphones. In the planned near-field measurements, the wind created by the rotor itself had to be taken into account, and the resulting wind direction at the microphone positions might not be well-defined. It was therefore necessary to use windscreens rather than nose cones on the microphones.

The acoustical instrumentation consisted of a sound intensity probe mounted on a microphone positioning system. The sound intensity probe type B&K 3519 was based on 0.25-inch microphones and had an acoustical separation of 12 mm, which gave a useful frequency range from approximately 100 Hz to 6 kHz.

The two microphone signals were analysed with a two-channel real-time frequency analyser type B&K 2133. The signals were analysed in one-third

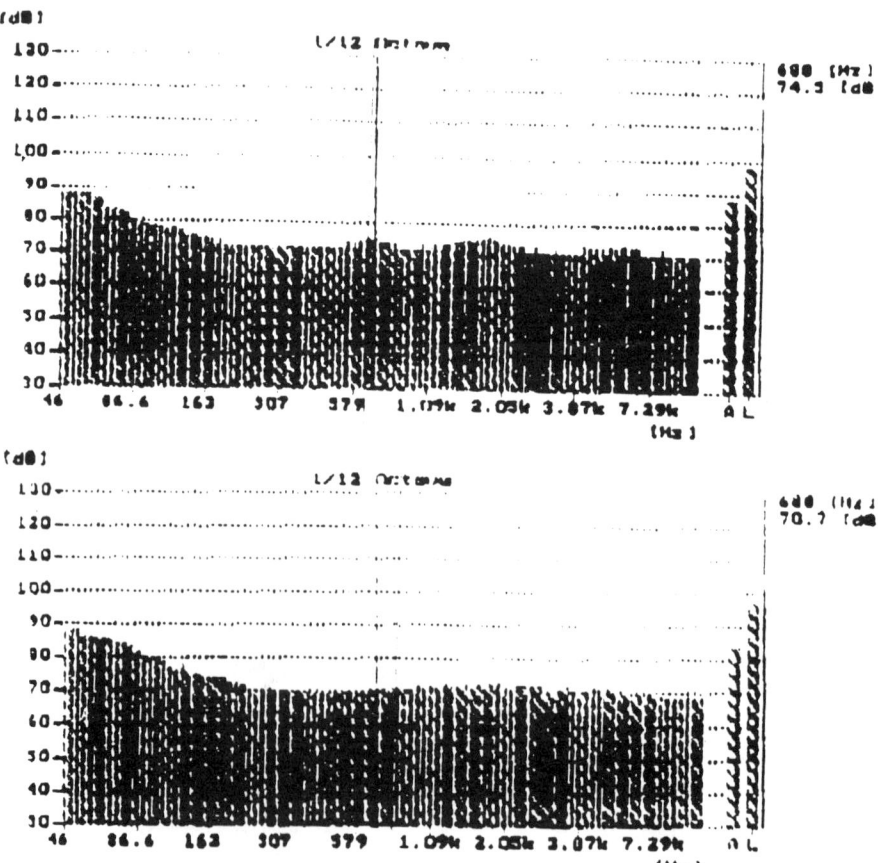

**Figure 3.12**  Wind-induced noise using (top) a small windscreen and (below) a medium-size windscreen (B&K data)

octave, one-twelfth octave and as enhanced time signals. All measurements were stored on floppy disks as spectra or time functions for later analysis. A personal computer, type Olivetti M211V was connected to the real-time frequency analyser via a GPIB interface.

The intensity probe was mounted on a microphone-positioning system with telescopic vertical movement and horizontal traversing mechanism. The distance from the front of the intensity probe to the rotor centre was between 50 mm and 200 mm during the measurements.

Measurements with wind-induced noise only (stationary rotor) (Figure 3.12) showed that the difference between the small and medium-size windscreens was very little, so that a fairly small windscreen could be selected, minimising the static load.

The measurement with the rotor running with rotational speed of 92 Hz and 3° pitch (Figure 3.13) shows that the signal from the rotor itself is at least 25 dB higher than the wind-induced noise from the small windscreen. This indicated that the small windscreen was suitable in the windspeed range to be tested (i.e. up to 35 m/s).

For application of the measurement method in a large wind tunnel such as the DNW, it would be necessary to use a remotely controllable support system (Figure 3.14). This system must be attached to the DNW sting to carry the necessary number of intensity-measuring microphone systems.

## 3.4 Code development activities

### 3.4.1 Aerodynamics

*Boundary-element methods*

**Doublet method**   ECD adopted a panel method [3] developed at the University der Bundeswehr, München, for the evaluation of the blades' pressure

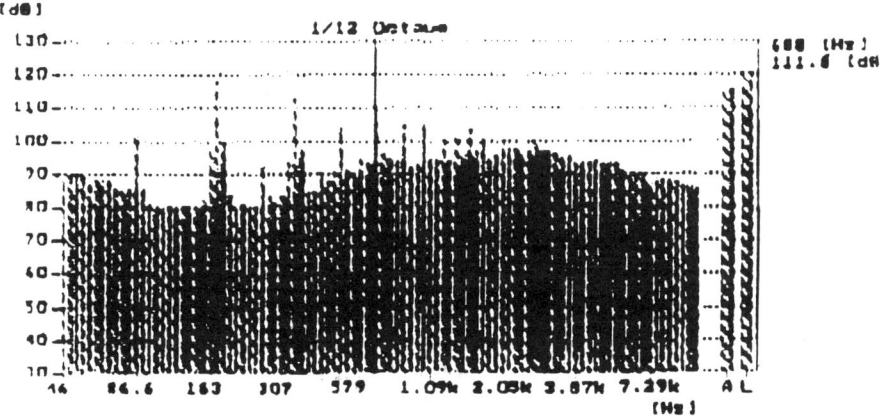

**Figure 3.13**   Signal and wind-induced noise using a medium-size windscreen (B&K data)

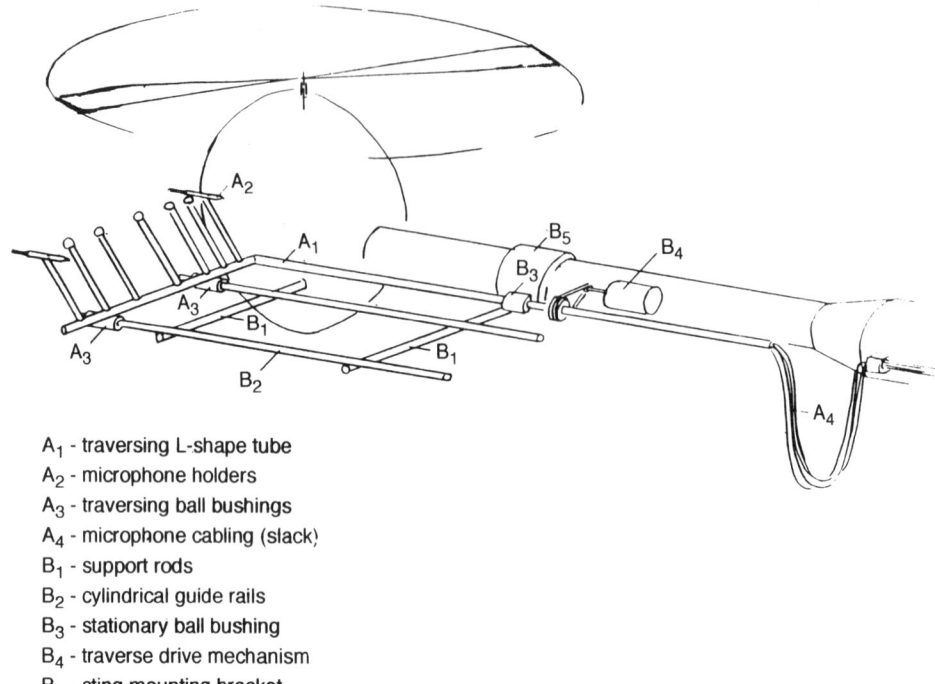

A$_1$ - traversing L-shape tube
A$_2$ - microphone holders
A$_3$ - traversing ball bushings
A$_4$ - microphone cabling (slack)
B$_1$ - support rods
B$_2$ - cylindrical guide rails
B$_3$ - stationary ball bushing
B$_4$ - traverse drive mechanism
B$_5$ - sting mounting bracket

**Figure 3.14**   Proposed support system for intensity measuring microphone array

distribution as input for the far-field noise calculation. The theory is based on incompressible potential theory and free wake analysis, allowing particularly the simulation of blade vortex interaction [3]. ECD extended the code to accomplish high panel numbers in the chordwise direction, necessary for a reliable incorporation of the blade pressure distribution.

The idealisation of the flow field on and behind the rotor blades was accomplished by panels corresponding to vortex or dipole rings. These panels describe the vorticity on the blade and along the rotor wake for arbitrary helicopter or tilt rotor flight conditions. Knowledge of the circulation at the position of the blades allows calculation of the spanwise and chordwise distributions of the pressure difference of upper and lower sides of the blade represented by a plate.

With the blades' pressure distribution known, existing far-field noise codes can be applied which use the pressure distribution as the source term. Owing to the incompressible formulation of the algorithm applied, only flight conditions at moderate flight velocity can be modelled. Amongst these, impulsive noise generating flight conditions such as descent and manoeuvres are of particular interest. For the future, an extension of the code towards a description of compressible fluids is planned.

In order to save CPU time, the rotor wake of a forward-flying helicopter or tilt rotor was calculated for one to three panels in the chordwise blade direction. (Three panels allow for a rough representation of blade camber.) After a stable periodic blade circulation had been established, the number of panels in the chordwise direction was increased to about 25. Figure 3.15 depicts such a periodic circulation and Figure 3.16 the corresponding stable wake configuration, calculated for the HELINOISE model rotor in a moderate descent flight (descent angle about 4°, cruise speed about 32 m/s).

The code is divided into three single routines. In a first step, the rotor geometry is described. Then, the wake development and the dipole intensities are calculated from the velocities existing at the blade positions. Finally, the aerodynamic characteristics including rotor forces are defined.

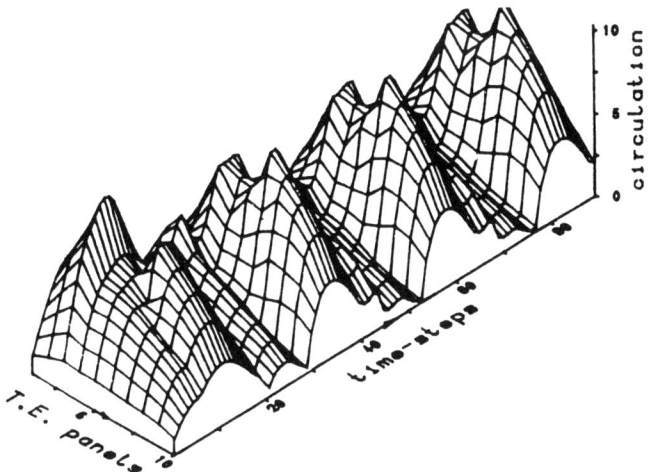

**Figure 3.15** Time history of spanwise blade circulation for a BO105 model rotor in moderate descent flight ($v = 32$ m/s, descent angle 4°)

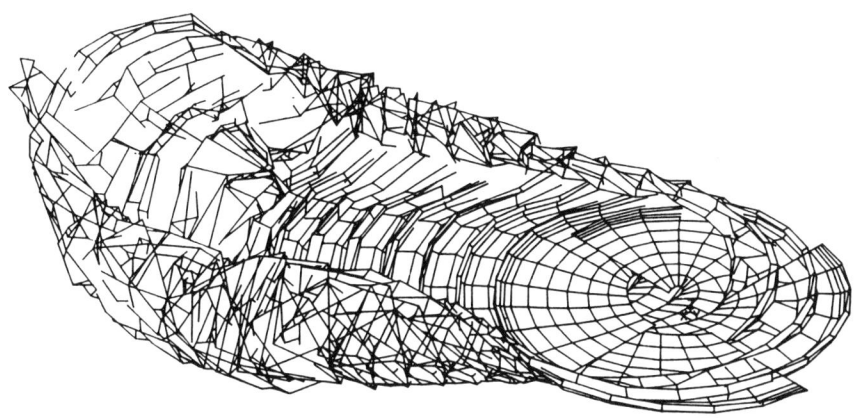

**Figure 3.16** Rotor wake corresponding to the flight conditions of Figure 3.15

The program uses a coordinate system, which rotates with the blades, having its origin at the rotor centre.

The following geometrical blade data have to be introduced:

Number of blades

Blade chord

Aerodynamically effective blade length

Sweepback angle of the leading edge

Number and borders of the panels in chordwise and spanwise directions

Linear or nonlinear twist

Collective and cyclic blade pitch angles

Position of the blade twist axis

Azimuthal flapping angle distribution (to be determined by preceding flight-mechanical computation)

To determine the aerodynamic parameters, the dipole intensity distribution on the blades is needed. In forward flight, this intensity varies with time or blade azimuth angle. Therefore, the flow conditions are evaluated within (constant) time steps. With every new time increment, one dipole row leaves the blade as a vortex sheet and a new dipole row is created along the blade's quarter line.

The program establishes the mutual influence of the panels by setting the dipole intensity $\Gamma = 1$ before the start of the first calculation step. With this configuration, the influence coefficient $A_{ij}$ from the $j$th vortex sheet on the $i$th control point is equal to the sum of the induced velocities generated by the different dipoles and calculated by Bio-Savart. The sum of all influence coefficients incorporates the 'influence coefficient matrix', being constant throughout the complete calculation.

The incident flow velocity at the separate control points is built up by rate of revolution, vehicle airspeed, blade flapping and pitching velocity, and induced velocities from all dipoles. Since the kinematic flow condition does not allow for flow components, vertical to the panels, the dipoles must induce a normal velocity component of the same amount as the ones of the other velocities but with reversed sign. As every new vortex sheet induces different velocity components at the different control points, the dipole forces are determined by a linear set of equations, which can be solved by a Gauss–Jordan procedure.

Since the theoretical induced velocity in the close vicinity of a vortex is unrealistic, it is damped by an exponential function along the mean distance between two individual vortex fragments.

For the simulation of the pressure distribution of tilt rotor blades, the aircraft's wing can be taken into consideration with respect to its aerodynamic influence on the rotor. For this, suitable vortex elements are placed at the position of the wing.

The actual blade representation in the form of a plate will be replaced in future by a realistic airfoil, so that the finite thickness of the blades can be taken into account and the surface pressure distribution can be determined.

In order to be able to predict high-speed noise, a compressibility approach has to be implemented in the program.

To validate the analytical model described above, four different experimental data sets were applied:

- DNW tests with a two-bladed Cobra (ASH-1/LOS) rotor [4].

- Wind-tunnel tests by Meyer and Fallabella with another two-bladed rotor [12].

- DNW model rotor wake measurements by DLR and ECD with a four-bladed BO 105 model rotor.

- The HELINOISE DNW tests performed during the described project period.

A comparison between results of the HELINOISE tests and recent theoretical results is shown in Figure 3.17 in the form of the pressure difference (lower side minus upper side) versus rotor azimuth angle for different radial stations at 3% chord. The agreement appears acceptable if one considers that the theory treats the rotor blades as flat plates in an incompressible flow. At the same time, the necessity of a pertinent extension of the theoretical approach must be emphasised.

**Velocity potential method**  As an appropriate method equivalent to the one described above, Morino and Genaretti [5] have introduced instead of the doublet method the direct velocity potential formulation. For this, an explicit treatment of the rotor wake and a particular formulation of the wake transport were developed. As in the doublet method, potential flow conditions are assumed with the exception of the space formed by the boundary elements (i.e. the rotor blade surface and the wake). Again, for the boundary elements themselves, it is required that all velocity components perpendicular to the boundary elements vanish (no penetration of blades and wake by the air flow).

The work started with incompressible flow conditions and was then extended to the potential flow formulation for compressible flows. The integral formulation for airplanes was further developed to bodies in arbitrary motion, including the rotation of the blades of a hovering helicopter and subsequently the superposition of rotational and translational motion of the blades of a forward-flying helicopter.

Description of the aerodynamic phenomena is normally based on a frame of reference moving in uniform motion with respect to the undisturbed air. Hence, in the case of rotors, the motion of the body (blade) with respect to the frame of reference is large and, as a consequence, the integral equation is not easy to use for the analysis of helicopter rotors and propellers. Therefore, Genaretti, Macina and Morino introduced for the analysis of unsteady com-

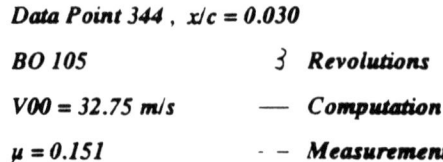

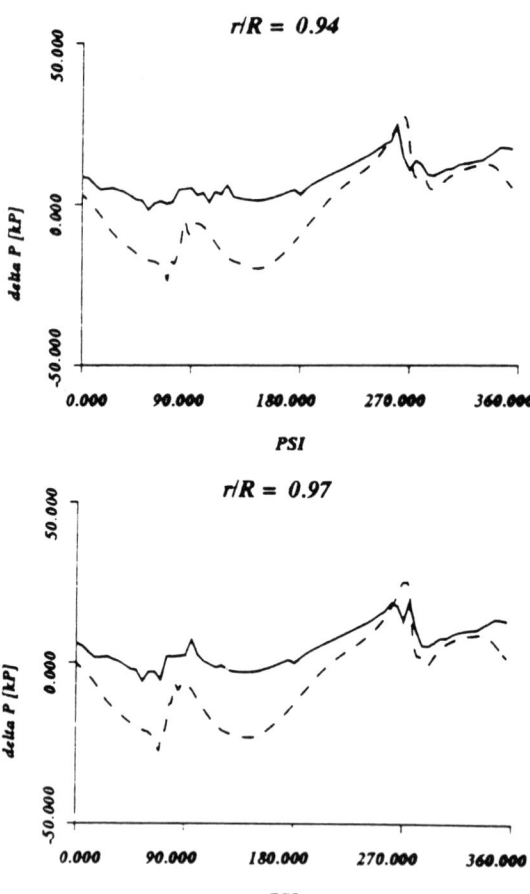

**Figure 3.17** Pressure difference of BO105 model rotor during DNW tests with $V_{tunnel} = 32.75$ m/s and $\alpha_{TPP} = -095°$ (inclination of tip path plane, corrected with respect to tunnel effects)

pressible potential flows a frame of reference assumed to move in arbitrary motion. As an example for an advancing rotor, Figures 3.18 and 3.19 show the thrust coefficient of a one-bladed rotor with a collective pitch of 5.7° as a function of the rotor azimuth angle for the incompressible and compressible case, compared with the work of Tai and Runyan.

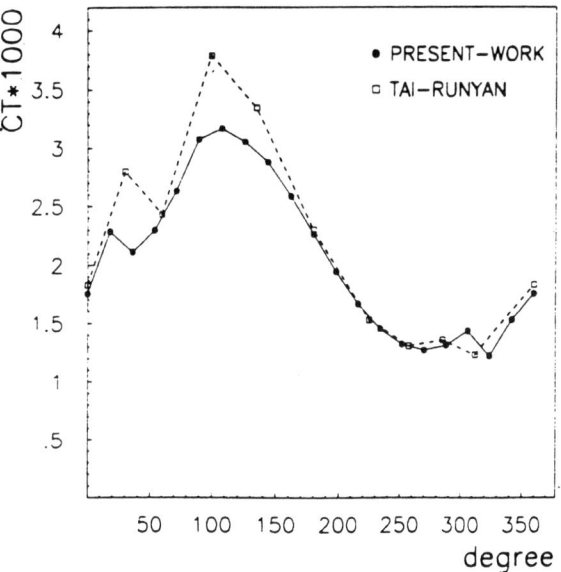

**Figure 3.18**   Thrust coefficient $c_T$ versus rotor azimuth angle $\alpha$ for incompressible flow, for a one-bladed rotor with collective pitch of 5.7°

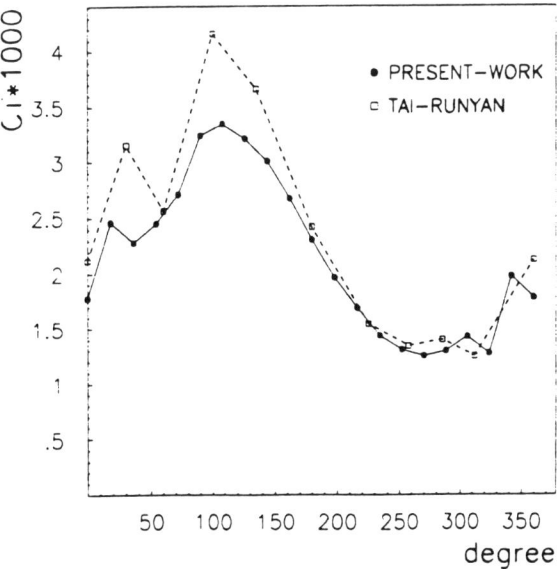

**Figure 3.19**   As for Figure 3.19 but with compressible flow

*Lifting-line methods*

**The ALFAPI approach** In the area of aerodynamic code development, ALFAPI addressed the following:

- Calculation of rotor free wake geometry and harmonic air loading, according to Scully's methodology.

- Nonuniform air loading calculation of a helicopter in forward flight.

- Blade vortex interaction (BVI) loading.

The reason for the reprogramming and development of the Scully code was the need for realistic prediction of air loading, especially that due to the BVI phenomena, and realistic blade motion calculation. As such, the code is to serve as a reference for the partners dealing with rotor noise prediction. For prescribed wake geometries, extended air-loading predictions based on rotor geometries related to the project have been performed. In the free wake analysis, the second main part of the code, numerical–computational improvements are necessary for achievement of realistic results and further code evaluation with theoretical and experimental data, as well as the code effectiveness improvement, are necessary. Work on this code has been continued and comparison with available DLR blade pressure data is due to be carried out.

For a standard test case, local lift, angles of attack and induced velocities have been evaluated. Figure 3.20 depicts a typical result for the azimuth distribution of local lift for different radial positions.

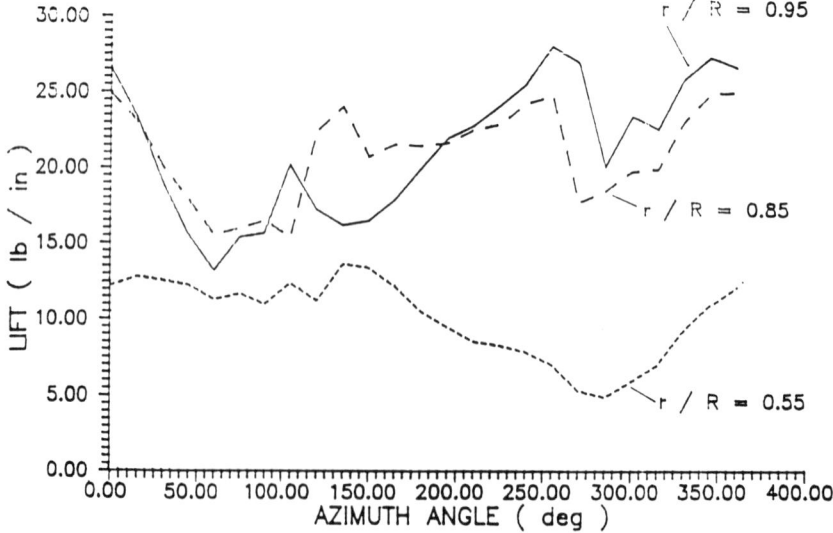

**Figure 3.20** HRACF computation of azimuthal distribution of local lift in the radial stations $r/R = 0.55$, $0.85$ and $0.95$ for a four-bladed rotor with $c_T = 0.0055$, $\mu = 0.18$, $R = 8.53$ m; NACA 0012

The code VEMRAC for the nonuniform air-loading calculation concerns the nonuniform inflow induced by the vortex wake. The induced velocities are calculated by integration of the Biot–Savart law over a rigid wake, or alternatively for a higher accuracy over a distorted wake geometry. The distorted wake geometry is used for the tip vortices, since errors in their location lead to significant discrepancies in the induced velocity and air-loading distribution.

In order to reduce the computational effort needed to describe the distorted wake geometry, a calculation procedure based on Beddoes' proposal concerning the vertical distortion of the rotor wake has been developed. Beddoes' wake model combines satisfactory results and low computing time in comparison with a 'free wake' model.

The calculation applied for the helicopter rotor wake assumes that the wake comprises a series of finite-length vortex segments, and the downwash induced by an arbitrary trailing vortex is derived by summing the individual contributions of the vortex segments at the point of interest. The relative positions of the rotor blades and tip vortices are also provided in order to perform an initial local calculation of the air-loading generated during blade/vortex interaction.

Assuming a large aspect ratio for the rotor blades, the solution of the integral equation of circulation is split into outer and inner problems, which are solved individually and then combined through a matching procedure. The lifting-line formulation is used to solve the inner problem where corrections for the compressibility and the viscosity are included using steady two-dimensional airfoil characteristics.

Typical results for the load distribution of a forward-flying rotor are shown in Figure 3.21 for a rigid wake and in Figures 3.22 and 3.23 for a distorted wake.

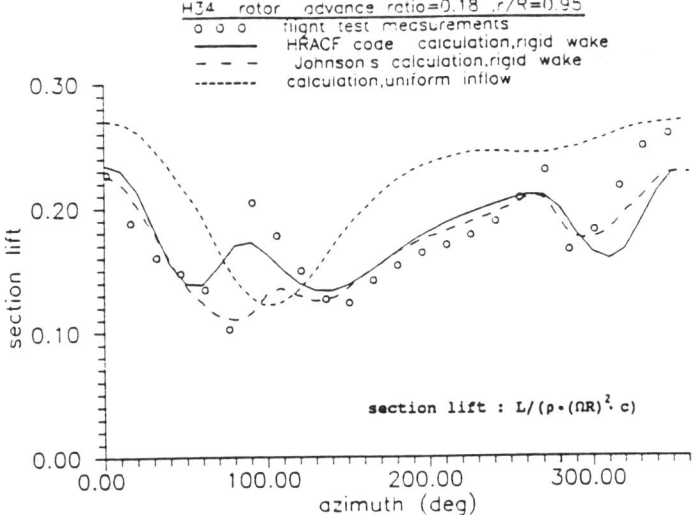

**Figure 3.21** Air-loading distribution of a forward-flying four-bladed rotor at 95% radius; $c_T = 0.0051$, $\mu = 0.18$, $R = 0.853$; NACA 0012 (computation with rigid wake)

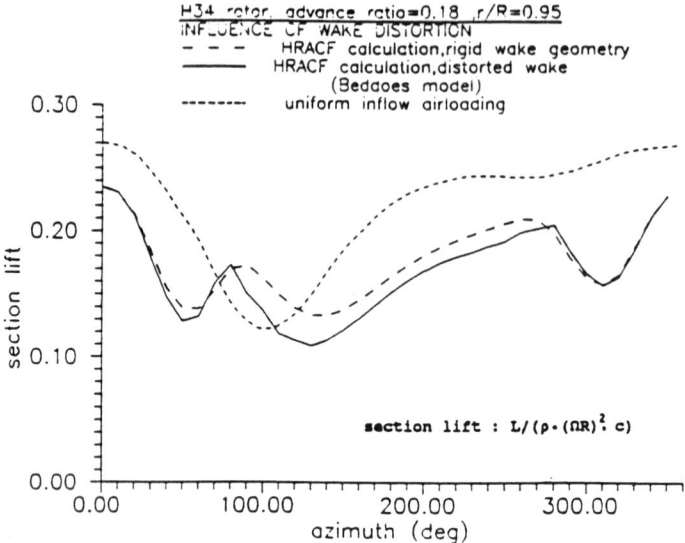

**Figure 3.22** Air-loading distribution according to Figure 3.21—HRACF result elaborated with distorted wake

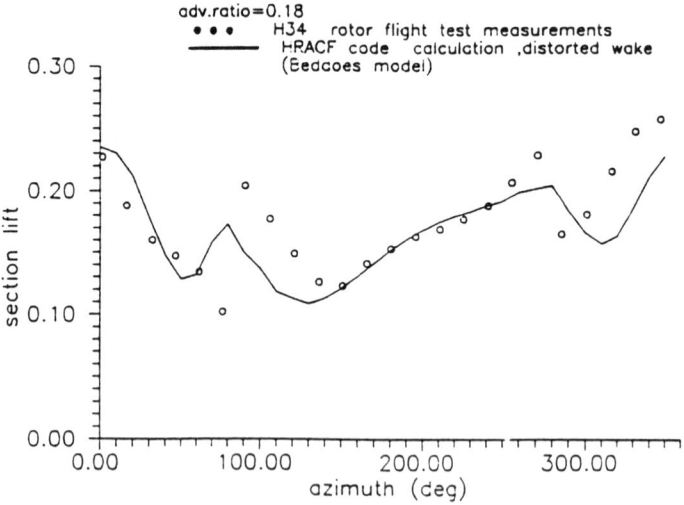

**Figure 3.23** Air-loading distribution according to Figure 3.22—compared with flight test results

The physical modelling of fixed-wing blade/vortex interactions in three-dimensional, turbulent and transient (unsteady) local flow fields has been approached in two distinct stages. In the first case, the airfoil is considered to be of infinite span and to interact with an infinite line vortex which is conveyed by a uniform free-stream and is parallel to the blade axis. This is the two-dimensional case. In the second case, the airfoil is considered to be of finite length and to interact with an infinite line vortex conveyed by a uniform free-stream but initially inclined to the blade axis, this being the three-dimensional model.

The developed module analyses the dynamic situation encountered at the section of the blade which is nearest to the tip vortex. The incorporated physico-mathematical model takes into account, in the 2-D case, the influence of the airfoil presence on the vortex passage, the airfoil shape, the variation of the airfoil angle of attack during the advance of the vortex over the blade surface; and it also analyses the variation of the bound circulation and the position of the free vortices in the wake (free wake analysis). The 3-D model, in addition to the above, also takes account of the direction of the vortex and the blade's geometrical twist, and hence accounts for the variation of the pitch angle over the length of the blade.

A vortex lattice method is employed for the representation of the airfoil and its wake. Owing to the transient (unsteady) nature of the problem, the time-dependent variations in circulation on the blade surface result in the shedding of vortices in addition to those trailing vortices which result from the nonuniformity of distribution of the blade circulation. Together, these vortices form the vortex lattice which describes the wake. Hence, by considering, in conjunction with the blade boundary conditions, the combined effect of the interacting external vortex, the shed vortices, and the bound and trailing vortices, the instantaneous associated blade circulation, and hence the aerodynamic coefficients of the blade, can be computed for distinct time intervals.

**The Bristol University approach**   A nonlinear lifting-line method for rotor aerodynamic performance prediction was developed at Bristol University. The method is validated by predicting the thrust and power for two propeller types that were used in acoustic tests by Dobrzynski et al. [6] in the German–Dutch wind tunnel DNW. The results of using the computed aerodynamic loads for these propellers for the prediction of acoustic pressure signals are described in Aston et al. [7]. The nonlinear lifting-line method described here is a natural extension of the work by Gould and Fiddes [8] on the prediction of power output from a wind turbine.

In the lifting-line method, the rotor blades are represented by lifting lines from which the shed vorticity forms a helical wake which induces a velocity at the propeller blades, and as a consequence the effective angle of attack of the wind is changed. Accurate modelling of the wake-induced velocity is therefore critical for accurate performance prediction.

The wake contracts over a distance of about one diameter downstream of the rotor. A lightly loaded wake model is assumed for the propeller, and the near-wake region is modelled using a prescribed contracting wake. An example of a near-wake treatment for a ship propeller is given by Greeley and Kerwin. Beyond the region one diameter downstream of the rotor, the wake is effectively of constant pitch and radius and the method described below for a constant-radius far wake may be used.

A general b-bladed unyawed propeller is considered. The trailing vorticity from each blade is represented by $(m - 1)$ semi-infinite horseshoe vortices which are distributed across the blade according to a Weber point distribution. This ensures that the highest density of vortices is near the tip and hub of the blade where the rate of change of circulation along the rotor blade is greatest.

It is possible to find an analytic expression for the far-wake influence $\int dv$ by approximating the integrand $dv$. In references [9] and [10], $dv$ is approximated by lower and upper bounds which can be integrated exactly. Wood and Gordon [11] obtained a very simple far-wake influence expression based on an approximation of the distance from a line vortex to a collocation point. Whilst the method of Wood and Gordon is simpler than the method described in this section and the methods given in references [9] and [10], it is also the most approximate; and as the majority of computational effort is involved in near-wake calculation, an accurate far-wake influence calculation is required.

In the new method described below, a power series in $1/x$ is obtained using a binomial expansion, valid for large $X$, where $X$ is the axial distance between the collocation point and wake truncation point.

For the prescribed wakestream model, the wakestream shape is defined as a function of axial distance by imposing the following four conditions:

1.  The wake stream edge is initially uncontracted.

2.  The degree of wake contraction is specified.

3.  The wakestream slope at the rotor plane is specified.

4.  A smooth transition to the noncontracting far-wake region is assumed.

For the application of the lifting-line method it is recommended to use at least 10 spanwise vortex filaments on each lifting line, and the near wake is modelled to complete the turn at one diameter downstream using 64 discrete line vortex sections per turn. The far-wake model is used from the start of the next complete turn.

Convergence studies for the circulation distribution and the angle of attack distribution for the propeller AN1 are shown in Figures 3.24 and 3.25, and it is clearly shown in Figure 3.26 that power coefficient $c_P$ and thrust coefficient $c_T$ have fully converged for $m = 14$.

### 3.4.2   The Farassat method [13, 15]

*The Agusta approach*

**Theory development**   Agusta activities started with the realisation of a simple aeroacoustic code, based on the acoustic analogy that allows the prediction of thickness and loading noise for near and far fields. The code is based on a formulation that can be called 'acoustic line' since it is supposed that all the dimensions of the blade can be neglected in respect of the chord (this is an acceptable hypothesis for common helicopter blades). It is possible to model twisted and swept blades, and all the rigid and elastic motions of the blade

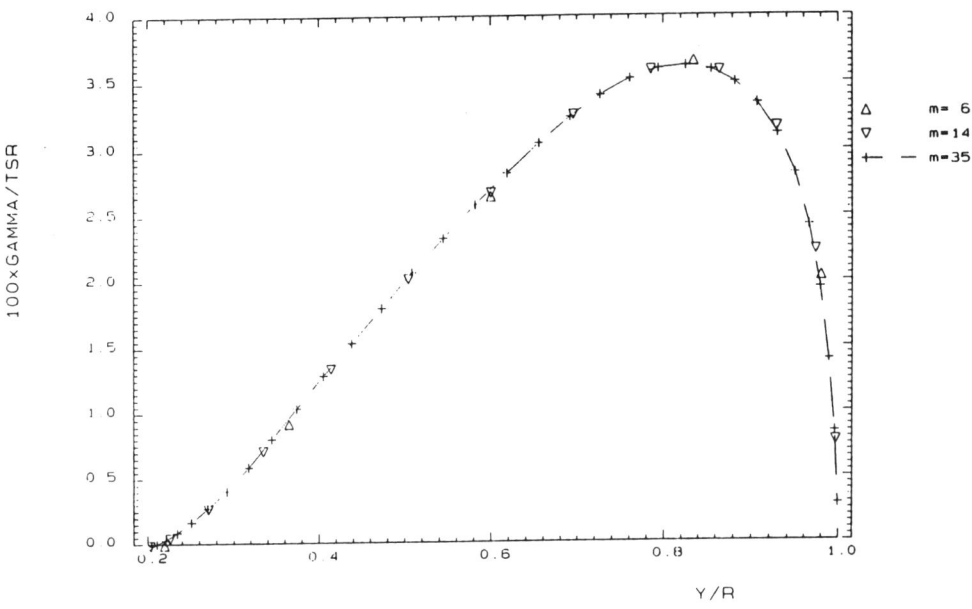

**Figure 3.24** Radial circulation distribution of test propeller ($m$ = number of spanwise filaments)

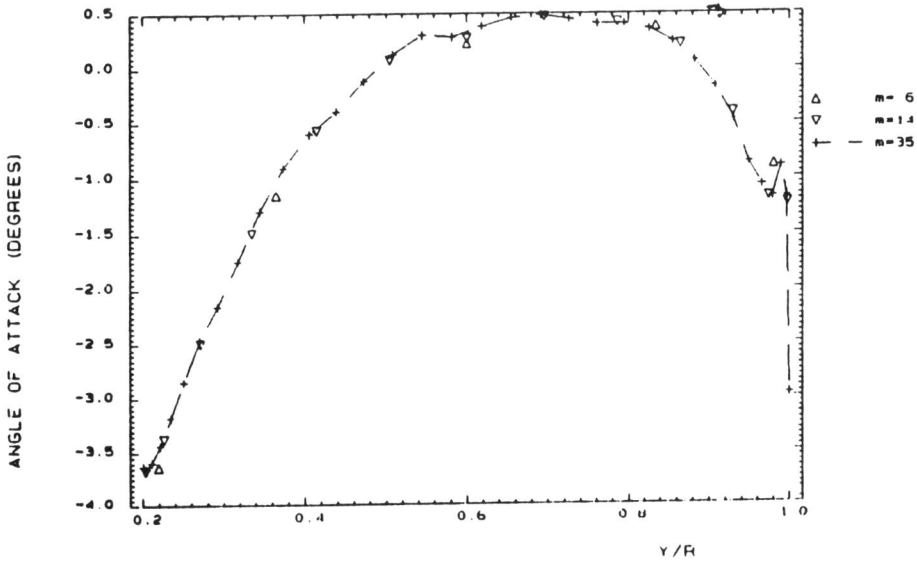

**Figure 3.25** Radial distribution of blade angle of attack of test propeller

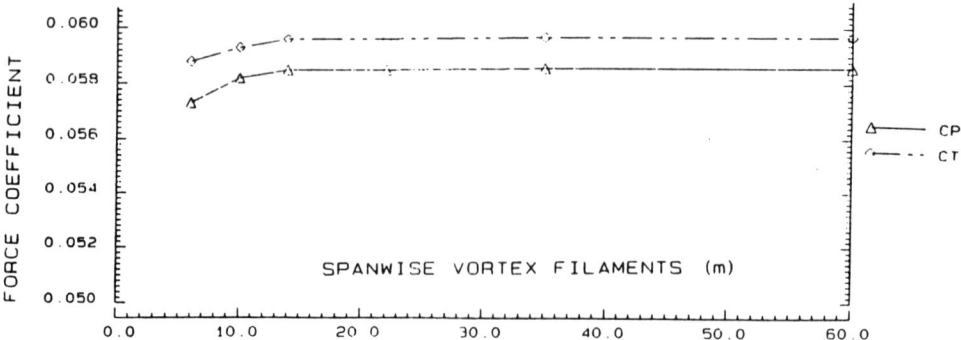

**Figure 3.26**   Lift and power coefficients ($c_T$, $c_P$) versus span vortex filaments ($m$)

(flapping, feathering, lagging) can be described. Furthermore, unsteady loads typical of forward flight can be modelled.

The required input data, such as the aerodynamic loads on the blade and the blade motion, are obtained using the aeroelastic code COSMIC (a precursor of CAMRAD), modified in order to obtain the desired quantities in the trimmed condition. Since the acoustic code is fully interfaced with COSMIC, it constitutes a handy instrument to obtain quickly a reasonable evaluation of radiated rotor noise.

Obviously, since the code takes into account only the linear terms and neglects the chordwise distributions of the aerodynamic loads, it can be used successfully only in cases in which the approximations can be considered reasonably true (such as hover or level forward flight at low tip Mach numbers). After realisation of the simplified approach described above, a more general formulation permits an analysis of the radiated noise also for more critical flight conditions, such as low-speed descending flight under BVI conditions or high-speed forward flight.

The starting point is always the acoustic analogy (the flow behaviour around the blades is supposed to be known), but now we directly integrate the Ffowcs–Williams–Hawkins (FWH) equation that provides an 'exact' description of the noise radiation:

$$\frac{1}{c_0^2}\frac{\partial^2 p_x}{\partial t^2} - \frac{\partial^2 p}{\partial x_i \partial x_j}$$

$$= \frac{\partial}{\partial t}[\rho_0 v_n \delta(f)] - \frac{\partial}{\partial x_i}[P_{ij}n_j\delta(f)] + \frac{\partial^2}{\partial x_i \partial x_j}[(\rho u_i u_j + P_{ij} - c_0^2\rho\delta_{ij})H(f)]$$

The first two terms on the right-hand side are surface terms that describe the thickness and loading components of the noise, while the last one is a volume term that describes the quadrupole component of the noise. During the first six months, Agusta focused on the thickness and loading terms, and worked on realisation of the code BENP for evaluation of both terms in the general case of a body (blade) with arbitrary motion and geometry.

Two different approaches were developed to execute the integrals. The first one was the well-known Farassat formulation 1, while the second was a formulation developed by Agusta that allows evaluation of the integrals on the acoustic surface. This approach has the advantage that the sonic singularity, due to the Doppler factor, disappears from the integrals in the analytical approach.

In order to execute the integrals, the surface is divided into panels. The panel geometry is described in terms of the positions of a certain number of points called 'nodes' of the panel, and a finite-element-like approach is used that allows an accurate integration and permits an efficient evaluation of the retarded value of all the required quantities.

At the moment, first-order panels (four nodes) and second-order panels (nine nodes) are available; furthermore, triangular panels can be analysed as a degeneration of four-node panels. Great care has been taken to keep as general as possible the geometric and kinematic descriptions of the body.

It is in fact possible to describe, without introducing any approximations, the geometry of modern rotor blades, and to include in a realistic way all blade motions, both rigid and elastic. Unsteady aerodynamic loads can also be described, and it has been demonstrated that the code is able to capture highly impulsive phenomena due to BVI.

In order to describe high-speed noise, the quadrupole term has been analysed and its surface component has been included in the code. Following Farassat [13, 15], it is possible to show that the quadrupole term can be described by two addends, being surface terms, and by a volume term usually called 'pure quadrupole' [14].

It is important to point out that the surface quadrupole terms have to be evaluated for all the surfaces across which the Lighthill stress tensor $T_{ij}$ is discontinuous, and so not only for the body surface, but also for the wake and eventually also for the shock waves.

There is thus the need to execute several surface integrals on surfaces like shock waves, that can greatly change their shape during the motion. Therefore, the method divides the deformable surface into panels, giving the positions of the nodes that define each panel for a certain number of time instants in such a way that, with an appropriate interpolation scheme, it is possible to reconstruct the position of each node for every time instant. Then a local mapping of the deformable surface to an undeformable one is defined for each panel, and the integration is executed, introducing the determinant of the Jacobian matrix of the transformation to the undeformable surface.

In addition to the above activities, efforts have been made to modify a two-dimensional full-potential aerodynamic code in order to provide the required input data for the surface quadrupole analysis. Subsequently, the same modification will be introduced into a three-dimensional code.

**Theoretical results**  In order to validate the codes, high-quality experimental data were applied. Results of this validation are described in the following.

Figures 3.27 and 3.28 analyse the effects of the chordwise distribution of the loads, for observers placed in the far field and near field. The plots refer to a

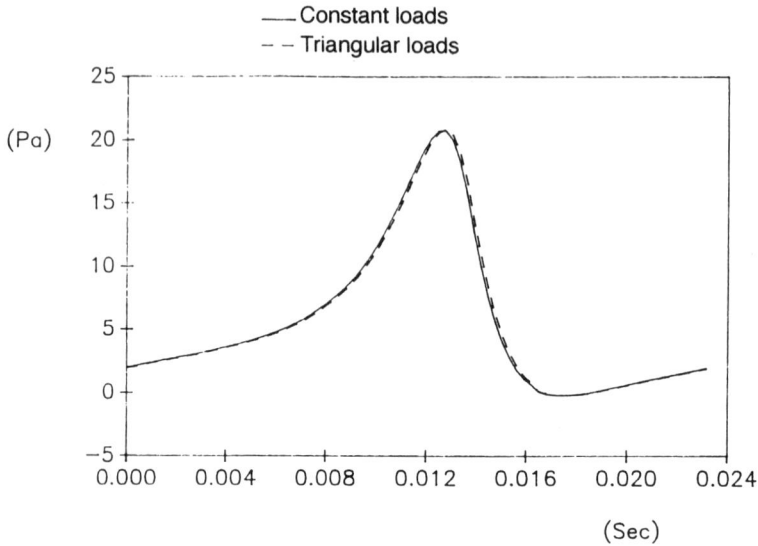

**Figure 3.27**   Effect of chordwise load distribution for observers placed in the far field

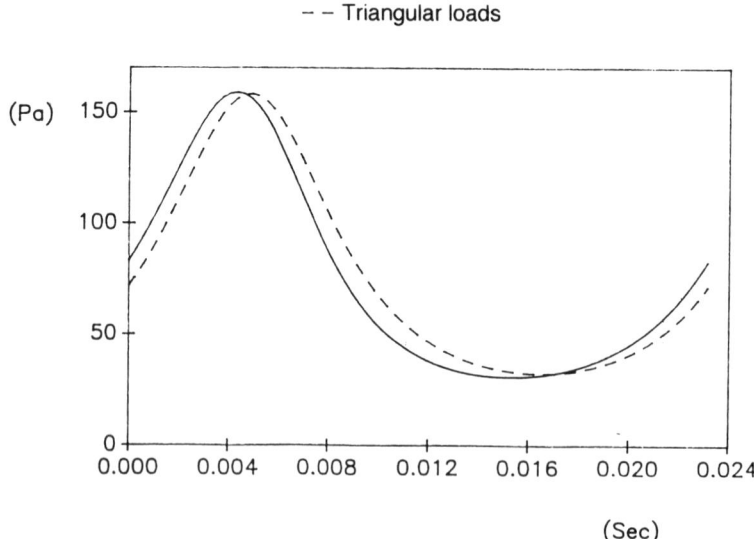

**Figure 3.28**   Effect of chordwise load distribution for observers placed in the near field

loading noise calculation for a two-bladed rotor in forward flight with the same spanwise load distribution that is, however, reconstructed in two different ways along the chord.

Figure 3.29 shows a comparison between the 'acoustic line' approach and the more sophisticated one (BENP) for a loading noise calculation of a two-bladed rotor in forward flight.

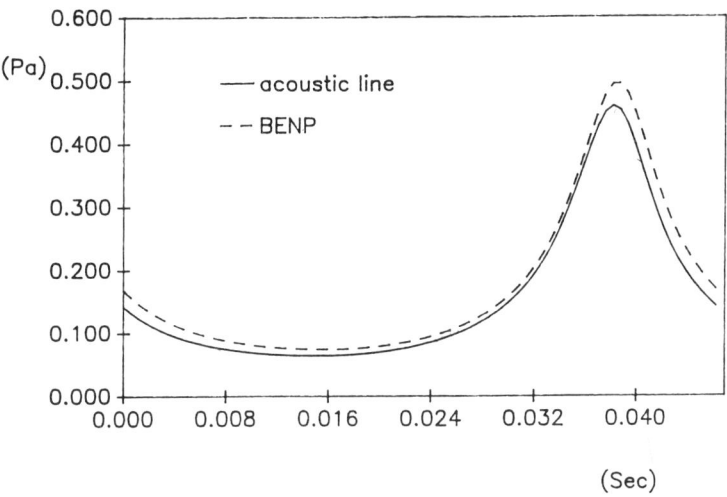

**Figure 3.29** Comparison of 'acoustic line' approach and BENP approach for a two-bladed rotor in forward flight

Figure 3.30 shows the effect of the thickness noise of the blade tip closure, for an observer placed in the rotor plane.

Figure 3.31 reveals the equivalence of the pressure history based on integration of the real and the acoustic (retarded) blade surface for a moderate Mach number.

In Figure 3.32 the BENP code has been used to compute the pressure disturbance produced by a fixed wing in steady flight when the observer is placed very near the surface of the blade. It should be possible, using such an

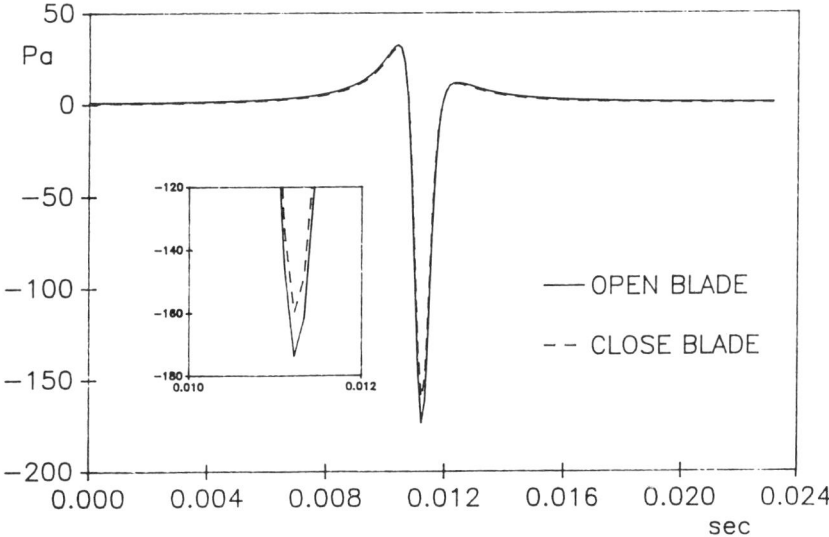

**Figure 3.30** Effect of thickness noise for an observer placed in the rotor plane

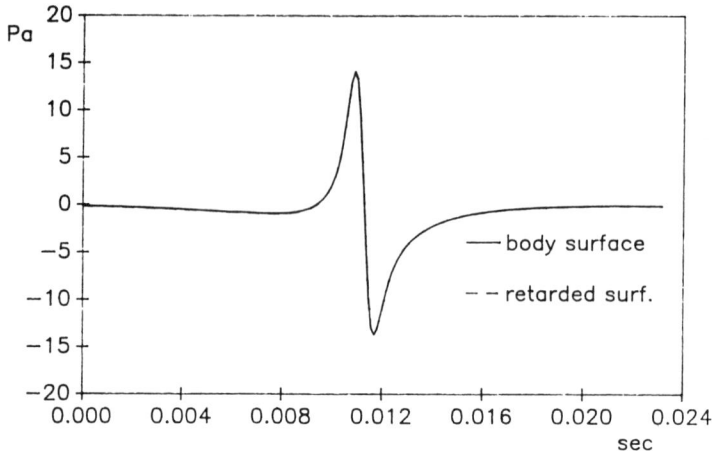

**Figure 3.31**   Comparison of body and acoustic surface integration

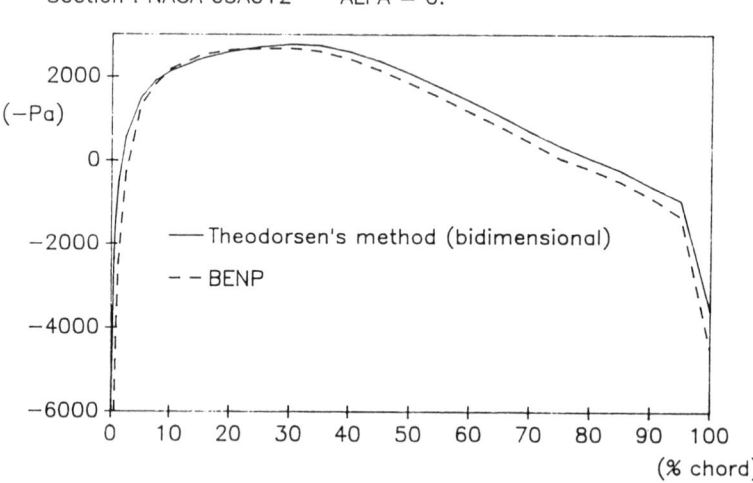

**Figure 3.32**   Pressure disturbance produced by a fixed wing in steady flight for an observer very near the blade surface

aeroacoustic approach, to reconstruct the aerodynamic loads along the wing section. The figure shows a comparison of the pressure computed with BENP and with the Theodorsen method for the central section of the wing where the flux can be considered bidimensional. The result refers to a wing with symmetrical section placed at zero incidence angle.

*The Cira approach*

During Cira's development work, the code HERNOP, originally designed to calculate only the linear terms of the FWH equation for a helicopter in hover, was modified. It represents actually an extremely articulated and complete program, which permits diversified analysis for the quadrupole terms and implements realistic blade motions, in hover and forward-flight conditions.

HERNOPS's structure allows the drafting of completely independent code blocks so that their connection with the main program is very easy; in this manner it is possible to separate the analysis of various terms of the equation, and to compare different methods of numerical solution, both for each integral term and for the complete equation, using different formulations.

The theoretical basis for the analysis of sound generated by a body moving in a fluid flow is the FWH equation. Several codes operating for the prediction of helicopter rotor noise (or, in general, of rotating machinery noise) are based on different forms of solution of this equation.

HERNOP solves the FWH equation in the time domain utilising two different approaches due to Farassat, formulations 1 and 1A [13]. They treat the first two source terms of the FWH equation. In HERNOP, the above formulations are properly modified to take into account the effect of the nonlinear quadrupole source term; then, upon application of the Green function method, the acoustic pressure is expressed as the sum of three integral terms, called thickness noise (or monopole term) $p_T$, loading noise (or dipole term) $p_L$ and quadrupole noise $p_Q$:

$$p(x, t) = p_T(x, t) + p_L(x, t) + p_Q(x, t).$$

**Thickness noise**  Determination of the monopole contribution in the FWH equation is, of course, the simplest task to be dealt with. In fact, thickness noise depends only on the blade geometry and the kinematics of blade motion. In formulation 1, thickness noise is expressed by a time derivative of a surface integral.

The local blade velocity, the unit normal vector to the blade surface, and the vector between source and observer are known for every point on the blade, so the surface integral can be evaluated at any observation time. Then the resulting signature is differentiated to obtain the pressure $p_T$.

Because of its dependence on geometric and kinematic quantities only, the thickness noise can be calculated for hover and forward-flight conditions with no significant differences. HERNOP is able to consider a blade with a linear variation of the section twist angle and a distribution of chord and thickness along the span. The blade can be described by approximating sections with several kinds of NACA airfoils; alternatively, the coordinates of the blade surface points may be supplied by suitable input files.

The code can also simulate realistic blade motions during the revolution period. In particular, it is able to consider periodic variations for flapping, feathering and lead–lag angles where the motion coefficients have to be inputted to the code.

**Loading noise**  For the determination of loading noise, knowledge of the pressure distribution on the blade is required. There is a substantial difference between helicopter rotors in hover and in forward-flight conditions. In fact, in the former case, the blade pressure distribution is time-independent, while in forward flight it becomes a function of the azimuth angle.

If the hover condition is to be studied, only the steady pressure distribution is required on the blade surface. On the other hand, in order to study forward flight a set of pressure distributions, corresponding to different values of the azimuth angle, must be provided. The most immediate form of solution of the FWH equation exhibits, as a loading noise contribution, the divergence of a vector integral; this has been turned by Farassat into a different expression, leading to a much simpler numerical treatment [15].

As in the case of thickness noise, in Farassat's formulation 1A the time derivative is taken inside the integral; nevertheless, the advantages of formulation 1A are not so evident in this case, unless an aerodynamic code giving directly the time derivative of pressure distribution is available.

The main issue in calculating loading noise is the possibility of exploiting a set of accurate aerodynamic data, based upon numerical or experimental results. For hover conditions, and for a blade with NACA profiles, HERNOP calculates the dipole term using standard aerodynamic data related to profile shape. Otherwise the code, which requires as an input the pressure coefficient $c_P$ on the blade or the lift coefficient $c_L$ on discrete stations along the span, interpolates and fits these data to its own grid; then, for any point and observation time, the azimuth position at the corresponding emission time is found, and the value of the blade pressure is determined.

**Quadrupole noise**   Both Farassat formulations 1 and 1A were written neglecting the quadrupole source term. This approximation in the solution of the FWH equation is certainly acceptable, as far as the velocities involved in the phenomenon are low. When the analysis concerns the prediction of noise from transonic tip speed rotors, neglecting the effect of the quadrupole source, large errors can be expected.

The quadrupole contribution is normally expressed by a volume integral which requires a complete knowledge of the flow field around the blade and increases the complexity of the numerical procedure.

Upon transforming the space derivatives into time derivatives, the volume integration can be simplified by specific kinds of approximations. The method proposed by Schultz and Splettstoesser [4] allows a good representation of the quadrupole source. Their 'momentum thickness' permits the volume integral to be split into two simpler integrals: the first one is carried out along the normal to the blade surface, where a surface integral is then performed. In general, it is necessary to know the history of velocity and pressure distributions in a volume around the blade, in order to perform a complete volume integration.

**Results from the current version of the HERNOP code**   As a first example, comparisons are established with results from the US noise prediction code WOPWOP developed by Brentner at Langley Research Center. Some results are considered here, which were proposed by the author [16] along with previously released experimental data. Figure 3.33 shows results for a test case with an advancing tip Mach number of 0.828; both microphones were located 40° from the upstream direction, on the advancing and the retreating side. The thickness signature from HERNOP is very close to the WOPWOP prediction. When a quadrupole contribution is added to the thickness term, good agreement is achieved with the main peak of the experimental data reported by Brentner. Similar results are obtained at higher forward speed: Figure 3.34 shows a comparison referring to a test case with an advancing tip Mach number of 0.866 and with the same microphone locations.

For validation of the quadrupole signature, a test case was selected from an extensive study conducted by Prieur [17] about the radiation from a hovering UH-1H model rotor with a tip Mach number of 0.88 at which no delocalisation effect is present (Figure 3.35). The delocalisation represents an extension of the supersonic areas beyond the blade tip up to the sonic cylinder—the radius station where the rotor blade tip would reach sonic speed.

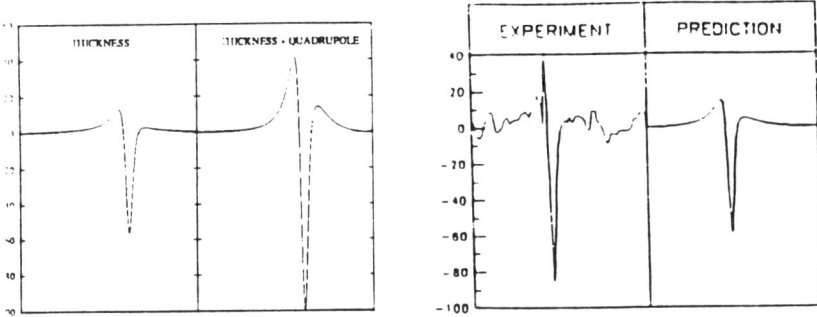

a) Microphone position: 40° from upstream direction, advancing side

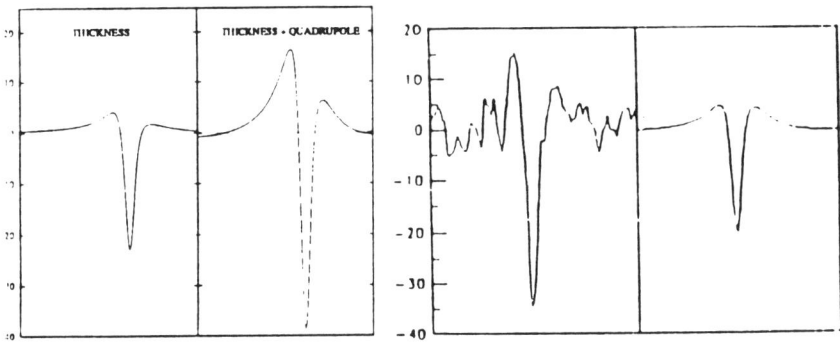

b) Microphone position: 40° from upstream direction, retreating side

**Figure 3.33**  HERNOP calculations (left pictures) and results from reference [16] (right pictures) at two different microphone locations, with advancing blade tip Mach number = 0.827)

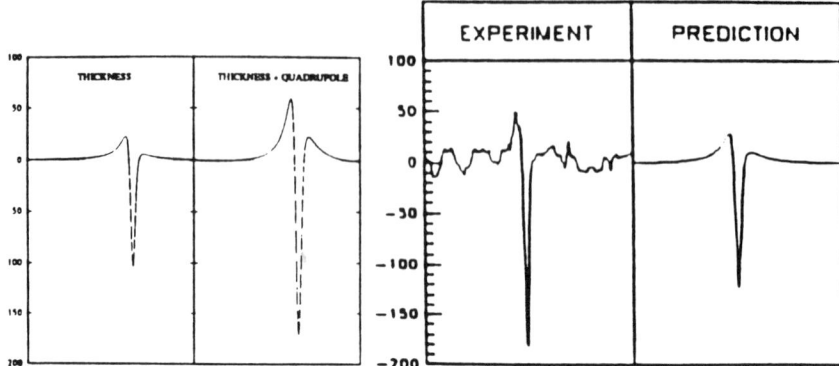

a) Microphone position: 40° from upstream direction, advancing side

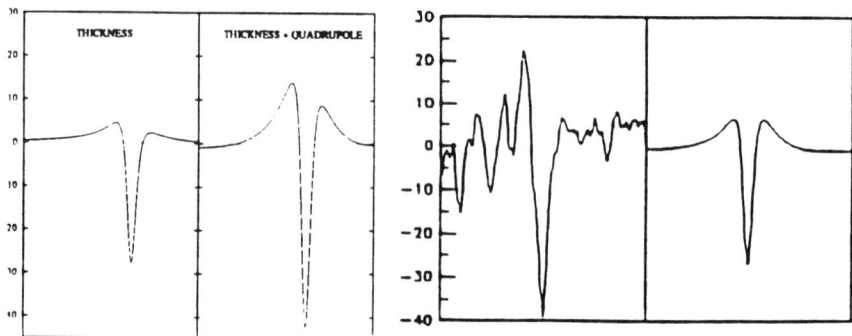

b) Microphone position: 40° from upstream direction, retreating side

**Figure 3.34**  HERNOP calculations (left pictures) and results from reference [16] (right pictures) at two different microphone locations, with advancing blade tip Mach number = 0.866

From Schmitz and Yu [18] some flight test data have been selected in order to check the ability of HERNOP code to predict the acoustic behaviour of the UH-1H rotor in forward flight. In particular, comparisons with in-flight measured data have been established, relating to the lateral directivity in level flight at an indicated airspeed of 115 knots, leading to an advancing tip Mach number of 0.9. Observer locations are near the tip path plane, 7° below the flight level (see Figure 3.36).

*The Lowson method*

**The Bristol University approach**  Bristol University addressed the theory of noise radiation of a moving point source following the work of Lowson and Ollerhead [19–21].

The Bristol University team demonstrated that the noise field for a nontranslating source can be obtained far more efficiently by use of Fourier space than

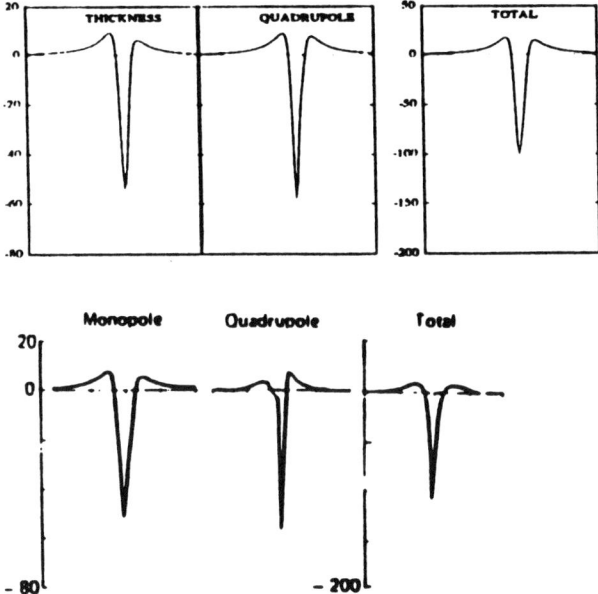

**Figure 3.35** HERNOP calculations (left pictures) and results from reference [17] (right pictures) for a hovering rotor, with blade tip Mach number = 0.88

by the conventional methods of iteration. Computed results agree well with the asymptotic theory of Lowson and Ollerhead [20].

A parametric study of the noise field as a function of tip Mach number revealed not only information as to the range of applicability of a point source model—in this case up to Mach numbers of about 0.9—but also led to the suggestion that standard subjective measures such as PNdB will not yield correct results at transonic tip speeds.

Predictions have been compared with the results of Dobrzynski et al. [6] for the noise radiation from a propeller.

Full details of the aerodynamic calculations are given in the report by Fiddes, Gould and Aston [22].

In reference [6], Dobrzynski et al. presented experimental pressure profiles and power spectra for two types of propeller operating under various conditions. The propellers were both of essentially Clark Y sections. One of the main differences between the propellers was their tip geometry, one having a square tip construction, the other a round tip.

For the purposes of comparing theory with experiment, the data sets AN-1 run 63 and BC-1 run 77 were chosen. The operational conditions for these data sets are detailed in Table 3.4. For computational purposes, the speed of sound was taken to be 340 m/s. In this report, only test case AN-1 run 63 is used.

Figures 3.37 and 3.38 show comparisons between the experimental pressure/time history and the total theoretical predictions, consisting of loading and thickness contributions, for the various microphones. Since the method employed to calculate the theoretical pressure/time history sets the phase of the

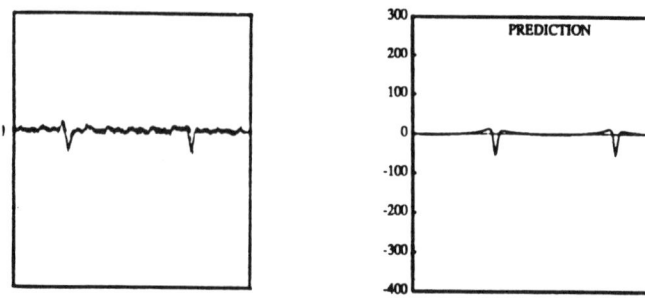

a) Microphone position: 72° from upstream direction, advancing side

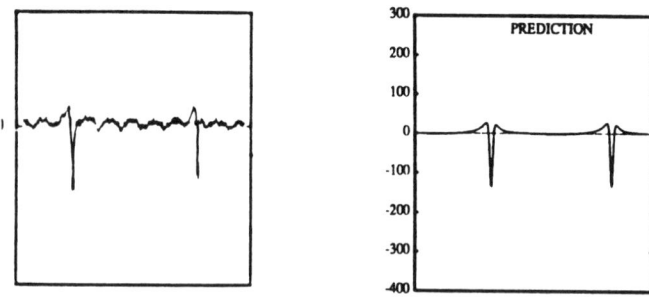

b) Microphone position: 53° from upstream direction, advancing side

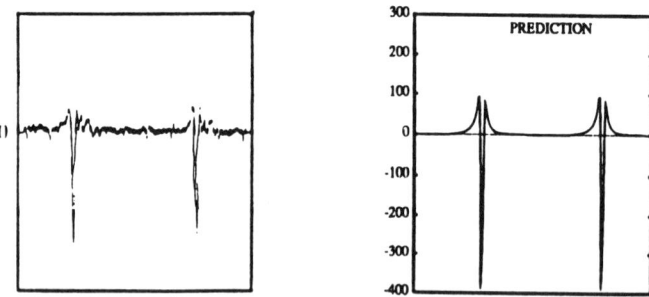

c) Microphone position: 29° from upstream direction, advancing side

**Figure 3.36** HERNOP calculations (right pictures) compared with in-flight measurements reported in reference [18] (left pictures), at three different (moving) microphone locations

Table 3.4   Propeller test data

| Operational conditions | Round tip | Square tip |
|---|---|---|
| | AN-1 run 63 | BC-1 run 77 |
| Rotational speed $\Omega$ (rads/s) | 219.0 | 188.49 |
| Tunnel speed (m/s) | 54.0 | 34.3 |
| Diameter (m) | 2.03 | 2.03 |

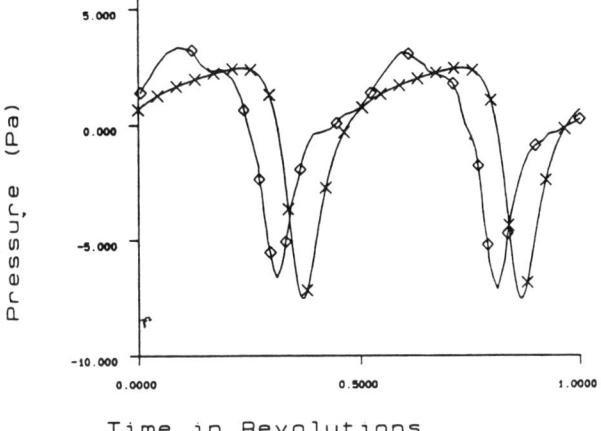

**Figure 3.37**   Acoustic pressure at 5 m distance; angle to propeller disc is 30°

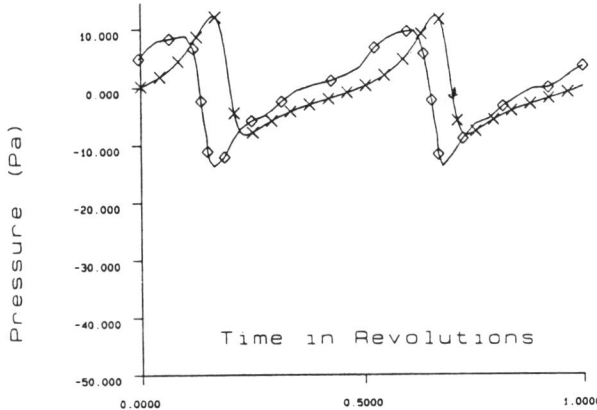

**Figure 3.38**   Acoustic pressure; microphone in propeller plane at 4 m distance

signature to be exactly zero, it seems reasonable to attribute the phase difference between theory and experiment to the processing of the experimental data.

Figure 3.39 shows a comparison between the experimental and theoretical power spectra. Figures 3.40 and 3.41 show how the theoretical pressure signal is composed of the loading and thickness contributions for different microphone positions.

In general, good agreement between acoustic pressure signals and experimental results have been found.

**The ALFAPI approach**   For the analysis of helicopter rotor noise in the frequency domain, the computer code HERONO-FDAJ valid for an open subsonic or supersonic rotor has been developed. The code utilises Johnson's

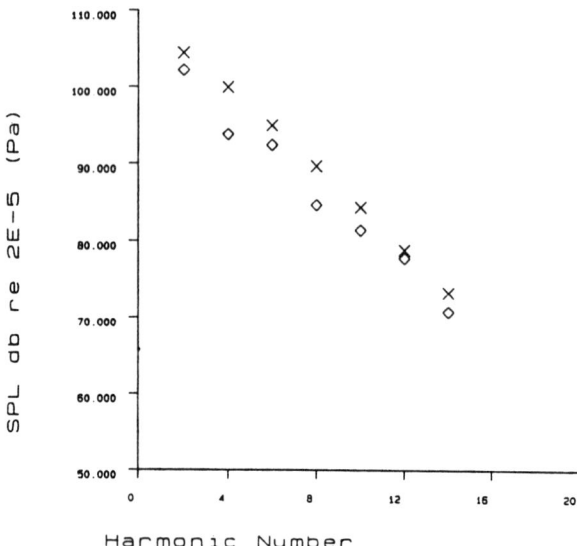

**Figure 3.39**   Power spectrum for microphone position of Figure 3.37

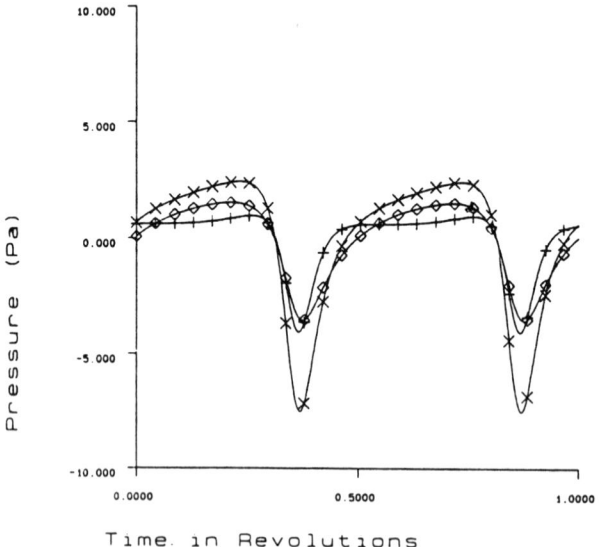

**Figure 3.40**   Acoustic pressure signal of Figure 3.37; contributions by different noise sources
× total, + thickness, ◇ loading

theoretical analysis applied for a rotor in forward flight based on Lowson's FWH linear acoustic formulas. According to the developed algorithm, the acoustic pressure/time histories are initially predicted for the frequency-domain calculations, which results in a significant reduction of computational time. The air-loading calculation is limited to inviscid steady flow, but the predicted

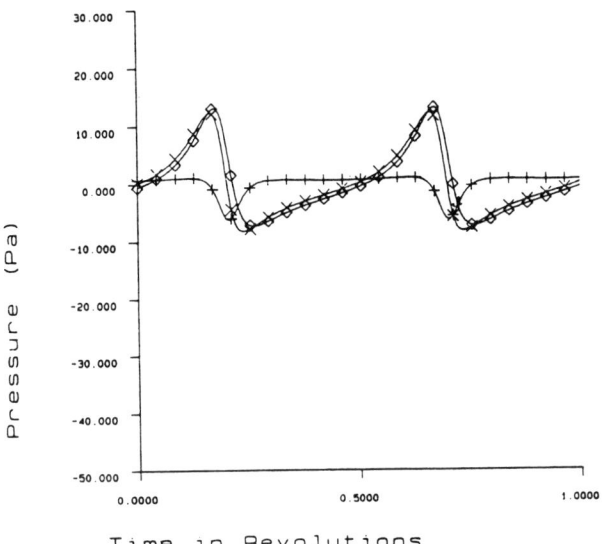

**Figure 3.41** Acoustic pressure signal of Figure 3.38; contributions by different noise sources

distribution can be replaced in the code by another one which is derived utilising more advanced calculations.

For the prediction of helicopter noise radiation in the time domain, the HERONO.TD code has been developed by applying the Farassat 1A acoustic formulation with the addition of the Brentner term concerning the derivative of normal blade velocity. This time-domain formulation is valid in both near-field and far-field domains. Aerodynamic data are calculated during the acoustic computational procedure using blade element theory. Flapping, feathering and lead–lag motion of the helicopter rotor are described by a truncated Fourier series that includes a constant term and the first and second harmonics. Blade section geometry and aerodynamic coefficients for all NACA sections are organised in a subroutine library. Calculation is performed on one blade rotation and acoustic pressure from blades is calculated using one blade's results and time-shifting. A preprocessing procedure is used for panelling the blade mean surface. Panelling subprograms are useful for several purposes since they are structured in such a way that they can be applied for the geometry segmentation on any type of rotor.

### 3.4.3 Broadband noise

The partner IST developed a method to predict broadband noise aspects [35]. There are at least four potential mechanisms of production of broadband noise—namely spectral broadening, convection of vortices, emission by turbulence and shear layer excitation.

Spectral broadening means that the acoustic energy in a signal is spread over the frequency band as it propagates through a random medium [23, 24]. For example, a tone emitted by a monochromatic source gives rise, after propagation through a turbulent jet, to a spike with sidebands [25], [33]. Vortices [26] and fluid inhomogeneities [27] in a nonuniform stream undergo hydrodynamic forces, which act as dipoles radiating sound [34]; although emission takes the form of 'pulses' (e.g. like for BVI), there is broadband noise as well. There are also essential broadband sources of sound, such as turbulence quadrupoles [28], represented by Reynolds or other stresses [29]. As a fourth potential mechanism, it has been demonstrated theoretically [30, 31] and experimentally [32] that a tone (e.g. due to BVI) transmitted through a jet can excite flow instabilities (e.g. in a shear layer), which amplify the sound, including broadband noise emission.

The present report describes an evaluation of the first of the mechanisms of broadband noise generation mentioned, namely spectral broadening of spikes, as IST's contribution to the code development activities of the consortium.

If the noise radiated by a helicopter rotor is assumed to consist of discrete tones, then the total energy over the spectrum

$$E_0(\omega) = \sum_{n=1}^{N} I_n \delta(\omega - \omega_n)$$

is the sum of the energies of each tone. When the rotor noise is measured at some distance, the acoustic energy is no longer concentrated at each frequency $\varpi_n$, but rather spread over a band owing to random scattering effects such as:

- Propagation across the turbulent wakes of rotor blades.

- Interaction with vortices, which may act as irregularities in refracting sound.

- Coarse sampling of the spectrum which can also 'spread' acoustic energy.

Thus, the emitted spectrum might be received as:

$$E(\omega) = I_0 + \sum_{n=1}^{N} I_n \exp\left(-\alpha\omega_n^2\right) \exp\left(-\beta(\omega - \omega_n)^2\right)$$

This equation can be interpreted as follows. Let a helicopter noise spectrum have $n = 1, 2, \ldots, N$ impulsive components, of intensity $J_n$, at the frequencies $\varpi_n$. The complete spectrum, including the broadband component, is then given by only three new parameters: the background noise $J_0$, the attenuation $\alpha$ and the correlation $\beta$.

We thus proceed to consider the three aforementioned parameters, starting with the attenuation. When sound propagates through a moving medium it experiences a Doppler shift; when the medium is turbulent, the accumulated Doppler shift is random, and corresponds to an aleatory phase lead or lag. Since neighbouring wave components may have different phase shifts, they can interfere, attenuating the sound field. The maximum interference, or attenuation, is the variance of the random phase shifts, which is a quadratic function of wave and flow parameters; thus it is proportional to the frequency of the wave squared, as indicated by the factor $\exp(-\alpha\varpi_n^2)$. The coefficient $\alpha$ depends on the square of the Mach number of turbulence, on the thickness of the turbulent region and on the sound speed. The interference of two wave components depends on their correlation; i.e. if they are the source's frequency $\omega = \varpi_n$ and there is no interference, and this increases statistically according to a Gaussian law as the frequency of reception $\varpi$ differs from that of emission $\varpi_n$. Again this is a quadratic effect, expressed by the factor $\exp\{-\beta(\varpi - \varpi_n)^2\}$, where $\beta$ depends on the correlation time of the random acoustic phases; the latter may be determined from the spectrum of turbulence. When the turbulence data are not available (e.g. not measured together with the spectra) we can estimate empirically the parameters $\alpha$ and $\beta$ by curve-fitting. The third parameter $J_0$ (the background noise) is always empirical for it depends on measurement conditions.

For clarity, the parameters used are defined once more:

- $I_n$ is the acoustic energy in the spike radiated at frequency $\varpi_n$. It is a result of theoretical models of helicopter rotor noise, or of measurement close to the blade. If a spectrum measured at some distance is available, then what is recorded as a peak is

$$H_n = I_n \exp(-\alpha\omega_n^2)$$

- $\alpha$ is the attenuation of sound. If $\alpha$ is small and the frequency not too high, then $H_n$ is close to $I_n$, i.e. low frequency peaks are less attenuated than high frequency peaks.

- $\beta$ is determined from the correlation length; the latter can be calculated from the spectrum of turbulence. If this information is not available, we can obtain $\beta$ by the best fit between the spectrum and the measured spectrum.

- $J_0$ is the background noise, and assuming it is uniform (white noise), its level can be estimated looking a spikeless high-frequency regions of the spectrum.

The present method is illustrated by applying it to a set of measured helicopter noise spectra [36].

The present theory of special broadening of spikes of the noise of a helicopter rotor in hover uses only two parameters over the entire frequency range (Figure 3.42). These parameters are the attenuation $\alpha$ and correlation $\beta$, and their adjustment is relatively straightforward (Figures 3.43 and 3.45). The present theory could be used in another way, viz. having measured the helicopter rotor noise at some position, the theory would be used to predict the spectrum at another position, e.g. (i) if the noise had been measured close to the rotor, the modified spectrum received by a distant observer could be predicted; (ii) if the noise could be measured only at some distance from the

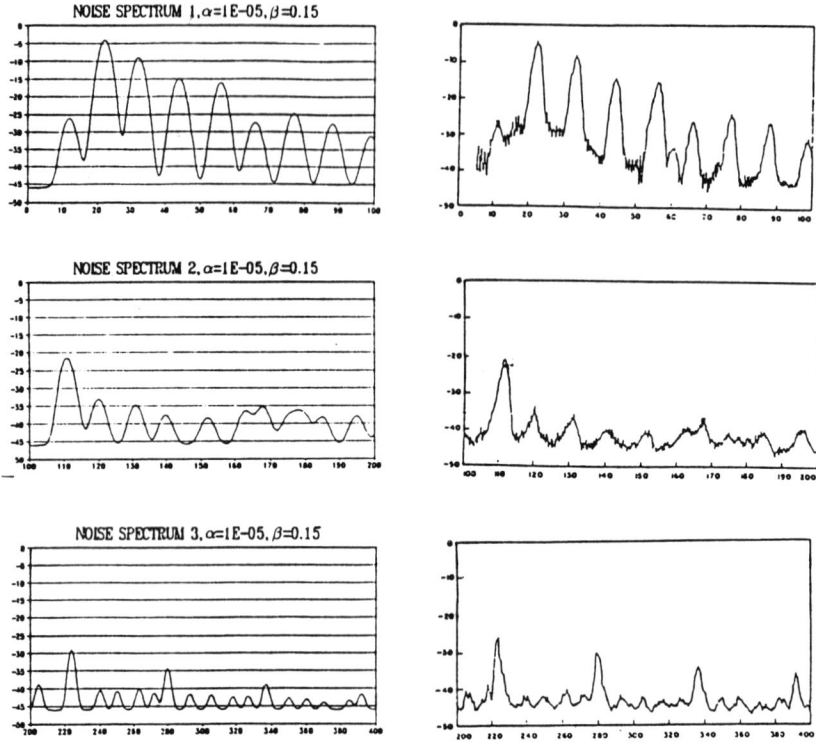

**Figure 3.42** Measured complete noise spectra of a Bell UH-1B Huey helicopter (r.h.s.) compared with the broadband spectrum (l.h.s.) predicted theoretically, from the empirical intensity and frequency of the spikes, with fitting of three parameters: attenuation $\alpha$, correlation $\beta$ and background noise

rotor, the theory could be used to predict the intensity of the spikes near the rotor. In both cases, the extrapolation is based on a change of attenuation and correlation parameters. The attenuation increases with the intensity of turbulence and extent of the turbulent region, and leads to a lowering of the spikes (right-hand side of Figures 3.43 to 3.45); the correlation decreases for strong turbulence well matched to the scattering of sound and leads to wider broadbands (left-hand side of Figures 3.43 to 3.45). The broadband component of the noise spectrum of a helicopter rotor can be due to at least four distinct physical processes: (i) scattering of impulsive sound or 'spikes' by the turbulent flow around the rotor, and also by atmospheric turbulence; (ii) the interaction of disturbances in the incident stream, e.g. vorticity or turbulence can produce not only spikes but also a broadband component; (iii) the turbulence in the boundary layer of the blades and in the wake of the rotor acts as a distributed quadrupole source with broadband emission of sound; (iv) the flow instabilities in boundary layers of blades and trailing shear layers can also emit or amplify noise spectra. The present theory concentrates on the first effect, namely scattering ([37] [38] [39]) and takes into account random physical processes, which lead to Gaussian statistics. It should be noted that the other three

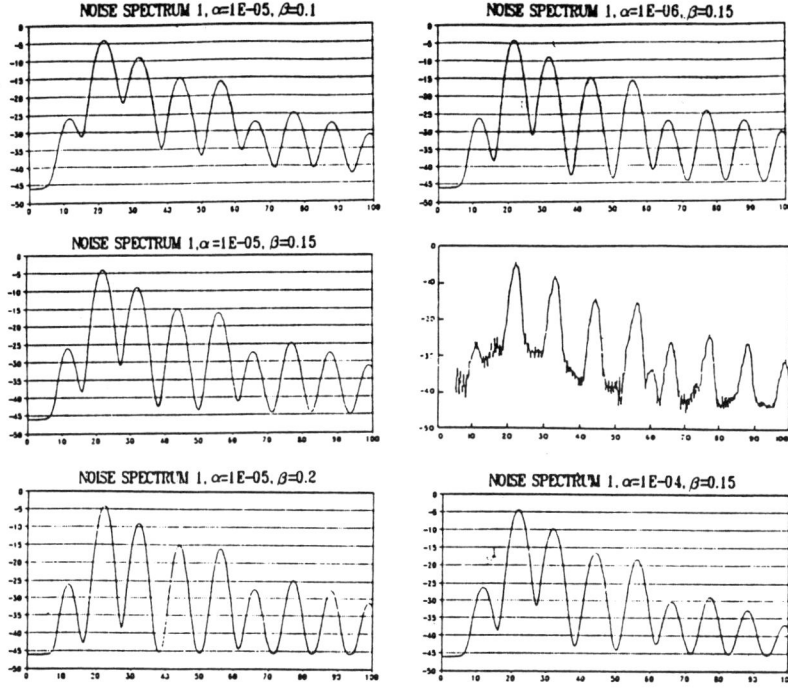

**Figure 3.43** Six noise spectra in the low-frequency band: (r.h.s. centre) measurement for Bell UH-1B Huey helicopter; (l.h.s. centre) predicted theoretically for optimal parameters, as in Figure 3.42; (r.h.s. top) with attenuation reduced, $\alpha = 10^{-5}$; (r.h.s. bottom) with correlation increased, $\beta = 0.2$ above optimum, $\alpha = 0.15$; (l.h.s. top) with correlation decreased, $\beta = 0.1$ below optimum.

processes (ii) to (iv) mentioned above also involve turbulence or random effects, so that Gaussian statistics may also apply with a modification of the attenuation and correlation parameters accounted for in the fitting procedure.

## 3.5 Experimental Results

The project HELINOISE provided a number of results relevant to the prediction of helicopter noise. The work comprised, besides the development of aerodynamic and aeroacoustic prediction tools, the performance of wind-tunnel tests for the validation of the aforementioned codes.

The results of the tests performed in the DNW with the BO 105 model rotor comprised the following:

- Radiated acoustics:
    sound pressure time histories
    narrowband spectra
    mid-frequency summary level (MFSL) contour plots
    low-frequency summary level (LFSL) contour plots.

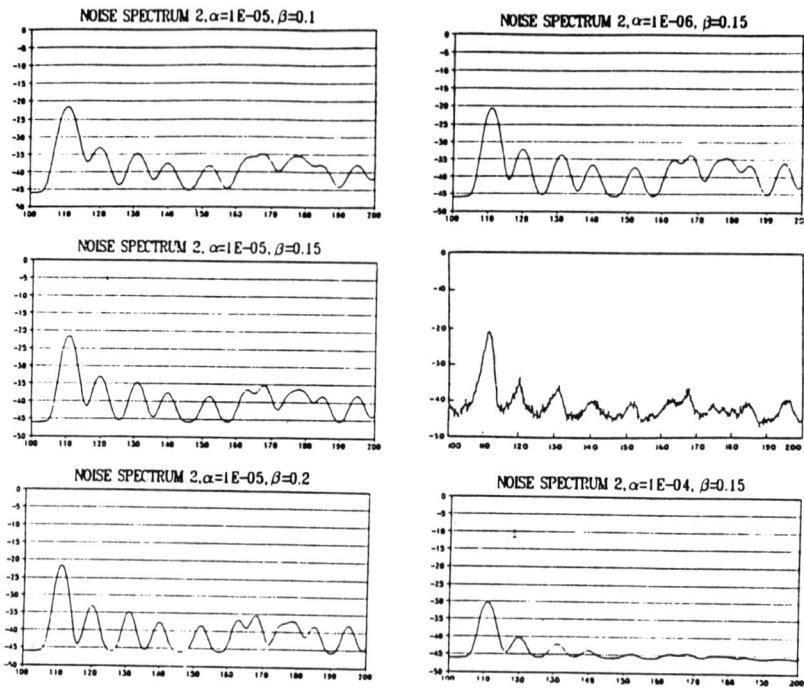

**Figure 3.44**   As Figure 3.43, for middle-frequency band

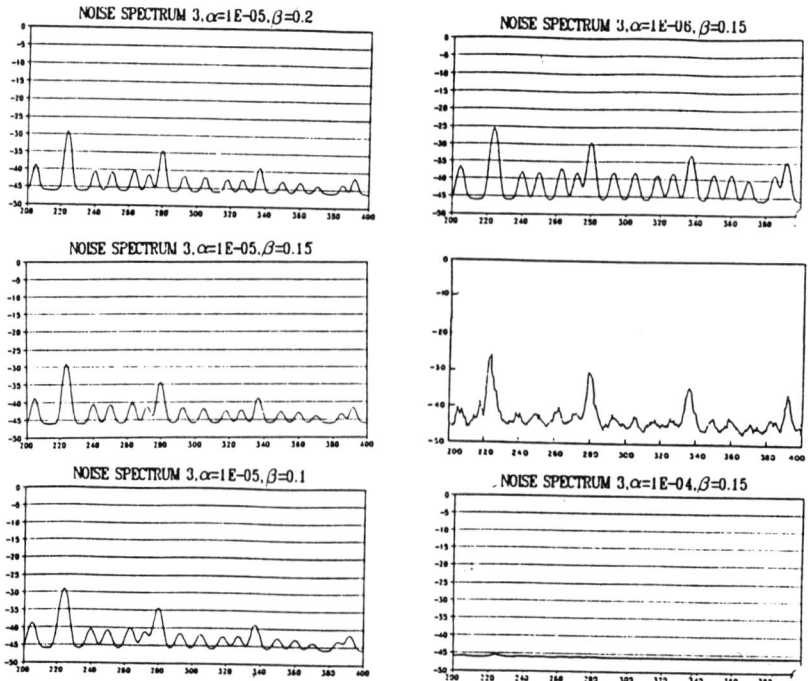

**Figure 3.45**   As Figure 3.43, for high-frequency band

- Blade surface pressures:
    time histories (in the azimuth)
    chordwise distribution
    (per one-revolution) azimuthal distribution.

- Vortex trajectories.

- Blade vortex miss distance.

- Blade deflection.

*Radiated acoustics*

Any particular test operational condition was defined by setting a tunnel speed and a rotor tip path plane inclination. Since both the rotor rotational speed and the thrust coefficient as such were kept constant, the former two parameters were the only ones to be varied. In combination, they in turn determine the climb or descent angle and speed. It will be recalled that flight ($\equiv$ tunnel flow) speed and the rate of climb or descent are the relevant operational parameters for the full-scale helicopter.

For each such defined test condition (e.g. 33 m/s flight speed, 6° descent angle), the complete set of both radiated acoustic data over the entire measurement area below the rotor as well as the blade surface pressure data were recorded simultaneously. Within each data point, the microphone array was continuously (rather than stepwise) moved at 45 mm/s; for online quicklook analysis, data were recorded 'on the fly' exactly every half metre in the streamwise direction for a time span corresponding to 30 rotor revolutions (1.7 s). Although during this short time interval the array was actually moving by about 75 mm, this was verified not to introduce any significant inaccuracy in the data. In addition, the acoustic signals for the complete microphone array sweep over typically 8–10 metres was recorded on analog tape for later offline analysis.

The initial outcome of measuring the radiated acoustics was a set of averaged time histories, one for each of the 11 microphones.

Figure 3.46 shows a typical 'baseline data' page, each box corresponding to one streamwise microphone array position. Typically 15 to 20 such pages were thus obtained for each test condition. These sound-pressure time histories can be readily converted into spectra (narrowband for example) such that the tonal content of the acoustic signal may be evaluated at each microphone position. A typical page, of which there are again 15 to 20 for each test condition, is shown in Figure 3.47.

One convenient way to reduce the vast amount of data is by means of their compression into a 'one number criterion' (e.g. a summary level) at each microphone location. A strategy along these lines is shown in Figure 3.48. Here any initial pressure time history is first converted into a narrowband spectrum. One may now add up the levels of the first few blade-passage-frequency (BPF) harmonics (say, up to the tenth or so). This 'low-frequency summary level'

| HELINOISE TEST | Series/Run: 10 17    Polar/DPt0: 99/1333    DPt:1341 | 19-APR-93 15:08 |
|---|---|---|
| $X_W$ : -0.99 m | $M_H$ :.644        $\mu$ :.149        $C_T$ :.0045        $\alpha_{TPP}$ : 5.3° | |
| $V_{TUN}$ : 32.63 m/s | SOUND PRESSURE TIME HISTORIES (30 AVG.) | Filter: 4 Hz |

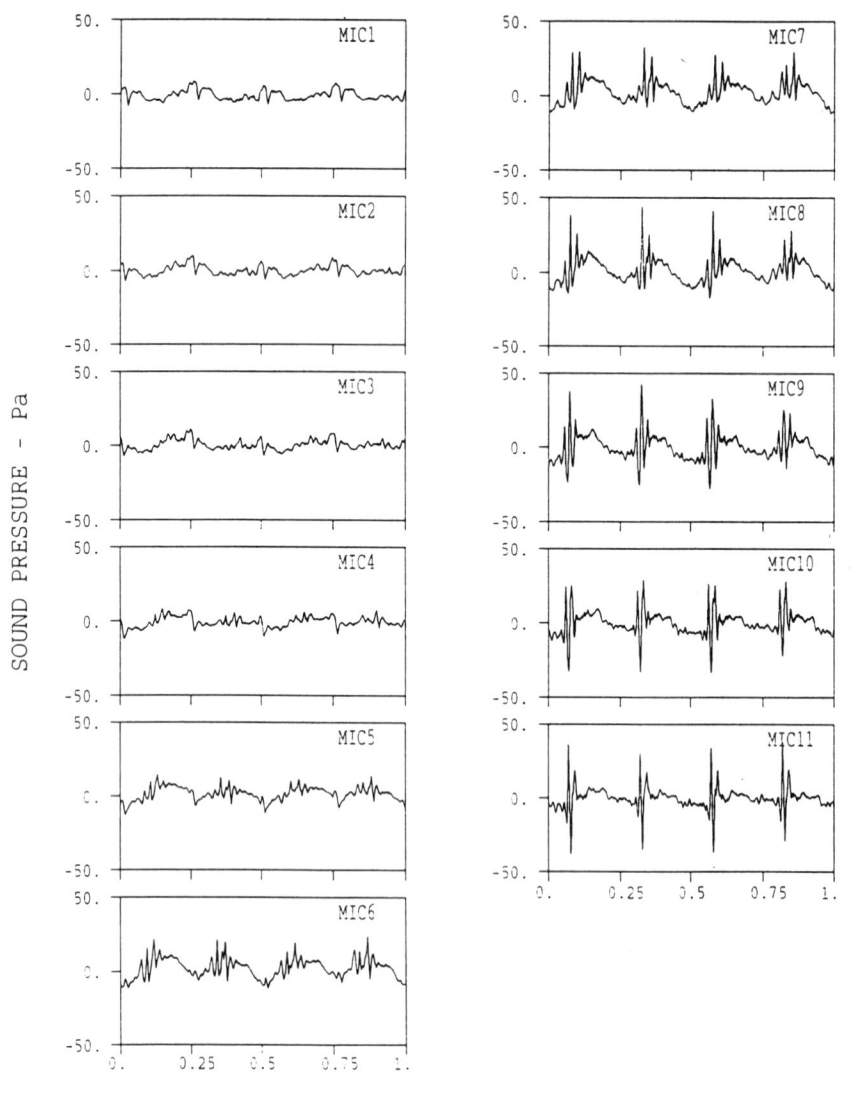

TIME - REV

**Figure 3.46**  Typical 'baseline data' page for a defined microphone array position

| HELINOISE TEST | Series/Run: 10 17  Polar/DPt0: 99/1333  DPt:1341 | 19-APR-93 15:11 |
|---|---|---|
| $X_W$ : -0.99 m | $M_H$ :.644   $\mu$ :.149   $C_T$ :.0045   $\alpha_{TPP}$ : 5.3$^0$ | |
| $V_{TUN}$ : 32.63 m/s | POWER SPECTRA (30 AVERAGES) | Filter: 4 Hz |

$$\Delta F = FROT = 17.35\,Hz$$

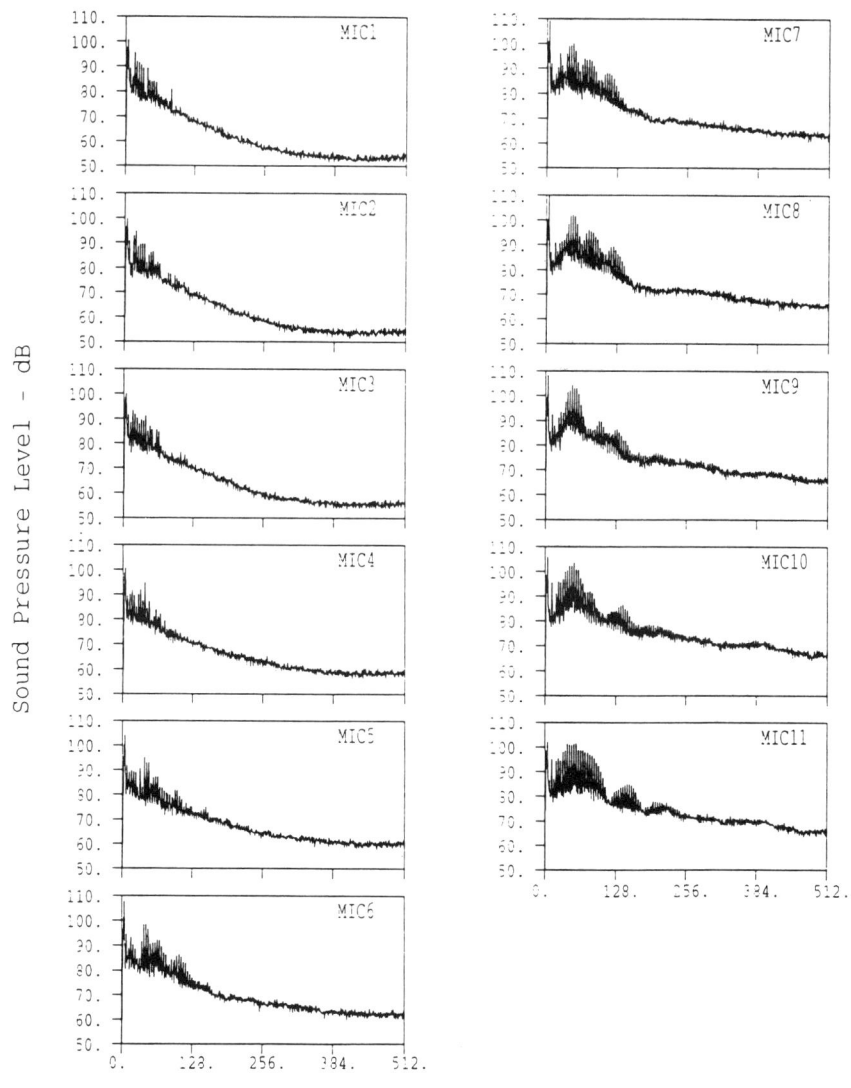

Sound Pressure Level – dB

Frequency/FROT

**Figure 3.47**  Narrowband power spectrum corresponding to the signals of Figure 3.46

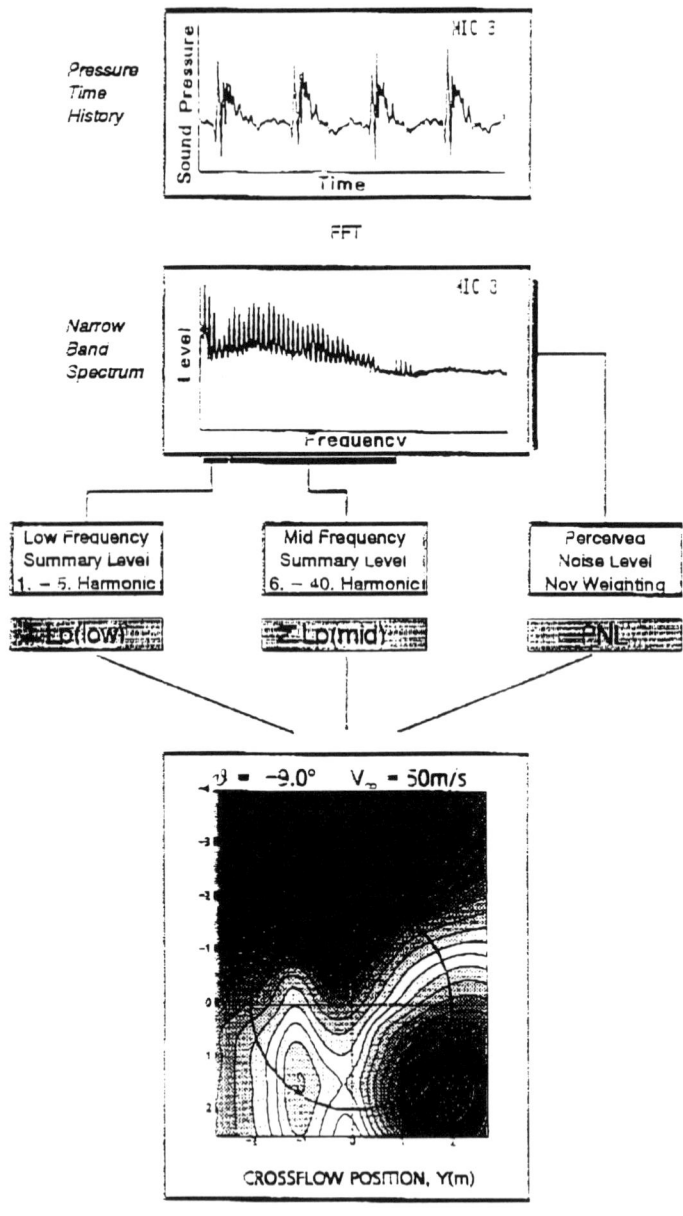

Summary Level Contour Plot

**Figure 3.48**   Method of data reduction

(LFSL) would be a measure of the long-distance sound propagation or of detectability; also, high-speed impulsive noise tends to be characterised by a rapid increase of the low-frequency spectral content.

Alternatively, one might add up the middle range of frequencies, say the levels of the 6th to the 40th rotor harmonics, to arrive at a 'mid-frequency summary level' (MFSL). This summary level could be considered a measure of the subjective acoustic perception of the human auditory system. Moreover, blade/vortex interaction noise is characterised by the mid-frequency spectral content of rotor noise.

Such characteristic summary levels may be plotted within the measurement area below the rotor to provide very instructive contour plots, which—at one glance—indicate critical noise regimes within an emitted radiation pattern. Several contour plots are shown here.

Figure 3.49 shows MFSL contour plots for a constant flight speed of 33 m/s and various descent angles, ranging from a 5.5° to a 12° descent. Clearly, at this speed the descent-angle regime from 6° to 7.5° is characterised by strong blade/vortex interaction, as is evident through the two conspicuous maxima under the first and the fourth rotor radiation quadrant.

Figure 3.50 shows MFSL contour plots for a constant descent angle of 6° and flight speeds ranging from 25 to 70 m/s. Again it is obvious that local maxima in the first and the fourth quadrants appear in a certain flight speed range from 25 to 40 m/s, characteristic for blade/vortex interaction.

Careful probing of the test matrix for especially intense blade/vortex interaction noise revealed critical combinations of descent angle and flight speed. Figure 3.51 indicates that BVI—contrary to earlier understanding—extends far into the high-speed and high-descent-angle regime.

At increasing forward speed and/or growing blade-tip speed (both causing higher advancing blade-tip Mach numbers), thickness noise and initial non-linear compressibility effects of high-speed impulsive noise become significant contributors to the total rotor noise radiation. An example of the typical negative pressure pulses of high-speed impulsive noise is shown in Figure 3.52 for an 80 m/s level flight at an upstream microphone position ($x/R = 2.5$) on the advancing side. Such pressure pulses do not exist at the downstream (retreating side) location. The LFSL contour plot (here comprising the 2nd to the 10th BPF harmonic) illustrates the typical high speed noise directivity pattern pointing upstream and towards the advancing side (with highest levels in the rotor plane). The corresponding power spectrum shows increased levels of the BPF harmonics in the lower frequency range, typical for the onset of high-speed impulsive noise.

### Unsteady blade surface pressures

Baseline data in terms of the ensemble averaged (60 averages) surface pressure time history during one rotor revolution for each of the 124 KULITE sensors on the instrumented blade was obtained simultaneously with the acoustic data for the respectively identical operational test conditions. A typical 'baseline data' sheet is reproduced in Figure 3.53. In this example, blade surface-pressure time histories are shown for several sensors on the upper blade surface for a case of high-speed level flight (80 m/s, corresponding to an advancing blade Mach number of 0.862). This figure is only intended to illustrate the general character of the data obtained.

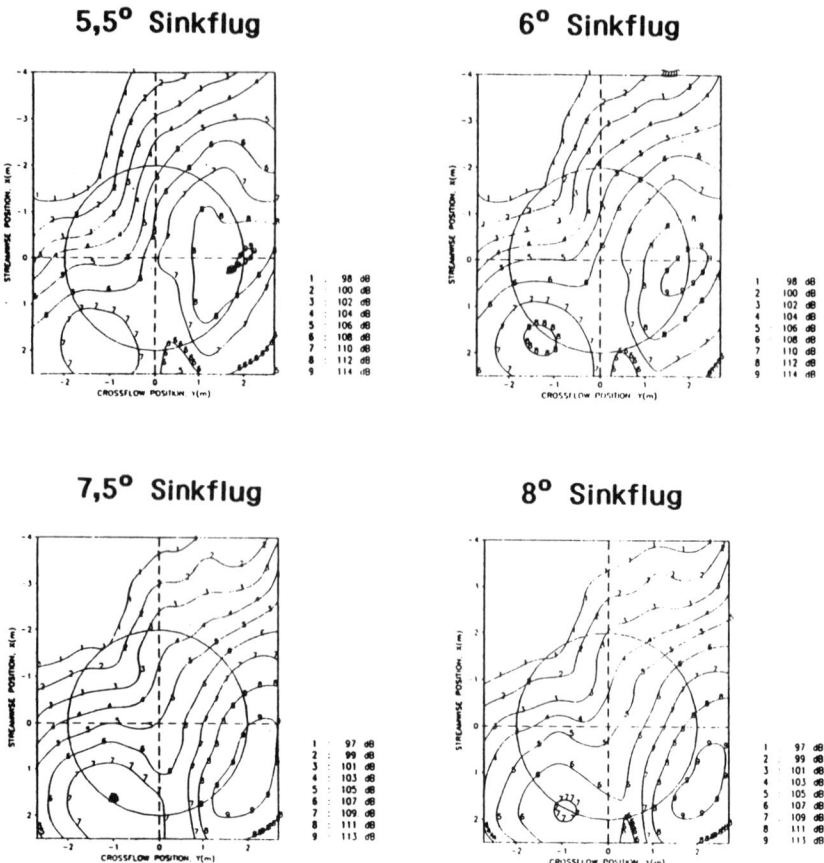

**Figure 3.49**   MFSL contour plots for varying descent angle at a flight speed of 33 m/s

These data can be further processed and utilised in a number of ways, as follows.

**Moderate-speed descent (strong BVI)**   The potential in helping to understand the physics of rotor noise is illustrated in Figure 3.54, where for a condition of moderate-speed descent (33 m/s flight speed, 6° descent) the blade surface-pressure time histories for several sensors close to the upper-surface blade leading edge (3% chord) at several radial stations are presented. The figure also shows a schematic of several vortex trails superimposed on the rotating blades. From the pressure time histories one may readily discern the interactions of blade tip vortices with the rotating blade in the first and the fourth rotor quadrants. Moreover, the relative strengthes of the interaction process at the various radial stations allows certain important conclusions (for example on the angle of interaction and its relative location).

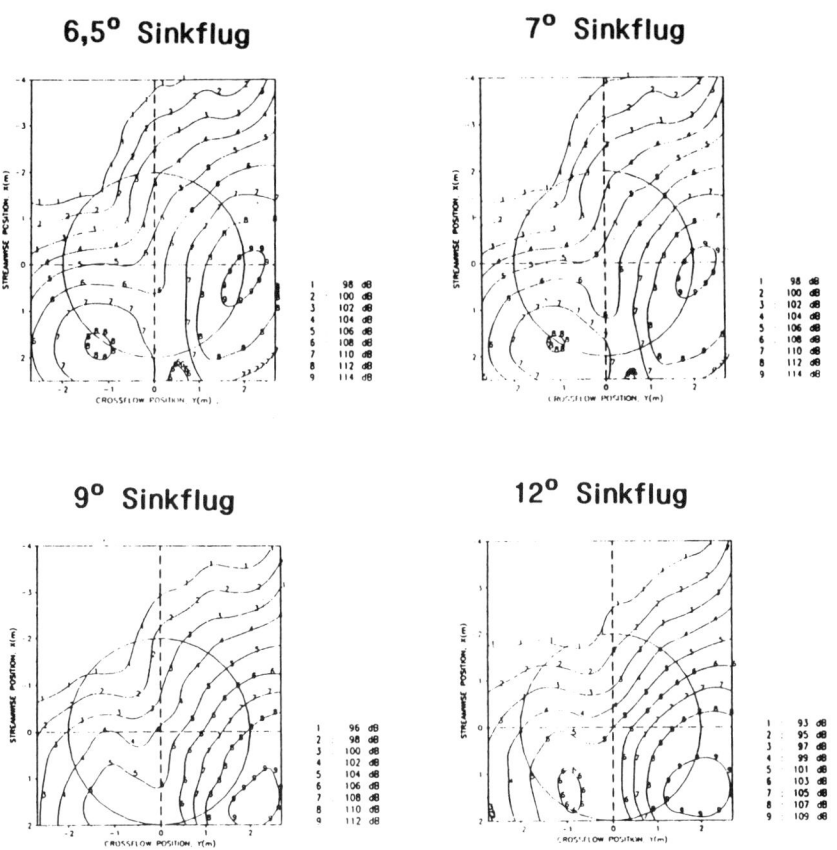

6,5° Sinkflug   7° Sinkflug

9° Sinkflug   12° Sinkflug

Stretching the time history (as in the upper left corner of Figure 3.54 where the fourth quadrant time history of the outermost sensor is shown) indicates the further interpretational potential, in that one may now actually determine quantitatively the vortex core dimensions.

Of course, rather than just providing physical insight into the various rotor aeroacoustic processes, these data are to be used to validate the mathematical models for surface pressure prediction, which in turn are a necessary input quantity into acoustic prediction schemes (utilising, for example, the FWH equation).

**High-speed flight (aerodynamic shocks)** Rather than following the pressure signal of a particular sensor for one rotor revolution, one might inspect the chordwise pressure distribution on the upper and lower surface at several azimuthally fixed rotor blade positions. Figure 3.55 illustrates a case of high-speed level flight, corresponding to an advancing blade tip Mach number of

## VT = 25 m/s

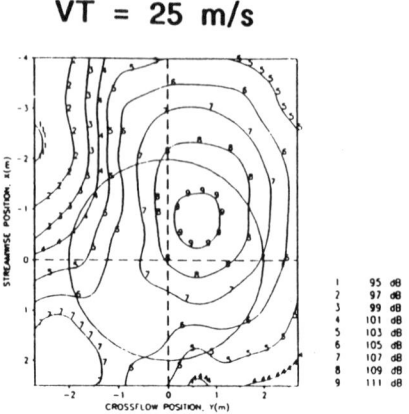

## VT = 50 m/s          ## VT = 60 m/s

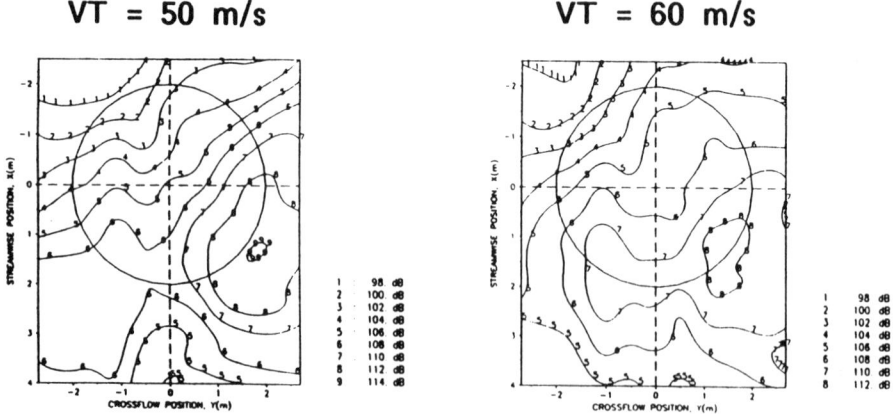

**Figure 3.50**  MFSL contours for varying flight speed at a descent angle of 6°

0.87, where at azimuthal positions from 45° to 90° strong aerodynamic shocks appear on the upper blade surface. These shocks—when they separate from the blade surface (typically at an advancing blade tip Mach number of about 0.9)—are the cause of extreme high-speed impulsive noise.

### Surface-pressure isobar plots

Again, the vastness of blade surface-pressure data may be compressed for easier interpretation of physical processes by plotting 'isobars' (i.e. lines of

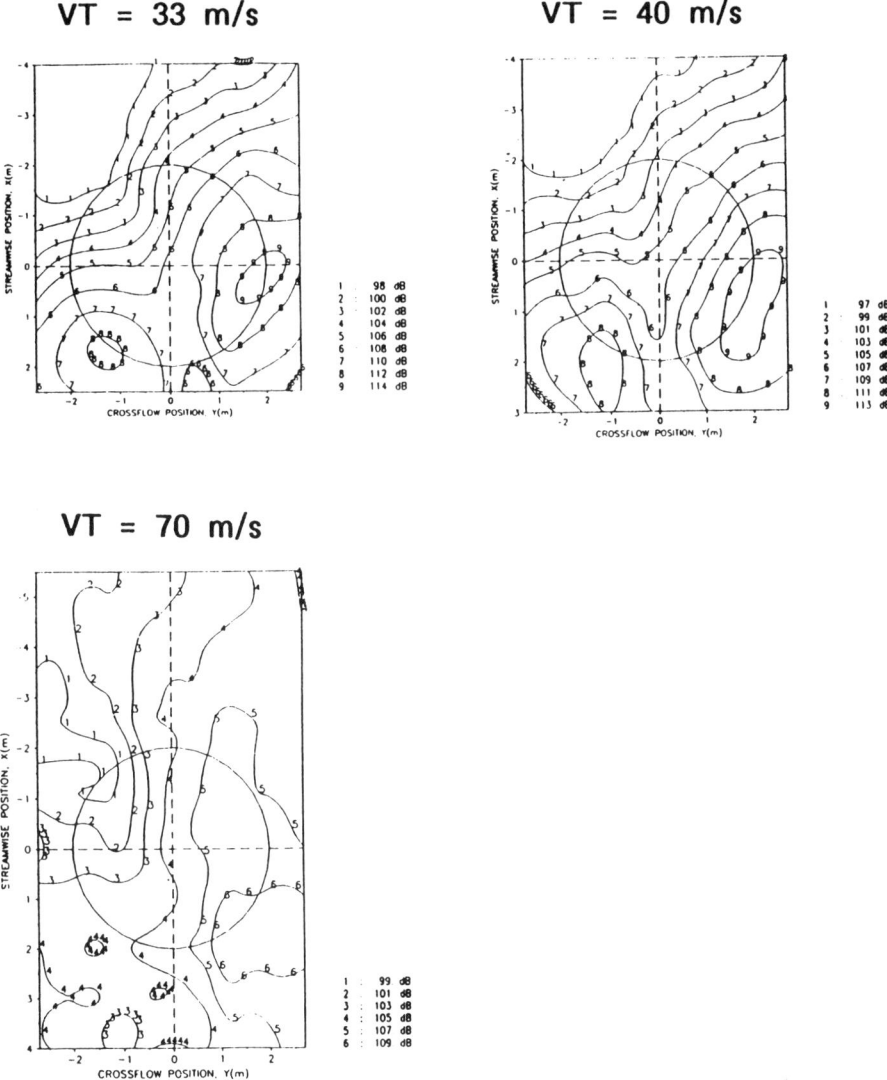

**Figure 3.50**   *(continued)*

equal pressure) for the entire rotor area instrumented, as it repeats for each revolution.

A typical plot is shown in Figure 3.56 for the case of a moderate descent speed, with the obvious appearance of strong pressure fluctuations due to BVI. Here in the first and the fourth rotor quadrants the tell-tale traces of several vortices interacting with the blade are clearly discernible. Vortex strength, vortex rotational direction, vortex/blade interaction angle etc. may be readily extracted from such illustrative representations.

This plot of the blade surface pressure characteristics might be related to the corresponding radiated acoustics field, as shown as one of the frames in Figure

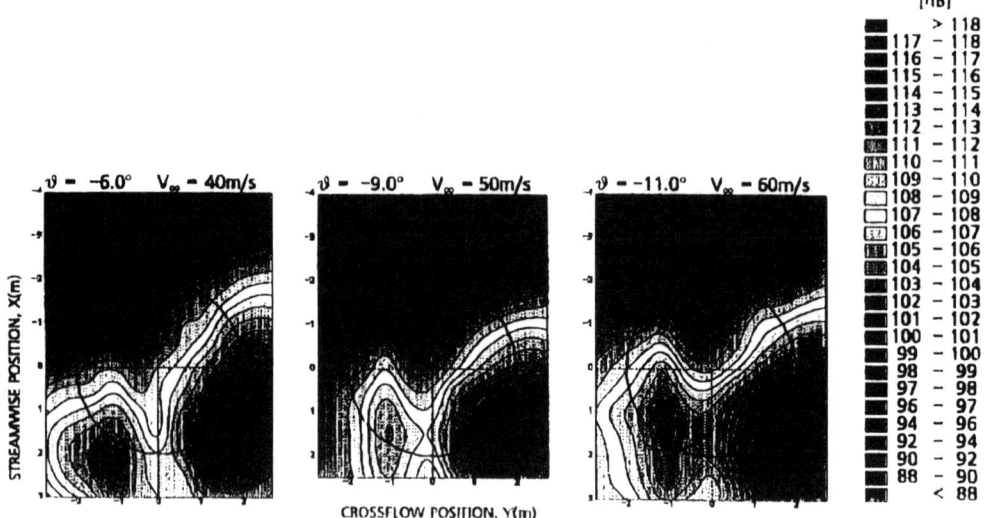

**Figure 3.51**   Development of BVI areas versus descent angle and flight speed

$$\mu = 0.347 \, , \quad \alpha_S = -13.0^\circ \, , \quad C_T = 0.0045 \, , \quad M_H = 0.640 \quad (79 \text{ m/s} - \text{LEVEL FLIGHT})$$

$$M_{AT} = 0.862$$

**Figure 3.52**   Typical negative pressure pulses for high-speed impulsive noise

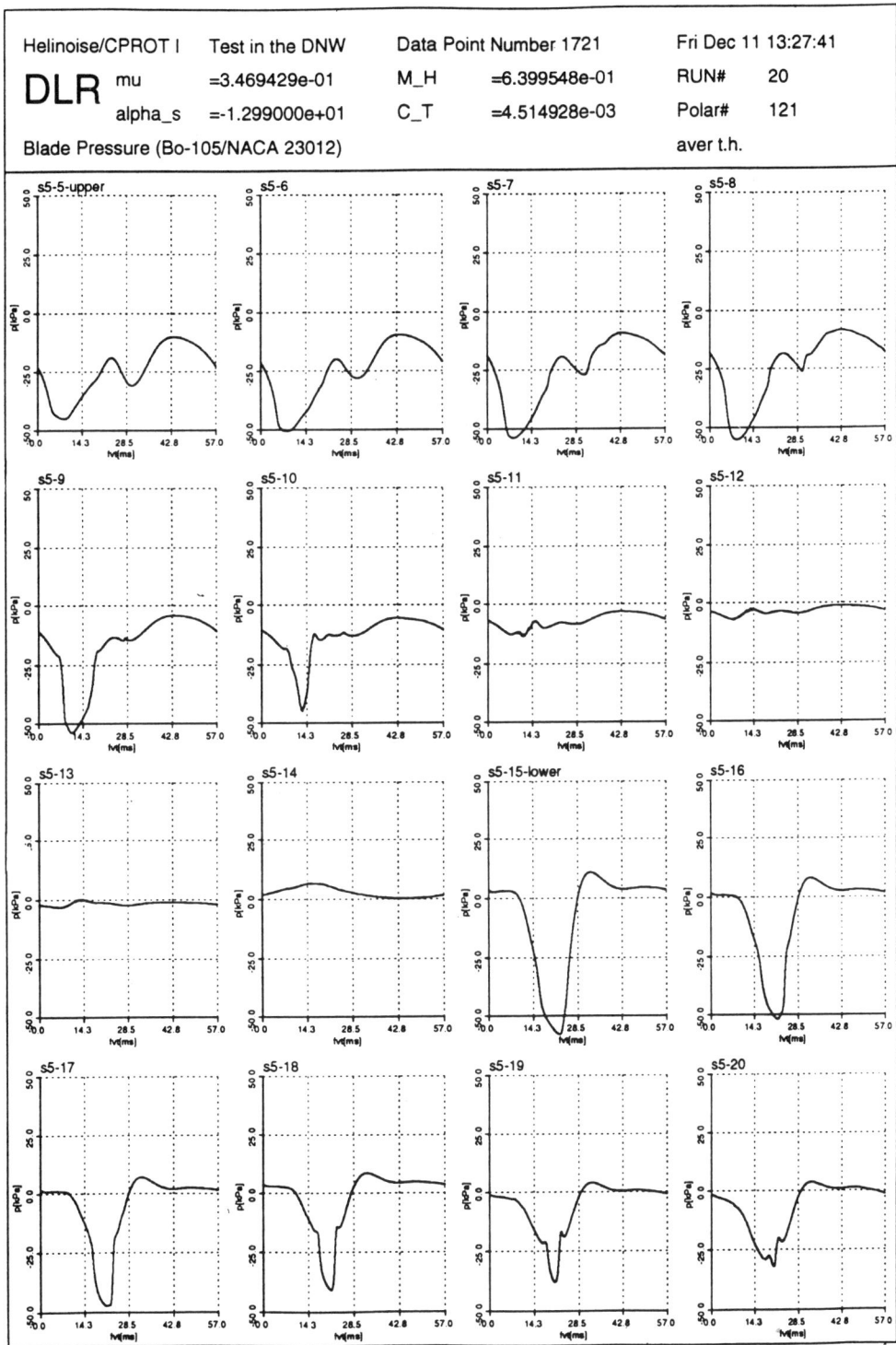

**Figure 3.53** Typical 'baseline data' sheet of pressure-probe time histories on upper blade surface for 80 m/s

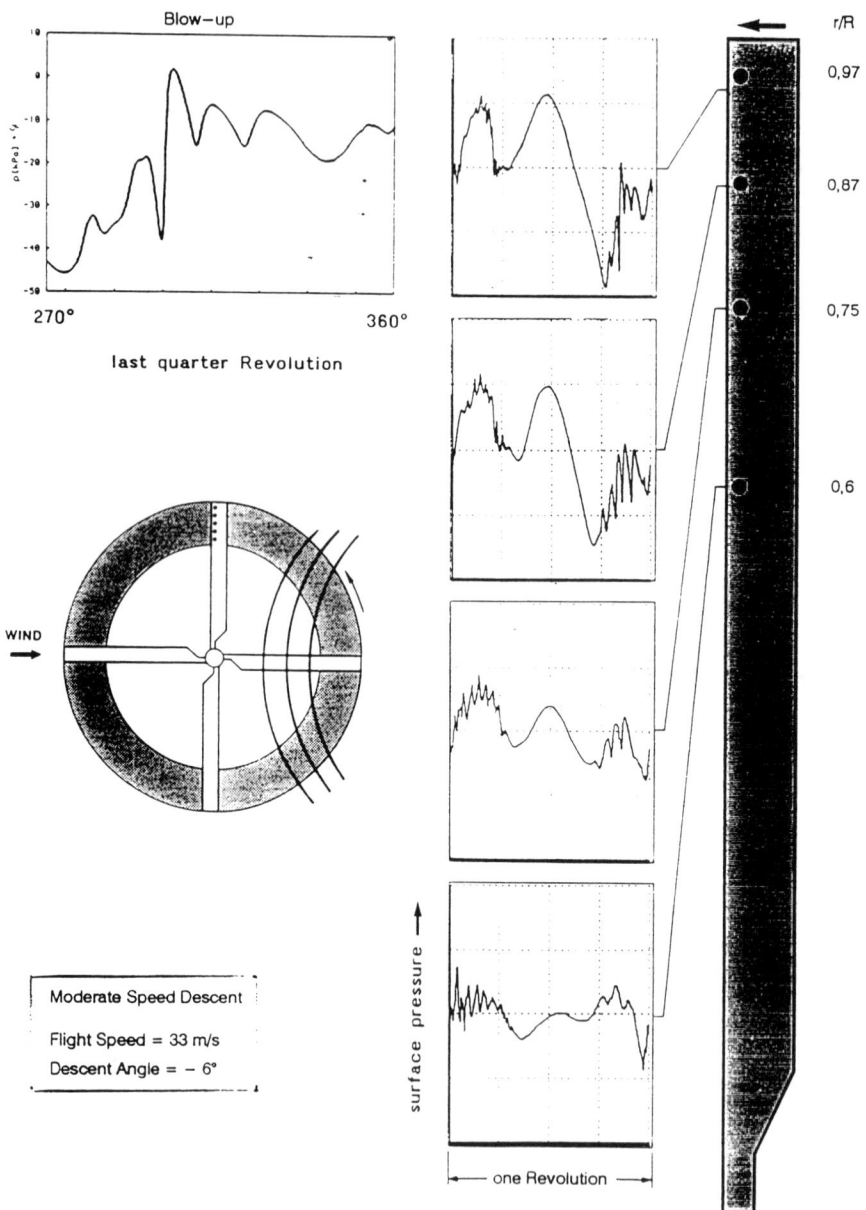

**Figure 3.54** Blade surface-pressure time histories for several sensors close to the upperblade leading edge (3% chord)

3.49 or Figure 3.50 (i.e. 33 m/s flight speed at 6° descent). One may readily relate the first and fourth quadrant blade/vortex interaction traces on the rotor with the two strong MFSL maxima in the first and fourth radiation quadrants under the rotor.

## Vortex visualisation

The loci of the vortices interacting with the rotor blades (misdistance and angle of vortex axis with respect to the blade) must be accurately predictable in order to guarantee a realistic description of the BVI noise-generation mechanism. Therefore, they were made visible with the aid of a laser light sheet system, in order to allow for a validation of the rotor wake prediction codes. Figure 3.57 shows two vortices with their misdistance to the blade being identified by the measurement grid. From the elliptic shape of the vortex cross-section, the angle between vortex axis and blade leading edge can be deduced.

### First comparisons between experimental and theoretical blade pressure distributions

First comparisons between experimental HELINOISE blade pressure distributions and theoretical results elaborated by ECD's vortex lattice method are shown in Figures 3.58 and 3.59. The figures show the azimuthal behaviour of

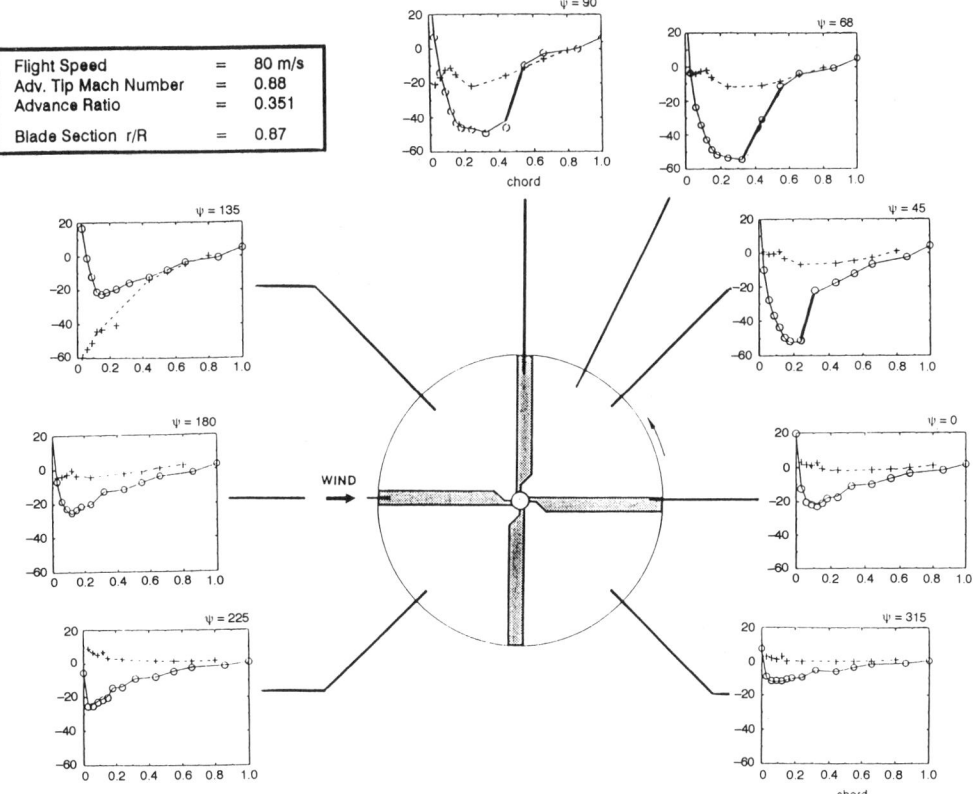

**Figure 3.55**  Azimuthal development of upper and lower side pressure distribution with appearance of strong shocks

Surface Pressure [kPa]

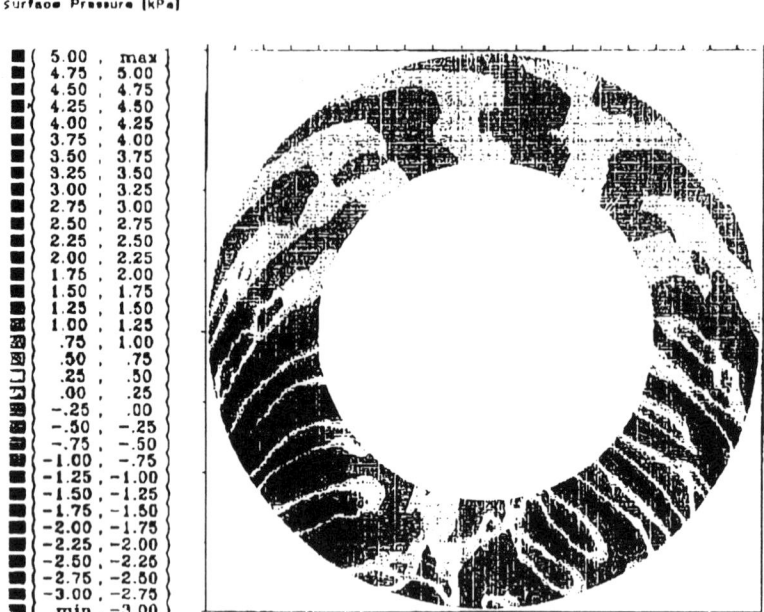

**Figure 3.56**   Isobars of upper blade surface near the blade leading edge (3% chord) along the outer blade radial stations for 6° descent and 33 m/s flight speed

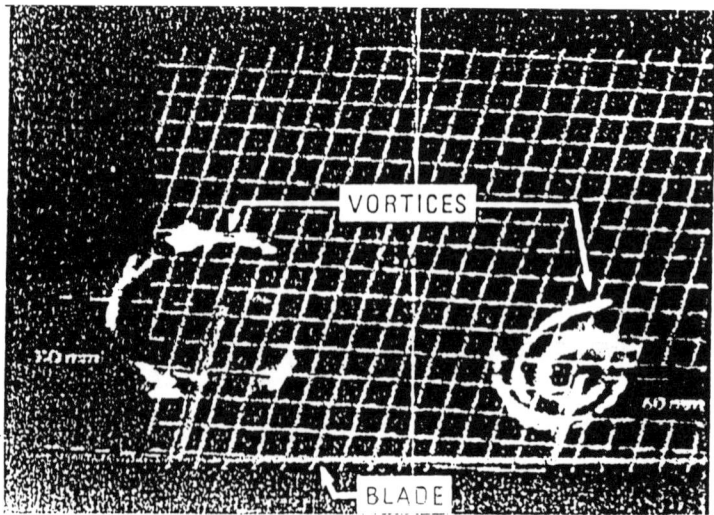

**Figure 3.57**   Blade tip vortices visualised by the laser light sheet method

the pressure difference between lower and upper blade sides for different radial stations, near the blade leading edge at 3% blade chord and at 6% blade chord. Figure 3.60 illustrates the corresponding computed wake of one single blade seen from above.

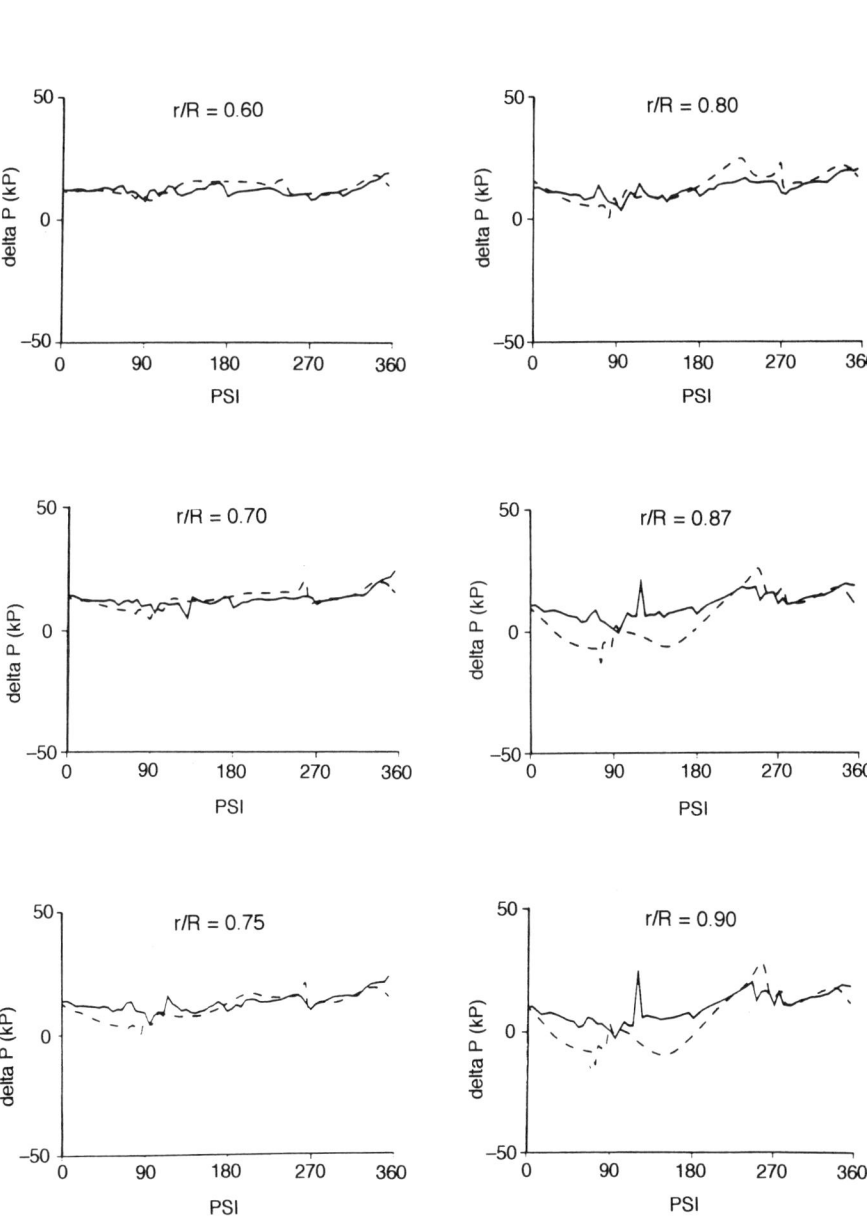

Figure 3.58  Comparison of measured and calculated blade pressure-difference distributions $\Delta p$ (lower side minus upper side pressure) at 3% chord

Pressure Difference

Data Point 344, x/c = 0.060

| | |
|---|---|
| BO 105 | 3 Revolutions |
| V00 = 32.75 m/s | — Computation |
| $\mu$ = 0.151 | - - Measurement |

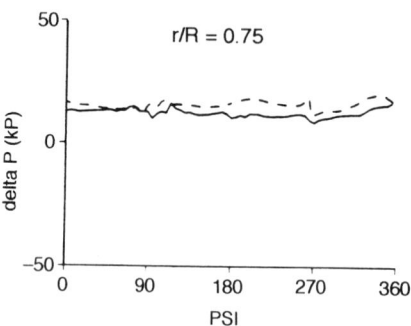

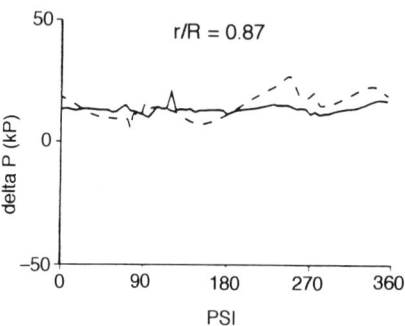

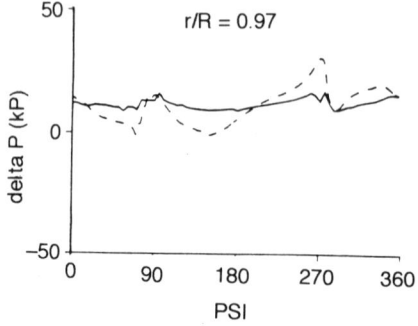

**Figure 3.59**  Comparison of measured and calculated pressure-difference distribution at 6% chord

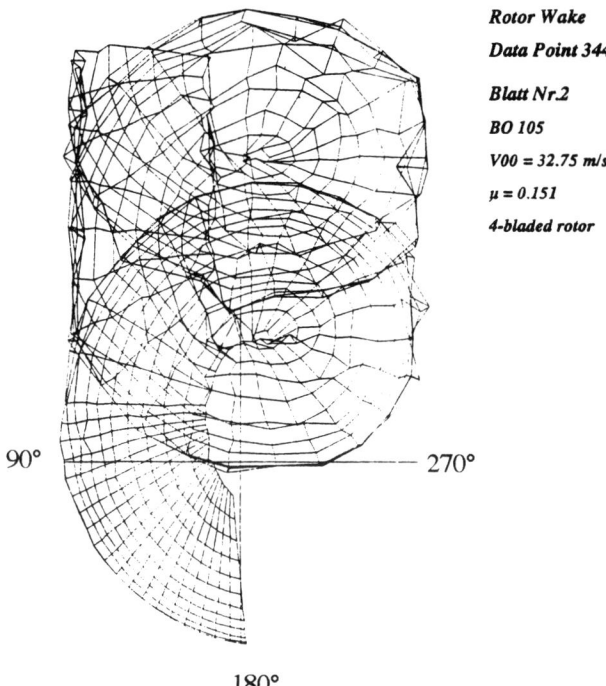

*Rotor Wake*

*Data Point 344*

*Blatt Nr.2*

*BO 105*

*V00 = 32.75 m/s*

*μ = 0.151*

*4-bladed rotor*

90°                                                                            270°

180°

**Figure 3.60**   Computed wake of one single blade for the test conditions of Figures 3.58 and 3.59

## 3.6   Conclusions

The project HELINOISE has turned out (in spite of early difficulties with the manufacturing of the model rotor) as a success. It provides a unique databank for the validation of aerodynamic and aeroacoustic prediction codes. The comparisons between first results of such codes and the experimental data are very encouraging.

The test data can also be used to study directly the phenomena of helicopter and tilt rotor aircraft noise. The noise footprints taken under different flight conditions reveal that blade/vortex interaction (BVI)—contrary to earlier under-standing—extends far into the high-speed and high-descent-angle regime.

Of particular relevance is the synchronous recording of blade pressure and acoustic signature below the rotor, since both signals are closely related, the pressure distribution being the dominant source of noise radiation.

The development of *different* aerodynamic/aeroacoustic prediction codes ad-dressing the same phenomenon proved to be very useful, as this allowed mutual control of the validity of the models involved. A final check will be the comparison with the experimental results, a very important aspect for realistic comparisons of different solution procedures.

Generally speaking, cooperative contact between prominent representatives of a specific scientific area can be expected to generate a substantial synergistic effect. HELINOISE can be taken as an example of this.

# References

[1] Mercker, E. *Test Documentation and General Description of the Measuring Techniques for the* HELINOISE *Test*. DNW-TR-93.01, 1992.

[2] Mercker, E. and Pengel, K. Flow visualization of helicopter blade tip vortices. 18th European Rotorcraft Forum, Avignon, 1992, ERF-1992-026.

[3] Lesching, A. and Wagner, S. Theoretical model to calculate aerodynamic interference effects between rotor and wing of tilt rotors. Presented at the 16th European Rotorcraft Forum, Glasgow, 1990.

[4] Schulz, K.J. and Splettstoesser, W. Prediction of helicopter rotor impulsive noise using measured blade pressures. Presented at the 43rd Annual Forum of the American Helicopter Society, St Louis, May 1987.

[5] Morino, L. and Genaretti, M. Boundary integral equation methods for aerodynamics. *AIAA Progress in Aeronautics & Astronautics — Computational Nonlinear Mechanics in Aerospace Engineering*, 1991/92.

[6] Dobrzynski, W.M., Heller, H.H., Powers, J.O. and Densmore, J.E. *Propeller Noise Tests in the German–Dutch Wind Tunnel DNW*. DFVLR-IB 129-86/3, FAA report AEE 86-3, 1986.

[7] Aston, J.A.G., Fiddes, S.P., Gould, J. and Lowson, M.V. *Propeller Noise: A Detailed Comparison between Experiment and Theory*. University of Bristol, Department of Aerospace Engineering, report 438, Feb. 1992.

[8] Gould, J. and Fiddes, S.P. Computational methods for the performance prediction of HAWTS. Presented at the European Wind Energy Conference, EWEC'91, Amsterdam, the Netherlands.

[9] Graber, A. and Rosen, A. Velocities induced by semi-infinite helical vortex filaments. *J. Aircraft*, May 1987, pp. 289–290.

[10] Chiu, Y.D. and Peters, D.A. Numerical solutions of induced velocities by semi-infinite tip vortex lines. *J. Aircraft*, Aug. 1988, pp. 684–694.

[11] Wood, D.H. and Gordon, G. Numerical evaluation of the velocities induced by trailing helical vortices. *AIAA Journal*, 1989, **28**(4).

[12] Meyer, J.R. and Fallabella, G. *An Investigation of the Experimental Aerodynamic Loading on a Model Helicopter Rotor Blade*. NASA report TN2953, 1953.

[13] Farassat, F. Linear acoustic formulas for calculation of rotating blade noise. *AIAA Journal*, 1981.

[14] di Francescantonio, P. A numerical method for the prediction of quadrupole shock wave noise. Presented at the 14th DGLR–AIAA Aeroacoustic Conference, Aachen, Germany, 11–14 May 1992.

[15] Farassat, F. *Theory of Noise Generation from Moving Bodies with an Application to Helicopter Rotors*. NASA technical report R-451, Dec. 1975.

[16] Brentner, K.S. *Prediction of Helicopter Rotor Discrete Frequency Noise*. NASA technical memorandum 87721, Oct. 1986.

[17] Prieur, J. Calculation of transonic rotor noise using a frequency domain formulation. *AIAA Journal*, 1988, **26**(2), 156–162.

[18] Schmitz, F.H. and Yu, Y.H. Helicopter impulsive noise: theoretical and experimental status. *J. Sound Vib.*, 1986, **109**, 361–422.

[19] Lowson, M.V. The sound field for singularities in motion. *Proc. Roy. Soc. Lond.*, 1965, **286A**, 559–572.

[20] Lowson, M.V. and Ollerhead, J.B. *Studies of Helicopter Noise*. USAAVLABS technical report 68–60, Jan. 1969.

[21] Lowson, M.V. and Ollerhead, J.B. A theoretical study of helicopter noise. *J. Sound Vib.*, 1969, **9**, 197–222.

[22] Fiddes, S.P., Gould, J. and Aston, J.A.G. *A Non-Linear Lifting Line Method for Propeller Aerodynamic Performance Prediction*. University of Bristol, Department of Aerospace Engineering, report BU/AERO/437, Feb. 1992.

[23] Candel, S.M., Guedel, A. and Julienne, A., Reffraction and scattering of sound in an open wind tunnel flow, *Proc. 6th Int. Cong. Instrum. in Aerospace Simulation Facilities* (Ottawa, 1975) p. 288.

[24] Campos, L.M.B.C., Sur la propagation du son dans les écoulements non-uniformes et non stationaires, *Rev. Acoust.*, 1984, **67**, 217–237.

[25] Campos, L.M.B.C., On the spectral broadening of sound by turbulent shear layers. Part 1: Scattering by interfaces and reffraction in turbulence. *J. Fluid Mech.*, 1978, **81**, 723–749.

[26] Powell, A., Vortex sound, *J. Acoust. Soc. Am.*, 1968, **36**, 117–185.

[27] Howe, M.S., Contributions to the theory of aerodynamic sound, with applications to excess jet noise and the theory of the flute, *J. Fluid. Mech.*, 1975, **71**, 625–673.

[28] Lighthill, M.J., On sound generated aerodynamically. I: general theory, *Proc. R. Soc.*, 1952, **A211**, 564–587.

[29] Campos, L.M.B.C., On the generation and radiation of magneto-acoustic waves, *J. Fluid Mech.*, 1977, **81**, 529–549.

[30] Crighton, D.G. and Leppington, F.G., Radiation properties of a semi-infinite vortex sheet. The initial value problem, *J. Fluid Mech.*, 1974, **64**, 393–414.

[31] Bechert, D. and Michel, U., The control of a thin free shear layer with and without thin semi-infinite plate by a pulsating flow field, *Acustica*, 1975, **33**, 287–307.

[32] Bechert, D. and Pfizenmaier, E., On the amplification of broad band jet noise by a pure tone excitation, *J. Sound Vib.*, 1975, **43**, 581–587.

[33] Campos, L.M.B.C., On the spectral broadening of sound by turbulent shear layers. Part 2: Comparison with experimental and aircraft noise, *J. Fluid Mech.*, 1978, **81**, 751–783.

[34] Campos, L.M.B.C., On the emission of sound by an ionized inhomogeneity. *Proc. Roy. Soc.*, 1978, **A351**, 65–91.

[35] Campos, L.M.B.C., and Macedo, C.M., On the prediction of the broadband noise of helicopter from the impulsive component, *J. Acoustique*, 1992, **5**, 531–542.

[36] Cox, C.R. and Lynn, R.R., A study of the origin and means of reducing helicopter noise, *U.S. Army Transportation Research Command Tech.* Rep. 1972, 62–73.

[37] Lighthill, M.J., On the energy scattered from the interaction of turbulence with sound and shock waves. *Proc. Camb. Philos. Soc.*, **44**, 531–551.

[38] Campos, L.M.B.C., On the fundamental acoustic mode in variable-area low-Mach number nozzles, *Progr. Aerosp. Sci.*, 1985, **22**, 1–26.
Campos, L.M.B.C., On waves in gases. Part 1: Acoustics of jets, turbulence and ducts, *Rev. Mod. Phys.*, 1986, **57** (1986) 117–182.

[39] Campos, L.M.B.C., On the generalizations of the Doppler factor local frequency, wave invariant and group velocity, *Wave Motion*, 1988, **10**, 193–207.

# *Index*

*Index compiled by Geoffrey C. Jones*